Multi-Criteria Decision-Making and Optimum Design with Machine Learning

As multicriteria decision-making (MCDM) continues to grow and evolve, machine learning (ML) techniques have become increasingly important in finding efficient and effective solutions to complex problems. This book is intended to guide researchers, practitioners, and students interested in the intersection of ML and MCDM for optimal design.

Multi-Criteria Decision-Making and Optimum Design with Machine Learning: A Practical Guide is a comprehensive resource that bridges the gap between ML and MCDM. It offers a practical approach by demonstrating the application of ML and MCDM algorithms to real-world problems. Through case studies and examples, it showcases the effectiveness of these techniques in optimal design. The book also provides a comparative analysis of conventional MCDM algorithms and machine learning techniques, enabling readers to make informed decisions about their use in different scenarios. It also delves into emerging trends, providing insights into future directions and potential opportunities. The book covers a wide range of topics, including the definition of optimal design, MCDM algorithms, supervised and unsupervised ML techniques, deep learning techniques, and more, making it a valuable resource for professionals and researchers in various fields.

Multi-Criteria Decision-Making and Optimum Design with Machine Learning: A Practical Guide is designed for professionals, researchers, and practitioners in engineering, computer science, sustainability, and related fields. It is also a valuable resource for students and academics who wish to expand their knowledge of machine learning applications in multicriteria decision-making. By offering a blend of theoretical insights and practical examples, this guide aims to inspire further research and application of machine learning in multidimensional decision-making environments.

Multi-Criteria Decision-Making and Optimum Design with Machine Learning

A Practical Guide

Edited by Van Thanh Tien Nguyen,
Nhut T.M. Vo, Van Chinh Truong, and
Van-Thuc Nguyen

CRC Press
Taylor & Francis Group
Boca Raton London New York

CRC Press is an imprint of the
Taylor & Francis Group, an **informa** business

Designed cover image: © Shutterstock—Summit Art Creations

First edition published 2025
by CRC Press
2385 NW Executive Center Drive, Suite 320, Boca Raton FL 33431

and by CRC Press
4 Park Square, Milton Park, Abingdon, Oxon, OX14 4RN

CRC Press is an imprint of Taylor & Francis Group, LLC

© 2025 selection and editorial matter, Van Thanh Tien Nguyen, Nhut T.M. Vo, Van Chinh Truong, and Van-Thuc Nguyen; individual chapters, the contributors

ISBN: 978-1-032-63508-8 (hbk)
ISBN: 978-1-032-63516-3 (pbk)
ISBN: 978-1-032-63517-0 (ebk)

DOI: 10.1201/9781032635170

Typeset in Times
by Apex CoVantage, LLC

To the pioneers and innovators in engineering and artificial intelligence, whose unwavering dedication and groundbreaking research have paved the way for transformative advancements. Your tireless pursuit of knowledge and commitment to excellence inspires us all. This book is dedicated to:

Ganesh Khekare, Ganesh Yenurkar, Anil V Turukmane, Gaurav Kumar Ameta, Pooja Sharma, Ajay Kumar Phulre

Hai-Ninh Do

Kannan A, Meenakshi Annamalai, Rajkumar. P, Kujani T

Kala Raja Mohan, R. Narmada Devi, Nagadevi Bala Nagaram, Regan Murugesan, Sathish Kumar Kumaravel

Nagadevi Bala Nagaram

R. Narmada Devi, Regan Murugesan, Nagadevi Bala Nagaram, Kala Raja Mohan, Sathish Kumar Kumaravel

Regan Murugesan, Sathish Kumar Kumaravel, Kala Raja Mohan, Nagadevi Bala Nagaram, R. Narmada Devi

Sathish Kumar Kumaravel, Regan Murugesan, Nagadevi Bala Nagaram, Kala Raja Mohan, R. Narmada Devi

Sathvik V Koushik, Shamanth Showri N R, Shreya C R, Urjitha P, Divya C D

Wasswa Shafik

Yang Minghai, Fluturim Saliu, Waqar Akbar Khan

Meenakshi, O. Mythreyi, S. Dhanushiya, Shivangi Mishra

Anita Mohanty, Ambarish G. Mohapatra, Subrat Kumar Mohanty, Abhijit Mohanty

J. Senbagamalar, V. Gomathi

Ganesh Khekare, Shivani Kerai, Anil V Turukmane, Urvashi Khekare, Rahul Sharma, Rahul Agrawal

Mousumi Karmakar, Subhajit Bhattacharyya

Pham Thi Hong Nga, Van Tron Tran, Thanh Tan Nguyen, Xuan Tien Vo, Van-Thuc Nguyen

Nguyen Cong Dat, Nguyen Tran Dan Truong, Nguyen Quang Tuan, Pham Thi Hong Nga, Van-Thuc Nguyen, Nguyen Vinh Tien, Pham Quan Anh, Nguyen Thanh Tan, Ho Thi My Nu

Trieu Khoa Nguyen, Nguyen Thy Ton That, Binh-Duong Nguyen

Truong Hoang Phuc, Huynh Nguyen Vinh Phuc, Le Nguyen Trung Nam, Pham Thi Hong Nga, Nguyen Thanh Tan, Nguyen Vinh Tien, Le Duong, Van-Thuc Nguyen, Ho Thi My Nu

Van-Canh Nguyen, Dung Hoang Tien, Thuy-Duong Nguyen, Van Hung Pham, Ba Nghien Nguyen, Duy Trinh Nguyen, Van Que Nguyen

Cong-Truyen Duong, Van-Chinh Truong, among others.

Your contributions have enriched our understanding and inspired future engineers and researchers. We express our deepest gratitude for your invaluable insights and dedication to advancing the frontiers of knowledge.

You are all highly appreciated.

Thank you!

Editorial Team

Contents

Chapter 3   Optimizing Ti-6Al-4V Milling under MQL Conditions Using
                SVR, NSGA-II, and TOPSIS 47

*Van-Canh Nguyen, Dung Hoang Tien, Thuy-Duong Nguyen,
Van Hung Pham, Ba Nghien Nguyen, Duy Trinh Nguyen,
and Van Que Nguyen*

Chapter 9 Enhancing Underwater Imagery Using Multicriteria
Decision-Making with Machine Learning Techniques 129

*Ganesh Khekare, Shivani Kerai, Anil V Turukmane,
Urvashi Khekare, Rahul Sharma, and Rahul Agrawal*

Chapter 10 Selecting Optimal Electric Vehicle Charging Station Sites
Based on Analytic Hierarchy and VIKOR...................................... 139

Mousumi Karmakar and Subhajit Bhattacharyya

Chapter 14 Locating Electric Vehicle Power Stations Using Neutrosophic TOPSIS 195

Regan Murugesan, Sathish Kumar Kumaravel,
Kala Raja Mohan, Nagadevi Bala Nagaram,
and R. Narmada Devi

Chapter 15 Multicriteria Decision-Making Modeling Using Spherical Neutrosophic Similarity Measures 203

R. Narmada Devi, Regan Murugesan, Nagadevi Bala
Nagaram, Kala Raja Mohan, and Sathish Kumar Kumaravel

Chapter 19 Artificial Intelligence Algorithms for Better Decision-Making 252

Ganesh Khekare, Ganesh Yenurkar, Anil V Turukmane, Gaurav Kumar Ameta, Pooja Sharma, and Ajay Kumar Phulre

Contents xvii

Chapter 24 Predicting Lumpy Skin Disease Using Machine Learning 317

Nagadevi Bala Nagaram, R. Narmada Devi, Kala Raja Mohan, Sathish Kumar Kumaravel, and Regan Murugesan

Preface

In the dynamic landscape of engineering and decision-making, integrating multi-criteria analysis with machine learning technologies has emerged as a pivotal paradigm for optimizing solutions across diverse domains. This book, *Multi-Criteria Decision-Making and Optimum Design with Machine Learning: A Practical Guide*, is a testament to the relentless pursuit of innovation and excellence within this evolving field.

This comprehensive volume, spanning twenty-four chapters authored by esteemed researchers and practitioners, encapsulates a wealth of knowledge and practical insights aimed at advancing decision-making methodologies. The authors of each chapter delve into unique facets of decision-making processes, from the intricate exploration of fuzzy systems for multicriteria optimization to the application of machine learning in optimizing technological parameters. The collaborative efforts of the contributing authors, under the editorial guidance of Van Thanh Tien Nguyen, Van-Thuc Nguyen, Nhut T.M. Vo, and Van-Chinh Truong, have resulted in a panoramic view of research that traverses diverse disciplines, from engineering design to sustainable development. This interdisciplinary approach underscores the significance of leveraging innovative methodologies to address complex decision-making challenges in real-world scenarios.

As readers embark on this intellectual journey, they will encounter a diverse array of topics, ranging from optimizing surface roughness in machining processes to identifying best teacher awardees using advanced algorithms. Each chapter is a testament to the authors' ingenuity and dedication to pushing the boundaries of knowledge and practice. Moreover, including practical case studies and real-world applications ensures that this book transcends theoretical discourse, offering actionable insights that can inform decision-making processes in various contexts. Whether navigating the intricacies of destination decision-making dilemmas or predicting the onset of lumpy skin disease using machine learning, the methodologies elucidated in these chapters hold immense potential for driving tangible outcomes and fostering sustainable development.

In conclusion, *Multi-Criteria Decision-Making and Optimum Design with Machine Learning: A Practical Guide* represents a seminal contribution to the burgeoning field of decision science. We sincerely hope this volume will inspire researchers, practitioners, and students alike to embark on their own explorations, armed with the knowledge and tools necessary to tackle the complex challenges of the modern world, and we highly appreciate all our readers.

The Editorial Team

About the Editors

Van Thanh Tien Nguyen (member, IEEE) received master's degrees in mechanical engineering and linguistics from Viet Nam National University Ho Chi Minh City, Bach Khoa University, and HCMC University of Social Sciences and Humanities in 2012 and 2020, respectively. He holds a Ph.D. in industrial engineering and management from the National Kaohsiung University of Science and Technology, Taiwan. He has published over 66 journal papers and conference papers, been a reviewer for over 75 SCI/Scopus journals with over 1010 review reports, and been an academic editor for several Q1 journals with over 65 scientific manuscripts. He has experience studying and working in labs as a researcher/professional in many countries, such as South Korea, Thailand, Russia, and Taiwan. He is a lecturer at the Industrial University of Ho Chi Minh City, Vietnam. His areas of interest are machine learning (AI), compliant mechanisms optimization design, numerical computation, multicriteria decision-making, and supply chain management. Dr. Nguyen's works have significantly influenced his field, as evidenced by his Scopus H-index of 19 and 748 citations (as of August 2024).

Nhut T.M. Vo (Member, IEEE) is a Lecturer at the School of Business, University of Management and Technology, Ho Chi Minh City, Vietnam. Her M.Sc. and Ph.D. degrees are from the National Kaohsiung University of Science and Technology, Taiwan, specializing in Industrial Engineering and Management. Her diverse professional background spans multiple sectors, including banking, the jewelry industry, information technology, and e-commerce, providing her comprehensive insights across various industries. She has contributed books on lean management and related fields. Her research interests encompass contemporary topics, including the Internet of Things, blockchain, cloud computing, machine learning (AI), green energy, logistics, e-commerce, and numerical computation.

Van Chinh Truong is not just a member of the Faculty of Mechanical Engineering at the Industrial University of Ho Chi Minh City, Vietnam, but a dedicated educator. Truong has also been actively involved in research and academia, having participated in several research projects. He has successfully developed and implemented various technologies, significantly contributing to the industry, but his true passion lies in inspiring and educating future generations of engineers, a commitment that shines through his work and contributions to the field of mechanical engineering.

Van-Thuc Nguyen is a lecturer at Ho Chi Minh University of Technology and Education in Vietnam. He has a Ph.D. from the National Kaohsiung University of Science and Technology, Taiwan, and has published over 50 SCIE journal papers. His areas of expertise include manufacturing material science and mechanical processing. He is a highly respected researcher and educator in his field.

Contributors

Kannan A
Dr. Rangarajan Dr. Sakunthala
  Engineering College
Vel Tech Multi Tech
Chennai, India

Rahul Agrawal
G H Raisoni College of Engineering
  Nagpur
Maharashtra, India

Gaurav Kumar Ameta
Parul Institute of Technology
Parul University
Gujarat, India

Pham Quan Anh
Ho Chi Minh City University of
  Technology and Education
Ho Chi Minh City, Vietnam

Meenakshi Annamalai
Vel Tech Rangarajan Dr. Sagunthala
  R&D Institute of Science and
  Technology
Chennai, India

Subhajit Bhattacharyya
Mallabhum Institute of Technology
West Bengal, India

Divya C D
Vidyavardhaka College of Engineering
  (affiliated with VTU, Belagavi)
Karnataka, India

Nguyen Cong Dat
Ho Chi Minh City University of
  Technology and Education
Ho Chi Minh City, Vietnam

R. Narmada Devi
Vel Tech Rangarajan Dr. Sagunthala
  R&D Institute of Science and
  Technology
Tamil Nadu, India

S. Dhanushiya
Vel Tech Rangarajan Dr. Sagunthala
  R&D Institute of Science and
  Technology
Tamil Nadu, India

Hai-Ninh Do
University of Economics Ho Chi Minh
  City
Ho Chi Minh City, Vietnam

Cong Truyen Duong
Industrial University of Ho Chi Minh City
Ho Chi Minh City, Vietnam

Le Duong
Ho Chi Minh City University of
  Technology and Education
Ho Chi Minh City, Vietnam

Pham Thi Hong Nga
Ho Chi Minh City University of
  Technology and Education
Ho Chi Minh City, Vietnam

Shivangi Mishra J
Vel Tech Rangarajan Dr. Sagunthala
  R&D Institute of Science and
  Technology
Tamil Nadu, India

Mousumi Karmakar
Mallabhum Institute of Technology
West Bengal, India

Shivangi Kerai
School of Computer Science and
  Engineering
Vellore Institute of Technology
Vellore, India

Waqar Akbar Khan
Shandong University of Finance &
  Economics
Jinan, China

Ganesh Khekare
School of Computer Science and
  Engineering
Vellore Institute of Technology
Vellore, India

Urvashi Khekare
Vellore Institute of Technology
Vellore, India

Sathvik V Koushik
Vidyavardhaka College of Engineering
  (affiliated with VTU, Belagavi)
Karnataka, India

Sathish Kumar Kumaravel
Vel Tech Rangarajan Dr. Sagunthala
  R&D Institute of Science and
  Technology
Tamil Nadu, India

Meenakshi Annamalai
Vel Tech Rangarajan Dr. Sagunthala
  R&D Institute of Science and
  Technology
Tamil Nadu, India

Yang Minghai
Shandong University of Finance &
  Economics
Jinan, China

Kala Raja Mohan
Department of Mathematics
Panimalar Engineering College
Poonamalee, Chennai-600 123

Abhijit Mohanty
Silicon University
Odisha, India

Anita Mohanty
Silicon University
Odisha, India

Subrat Kumar Mohanty
Einstein Academy of Technology and
  Management
Odisha, India

Ambarish G. Mohapatra
Silicon University
Odisha, India

Regan Murugesan
Vel Tech Rangarajan Dr. Sagunthala
  R&D Institute of Science and
  Technology
Tamil Nadu, India

O. Mythreyi
Vel Tech Rangarajan Dr. Sagunthala
  R&D Institute of Science and
  Technology
Tamil Nadu, India

Nagadevi Bala Nagaram
Vel Tech Rangarajan Dr. Sagunthala
  R&D Institute of Science and
  Technology
Tamil Nadu, India

Le Nguyen Trung Nam
Ho Chi Minh City University of
  Technology and Education
Ho Chi Minh City, Vietnam

Ba Nghien Nguyen
Hanoi University of Industry
Hanoi, Vietnam

Binh-Duong Nguyen
Industrial University of Ho Chi Minh City
Ho Chi Minh City, Vietnam

Duy Trinh Nguyen
Hanoi University of Industry
Hanoi, Vietnam

Thuy-Duong Nguyen
Hanoi University of Science and
    Technology
Hanoi, Vietnam

Van Thanh Tien Nguyen
Industrial University of Ho Chi Minh
    City
Ho Chi Minh City, Vietnam

Trieu Khoa Nguyen
Industrial University of Ho Chi Minh
    City
Ho Chi Minh City, Vietnam

Van-Canh Nguyen
Hanoi University of Industry
Hanoi, Vietnam

Van-Thuc Nguyen
Ho Chi Minh City University of
    Technology and Education
Ho Chi Minh City, Vietnam

Van Que Nguyen
Hanoi University of Industry
Hanoi, Vietnam

Ho Thi My Nu
Ho Chi Minh City University of
    Industry and Trade
Ho Chi Minh City, Vietnam

Mythreyi O.
Vel Tech Rangarajan Dr. Sagunthala
    R&D Institute of Science and
    Technology
Chennai, Tamil Nadu, India

Van Hung Pham
Hanoi University of Science and
    Technology
Hanoi, Vietnam

Huynh Nguyen Vinh Phuc
Ho Chi Minh City University of
    Technology and Education
Ho Chi Minh City, Vietnam

Truong Hoang Phuc
Ho Chi Minh City University of
    Technology and Education
Ho Chi Minh City, Vietnam

Ajay Kumar Phulre
VIT Bhopal University
Sehore, MP, India

Shamanth Showri N R
Vidyavardhaka College of
Engineering (affiliated with
    VTU, Belagavi)
Karnataka, India

Shreya C R
Vidyavardhaka College of
    Engineering (affiliated
    with VTU, Belagavi)
Karnataka, India

Fluturim Saliu
University of Tetova
Tetovo, North Macedonia

J. Senbagamalar
Vel Tech Rangarajan Dr. Sagunthala
    R&D Institute of Science and
    Technology
Tamil Nadu, India

Wasswa Shafik
University Brunei Darussalam,
    Gadong, Brunei
Dig Connectivity Research Laboratory
    (DCRLab)
Kampala, Uganda

Rahul Sharma
Parul Institute of Technology
Parul University
Gujarat, India

Kujani T
Vel Tech Rangarajan Dr. Sagunthala R&D
    Institute of Science and Technology
Chennai, India

Nguyen Thanh Tan
Ho Chi Minh City University of
    Technology and Education
Ho Chi Minh City, Vietnam

Nguyen Thy Ton That
Industrial University of Ho Chi Minh
    City
Ho Chi Minh City, Vietnam

Dung Hoang Tien
Hanoi University of Industry
Hanoi, Vietnam

Nguyen Vinh Tien
Ho Chi Minh City University of
    Technology and Education
Ho Chi Minh City, Vietnam

Van Tron Tran
Ho Chi Minh City University of
    Technology and Education
Ho Chi Minh City, Vietnam

Nguyen Tran Dan Truong
Ho Chi Minh City University of
    Technology and Education
Ho Chi Minh City, Vietnam

Nguyen Quang Tuan
Ho Chi Minh City University of
    Technology and Education
Ho Chi Minh City, Vietnam

Anil V Turukmane
VIT-AP University
Andhra Pradesh, India

Nhut T.M. Vo
School of Business
University of Management and Technology
Ho Chi Minh City, Vietnam

Xuan Tien Vo
Ho Chi Minh City University of
    Technology and Education
Ho Chi Minh City, Vietnam

Ganesh Yenurkar
Yeshwantrao Chavan College of
    Engineering
Nagpur, India

Acknowledgments

This book, *Multi-Criteria Decision-Making and Optimum Design with Machine Learning: A Practical Guide*, was completed through the collaborative efforts and unwavering support of numerous individuals whose contributions have enriched its content and brought it to fruition.

First and foremost, we extend our heartfelt gratitude to all the authors whose dedication and expertise have resulted in this volume's insightful chapters. Your commitment to advancing engineering, decision-making, and machine learning is truly commendable, and your contributions serve as a beacon of inspiration to researchers and practitioners alike. We would like to sincerely thank Van Thanh Tien Nguyen, Van-Thuc Nguyen, Nhut T.M. Vo, and Van-Chinh Truong for their editorial leadership and invaluable guidance throughout the publication process. Your vision, professionalism, and meticulous attention to detail have been instrumental in shaping the scope and direction of this book.

Special thanks are also due to the reviewers and editorial staff, whose expertise and constructive feedback have helped refine the chapters' quality and rigor. Your commitment to upholding academic standards and ensuring the scholarly integrity of this work is deeply appreciated. Furthermore, we acknowledge the support of our respective institutions and organizations, whose encouragement and resources have facilitated the realization of this collaborative endeavor. Your ongoing commitment to fostering research and innovation is fundamental to advancing knowledge and pursuing excellence.

Lastly, we thank our families, friends, and colleagues for their understanding, encouragement, and unwavering support throughout this project. Your patience, love, and belief in our endeavors have been a constant source of strength and inspiration.

In conclusion, we humbly acknowledge and appreciate the collective efforts of all individuals involved in creating this book. We sincerely hope that *Multi-Criteria Decision-Making and Optimum Design with Machine Learning: A Practical Guide* will be a valuable resource for researchers, practitioners, and students seeking to navigate the complexities of decision-making in today's dynamic and evolving landscape.

Thank you.

The Editorial Team

1 Innovations in Technical Methodologies: Advancing Decision-Making and Optimization

Cong Truyen Duong, Nhut T.M. Vo,
*and Van Thanh Tien Nguyen**

1.1 INTRODUCTION

In engineering, decision-making often involves navigating intricate systems with multiple competing goals. Traditional optimization methods can struggle to cope with uncertainties and imprecisions inherent in real-world engineering issues. As such, there is a growing need for advanced methodologies to address these challenges. In this introduction, we examine recent advances in optimization and multicriteria decision-making (MCDM) across various engineering domains. We overview the reviewed chapters, highlighting the significance of employing innovative techniques to tackle complex engineering problems effectively.

1.2 RECENT ADVANCES IN OPTIMIZATION AND MCDM

Many researchers employ fuzzy systems for multicriteria engineering optimization. Fuzzy logic provides a robust framework for handling the uncertainty and imprecision inherent in engineering decision-making. In this chapter, we explore various fuzzy techniques, including fuzzy goal programming, fuzzy inference systems, and fuzzy TOPSIS, demonstrating their effectiveness through case studies in engineering contexts. In Chapter 2, Anita Mohanty et al. determine that despite the potential of fuzzy systems, challenges remain in defining fuzzy rules and managing computational complexity.

In Chapter 3, Van-Canh Nguyen et al. describe integrating genetic algorithms, machine learning, and MCDM to optimize technological parameters for milling the Ti-6Al-4V alloy under minimum quantity lubrication conditions. This approach showcased the efficacy of combining advanced optimization techniques to achieve desired machining outcomes. However, selecting appropriate optimization algorithms and managing model complexity pose ongoing challenges in practical implementation.

*Corresponding author: nguyenvanthanhtien@iuh.edu.vn

DOI: 10.1201/9781032635170-1

For Chapter 4, Trieu Khoa Nguyen et al. conducted a deep study on selecting 3D printing parameters to optimize tensile strength using the Taguchi-based response surface method. By systematically investigating the influence of various printing parameters, the authors offered valuable insights into optimizing the mechanical properties of 3D-printed components. However, challenges persist in accurately predicting the relationships between printing parameters and mechanical properties due to the complex nature of additive manufacturing processes.

Annamalai Meenakshi et al. introduced an enhanced network MCDM optimization approach using the max product in Chapter 5. The authors aimed to optimize critical project management networks by integrating intuitionistic fuzzy graphs with MCDM. While this methodology shows promise in improving network efficiency, further research is needed to validate its effectiveness in practical applications.

In Chapter 6, Pham et al. investigated the optimization of surface roughness in H13 steel machined by wire electrical discharge machining. By exploring the effects of discharge time on surface roughness, the authors offer valuable insights into optimizing machining parameters for improved surface finish. However, challenges remain in balancing competing objectives and optimizing machining processes for complex materials.

Cong Dat Nguyen et al. examined the impact toughness of a PBT/PA6 composite reinforced with glass fibers in Chapter 7. Through experimental analysis, they highlighted the potential of glass fiber reinforcement for enhancing the mechanical properties of polymer composites. However, optimizing the reinforcement process to achieve desired mechanical properties poses challenges due to the complex interactions between materials and processing parameters.

In Chapter 8, Truong and fellow researchers explored the effect of chamber temperature on the flexural strength of thermoplastic polyurethane plastic via fused deposition modeling technology. This study underscores the importance of chamber temperature control in achieving desired mechanical properties in 3D-printed parts. However, challenges persist in optimizing printing parameters for specific materials and applications.

Ganesh Khekare et al. focused on enhancing underwater imagery using MCDM and machine learning in Chapter 9. By employing advanced image enhancement methodologies, the authors were aiming to improve the interpretability of underwater images for marine exploration. However, challenges remain in mitigating environmental factors and ensuring the reliability of underwater imaging systems.

In Chapter 10, Mousumi Karmakar and Subhajit Bhattacharyya address selecting optimal sites for electric vehicle charging stations using analytic hierarchy and the VIKOR method. By integrating MCDM with hybrid ranking, the authors offer a systematic approach to selecting optimal sites, but challenges persist in considering uncertainty and qualitative aspects of site selection criteria.

J. Senbagamalar and V. Gomathi introduce optimal indices on topological intuitionistic fuzzy graphs in Chapter 11. By defining various topological indices, the authors advance graph theory applications in computer science, but challenges remain in developing robust algorithms and applications for intuitionistic fuzzy graphs.

The latter chapters of this volume, serving as a discussion section, synthesize the essential findings and challenges that the earlier authors identified. They highlight common themes and opportunities for future engineering optimization and decision-making research. By critically examining the strengths and limitations of each methodology, this discussion section provides valuable insights into the ongoing efforts to address complex engineering problems.

1.3 CONCLUSIONS

In this first chapter, we have reviewed how some contemporary researchers are leveraging advanced optimization and MCDM methods to address complex engineering challenges effectively. While significant progress has been made in developing innovative, the contributing authors to this volume have concluded that ongoing research is needed to overcome existing limitations and realize full potential of these methodologies in practical engineering applications. By addressing the identified challenges, researchers can contribute to advancing the state of the art in engineering optimization and decision-making.

1.4 ACKNOWLEDGEMENTS

The authors would like to thank the Industrial University of Ho Chi Minh City, Vietnam, and the National Kaohsiung University of Science and Technology, Taiwan, for their assistance. Additionally, we would like to thank the reviewers and editors for their constructive comments and suggestions for improving our work.

1.5 CONFLICTS OF INTEREST

The authors state that they have no known competing financial interests or personal relationships that could have appeared to influence the work reported in this chapter.

Two authors (Van Thanh Tien Nguyen and Nhut T.M. Vo) are members of this book's editorial board but were not involved in the editorial review or the decision to publish this book. The authors state that the work has no potential conflict of interest. All authors have read and agreed to the published version of the book.

2 Fuzzy Systems for Multicriteria Optimization
Applications in Engineering Design

Anita Mohanty, Ambarish G. Mohapatra,
Subrat Kumar Mohanty, and Abhijit Mohanty

2.1 INTRODUCTION

Diverse and frequently incompatible criteria highlight the difficulties of decision-making in the context of contemporary engineering design. In this chapter, we offer an engaging examination of how fuzzy systems might address these issues. We explore the theoretical foundations of fuzzy logic as it progresses, clarifying ideas such as fuzzy sets and linguistic variables. After that, we move smoothly into the real world by showing how fuzzy systems can improve on the multicriteria optimization tools that are already in place. The chapter's strength is our vivid illustration of real-world applications in a variety of engineering areas, which highlights fuzzy logic's versatility and effectiveness in optimizing complex design spaces (Xu, C., 2022; Akram et al., 2021). Readers learn how to use fuzzy systems to negotiate the complexities of decision-making and provide realistic, nuanced solutions through engrossing case studies. As the chapter comes to a close, the reader will have a deeper understanding of this revolutionary intersection in engineering design as it considers future directions and possible developments in the symbiotic relationship between fuzzy systems and multicriteria optimization tools.

2.1.1 MOTIVATION FOR MULTICRITERIA OPTIMIZATION IN ENGINEERING

Multicriteria optimization in engineering has multiple benefits, such as the ability to manage real-world complexity; real-world engineering design problems seldom have a single, well-defined objective. Indeed, the ability to manage competing objectives is a second benefit of multicriteria optimization; engineering solutions often involve balancing competing objectives like cost-effectiveness, performance efficiency, and environmental sustainability. Multicriteria optimization also allows for managing complexity in a systematic manner; engineers use multicriteria optimization

DOI: 10.1201/9781032635170-2

methodologies to systematically manage the interconnectedness of design elements. The techniques also allow for exploring trade-offs between different design objectives.

Sustainability has a role in multicriteria optimization as well; the need for multiple criteria arises from the growing importance of sustainability in engineering, ensuring environmentally sustainable solutions, which are separately also what stakeholders increasingly want; consideration of stakeholder preferences is essential, and multicriteria optimization helps incorporate diverse perspectives. Along with environmental sustainability, multicriteria optimization goes beyond technical aspects to ensure that engineering solutions are socially responsible and aligned with ethical considerations. Finally, multicriteria optimization provides an integrated method of decision-making, considering various aspects simultaneously for well-balanced engineering decisions.

2.1.2 Challenges in Engineering Decision-Making

Making decisions in engineering is a complex process that requires overcoming many obstacles, reflecting the difficulty of creating and putting into practice solutions to issues that arise in the actual world. Some prominent challenges in engineering decision-making include:

- **Complexity of Systems:** Engineering systems are becoming increasingly complex, with interconnected components and dependencies. Understanding and optimizing these intricate systems pose challenges in terms of modeling and analyzing and predicting their behaviors under varying conditions.
- **Uncertainty and Variability:** Decision-makers often face uncertainties related to factors such as material properties, environmental conditions, and external influences. Variability in these factors can impact the performance and reliability of engineering solutions, making it challenging to make precise decisions.
- **Conflicting Objectives:** Engineering projects typically involve multiple, and sometimes conflicting, objectives. Balancing factors like cost, performance, sustainability, and safety requires trade-offs, and finding an optimal solution that satisfies all criteria is a complex and often elusive goal.
- **Limited Resources:** Resource constraints, whether in terms of budget, time, or manpower, present a constant challenge. Decision-makers must optimize resource allocation to meet project goals, and prioritizing certain aspects over others becomes a critical aspect of the decision-making process.
- **Rapid Technological Advancements:** The fast pace of technological innovation introduces the challenge of keeping up with evolving technologies. Decision-makers must carefully assess new technologies for their applicability, feasibility, and potential impacts on existing systems.
- **Regulatory Compliance:** Engineering projects must adhere to a myriad of regulations and standards. Ensuring compliance with these requirements

adds an extra layer of complexity to decision-making, as failure to meet regulatory standards can have legal, financial, and reputational consequences.

- **Interdisciplinary Collaboration:** Many engineering tasks necessitate interdisciplinary cooperation. It might be difficult to ensure effective communication and bridge the gaps between experts in different sectors, which can affect the effectiveness and caliber of decision-making processes.
- **Environmental and Ethical Considerations:** Putting more focus on sustainability and moral issues while making decisions in engineering creates a new level of complexity. Careful consideration of the wider effects of engineering decisions is necessary to strike a balance between economic objectives and those that affect society and the environment.
- **Data Overload:** The vast amounts of data available can be overwhelming. In a world of abundant information, there are obstacles in filtering relevant information, guaranteeing data quality, and making judgments based on data analytics.
- **Long-Term Impact Assessment:** It's critical to consider and evaluate the long-term effects of engineering decisions. This entails taking into account elements like upkeep, technological flexibility, and the engineered system's total lifespan.

To tackle these obstacles, a blend of specialized knowledge, multidisciplinary cooperation, sophisticated decision assistance instruments, and an all-encompassing method of problem-solving is needed. To make successful engineering decisions, technical factors must be optimized in addition to taking the larger context into account.

2.1.3 Role of Fuzzy Logic in Handling Uncertainty

Because fuzzy logic offers a flexible and useful framework for handling incomplete knowledge and unclear or ambiguous circumstances, it is crucial for controlling uncertainty (Diaz-Curbelo et al., 2020). The following are the main ways that fuzzy logic aids in uncertainty management:

- **Representation of Uncertainty:** Fuzzy logic introduces the ideas of fuzzy sets and linguistic variables to capture the delicate, nuanced transition between true and wrong, in contrast to classical logic, which functions in a binary true/false paradigm. This allows for a more realistic portrayal of uncertainty.
- **Linguistic Variables and Membership Functions:** Fuzzy logic employs linguistic variables and membership functions, such as "high," "low," "hot," and "cold." Membership functions define the extent to which a given value belongs to a linguistic variable. The utilization of linguistic strategy enables decision-makers to communicate vague preferences by facilitating the communication of subjective and qualitative information.
- **Rule-Based Reasoning:** Fuzzy systems use rule-based reasoning in which the relationship between input and output variables is modeled by a set of

IF–THEN rules. These principles make decision-making more intuitive and practical in real-world circumstances by utilizing language words to account for the inherent ambiguity in human decision-making.

- **Fuzzy Inference:** Using fuzzy logic to process input data by predefined rules, fuzzy inference produces output that is understandable and meaningful. With the use of this technique, systems can make decisions even in the face of ambiguous or lacking information, offering a more flexible and contextually aware method.
- **Handling Incomplete or Noisy Data:** Fuzzy logic is very good at handling noisy or incomplete data, which is frequently encountered in practical engineering applications. Fuzzy systems can handle scenarios where data may be missing or variable, reducing the need for exact inputs and enhancing the resilience of decision-making processes.
- **Quantification of Truth:** Thanks to fuzzy logic, which introduces the concept of degrees of truth, statements might be only partially true. This is particularly helpful in circumstances where a straightforward binary distinction between true and false is insufficient. With degrees of truth, it is feasible to comprehend the uncertainty that exists in real-world situations on a more sophisticated level.
- **Adaptability to Human Reasoning:** Fuzzy logic is consistent with human reasoning since it emulates how people reason in the face of uncertainty. This makes it the perfect fit for situations where human expertise and judgment are essential since it makes it possible to integrate human knowledge into automated systems more smoothly.
- **Control Systems and Decision Support:** Fuzzy logic is extensively employed in control systems and decision support tools due to its exceptional ability to handle dynamic and uncertain circumstances. Fuzzy logic is used in process management, robotics, and optimization to build more resilient and adaptable systems.

In conclusion, fuzzy logic is essential for managing uncertainty in engineering applications. By providing a framework for encoding and processing imprecise information, fuzzy logic enables decision-makers to navigate the complexities of real-world contexts and make well-informed, context-aware decisions in the face of ambiguity.

2.2 FUNDAMENTALS OF FUZZY LOGIC

A mathematical model acknowledged as fuzzy logic addresses uncertainty and vagueness in reasoning (Kose, U., 2012; Leitmann, G., 1996). Some vital ideas in fuzzy logic principles are the following:

- **Fuzzy Sets:** By allowing elements to have membership degrees (μ) ranging from 0 to 1, fuzzy sets develop on the knowledge of classical sets. A component in a classical set may have membership degree 0 (not a member) or membership degree 1 (a member), but fuzzy sets also permit partial membership, which captures the uncertainty and randomness inherent in real-world environments. For example, A is a fuzzy set with variable x, whose membership function ranges between 0 and 1 as shown in Figure 2.1.

- **Linguistic Variables:** The use of fuzzy logic demonstrates linguistic variable understanding. These variables stand for qualitative or nonnumeric terms such as "low," "medium," or "high," as in Figure 2.2, in which speed is the fuzzy set and these are the three possible values. These variables provide a sound mathematical foundation for representing approximate knowledge and reasoning similar to human thinking.
- **Membership Functions:** To measure an element's level of membership in a fuzzy set, membership functions are utilized. These functions provide a means of quantifying the degree of fuzziness inherent in linguistic concepts by establishing a relationship between the input variable and its membership in a fuzzy set. In Figure 2.1, the relationship between input variable x and its membership function is $\mu_A(x)$.
- **Fuzzy Rules:** Fuzzy logic systems employ fuzzy rules to make decisions. They consist of IF–THEN statements that define the relationship between the language variables used in the input and output. A system's behavior is captured by each rule, and the combined effect of several rules defines how the system reacts to different input conditions.
- **Fuzzy Inference System:** To produce meaningful output, a fuzzy inference system (FIS) processes the fuzzy rules. Fuzzification (converting

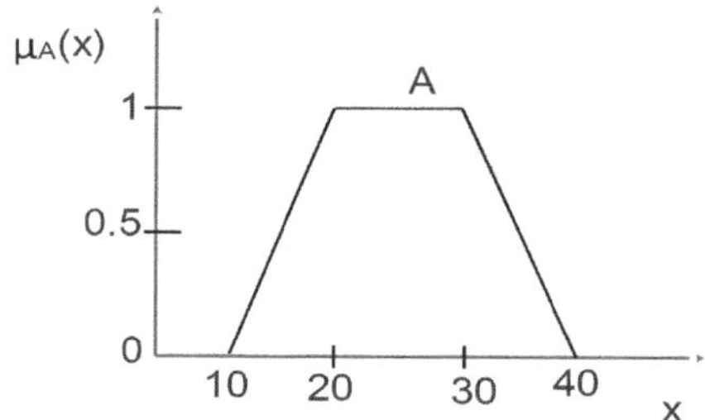

FIGURE 2.1   An example of a fuzzy set A.

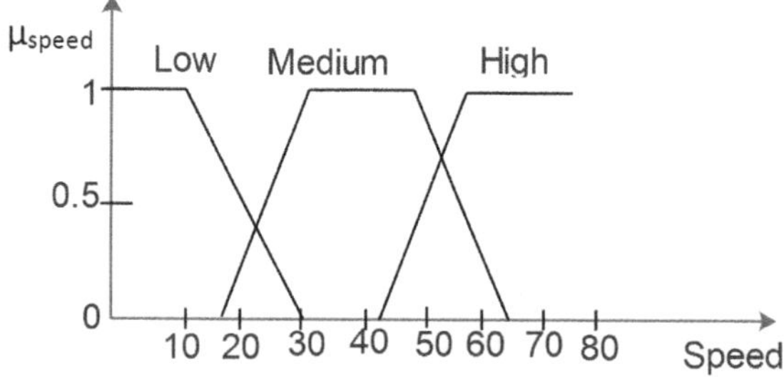

FIGURE 2.2   An example of speed as a linguistic variable speed.

crisp inputs to fuzzy sets), rule evaluation (applying fuzzy rules), and defuzzification (turning fuzzy output to crisp values) are its three primary components. Under uncertainty, human-like reasoning can be modeled thanks to the FIS.

- **Defuzzification:** The technique of turning the hazy output into a precise value is known as defuzzification. This stage is essential to getting a precise and useful outcome from the fuzzy inference procedure. Defuzzification can be accomplished using a variety of techniques, including weighted average, mean of maximum, and centroid.
- **Fuzzy Logic Controllers:** In control systems, fuzzy logic controllers (FLCs) mimic human decision-making by using fuzzy logic. Applications for FLCs can be found in industries such as consumer electronics, automotive systems, and industrial control. They perform well in situations for which it is challenging to develop exact mathematical models.
- **Applications:** Artificial intelligence, pattern recognition, control systems, and decision support systems are just a few of the fields in which fuzzy logic has been successfully used. Because of its capacity to manage uncertainty and imperfect information, it is especially well-suited for real-world applications where accurate modeling may be difficult.
- **Extends Multivalued Logic:** More complex representations of uncertainty are possible when using fuzzy logic to handle multivalued logic. When results are not strictly binary, this enhancement can take on multiple values.

Gaining an understanding of these core ideas lays a strong basis for using fuzzy logic in a variety of engineering and decision-making scenarios. Fuzzy logic is an effective technique for capturing the complexity of real-world systems and enabling more human-like thinking in computing processes because of its capacity to model and handle ambiguity.

2.2.1   Fuzzy Sets and Membership Functions

2.2.1.1   Fuzzy Sets

Developed by Lotfi A. Zadeh in 1965, fuzzy sets are an extension of classical set theory that take into account the natural imprecision and vagueness seen in real-world systems. An element in a classical set is either a part of the set or it is not. A fuzzy set, on the other hand, expresses the extent to which an element is a part of the set and permits partial membership. An element in a fuzzy set has a membership degree that falls between 0 and 1.

In mathematical terms, a membership function $\mu A(x)$, where x is an element of X, defines a fuzzy set A in a universe of discourse X. Each element x is given a degree of membership $\mu A(x)$ by the membership function, which indicates how much x belongs to the fuzzy set A. Take into consideration, for instance, a fuzzy set "tall" within the discourse universe "height." A tall person may be assigned a high value by the membership function $\mu Tall(x)$, a medium value by someone of ordinary height, and a low value by someone of modest stature.

2.2.1.2 Membership Functions

In determining an element's level of membership in a fuzzy set, membership functions are essential. These functions quantify the fuzziness associated with linguistic phrases by mapping the crisp input values to their corresponding membership degrees. There are several membership functions as shown in Figure 2.3, and each is appropriate in a particular situation:

- **Triangular Membership Function:** This straightforward, widely used membership function has a triangle form. Three characteristics describe it: the peak or center, the upper limit, and the lower limit.
- **Trapezoidal Membership Function:** This function is more flexible than the triangular function since it has a flat top. The lower limit, the left base, the right base, and the top limit are its four defining parameters.
- **Gaussian Membership Function:** The Gaussian function, which has a bell-like shape, is applied when the fuzziness is concentrated around a certain value. It is described by variables like the standard deviation and mean.
- **Sigmoidal (S-shaped) Membership Function:** This function is frequently employed to simulate slow transitions because of its sigmoidal shape. The steepness and midpoint are examples of parameters.
- **Generalized Bell Membership Function:** This function is more flexible than the Gaussian function; it lets you adjust the curve's width, slope, and center.
- **Piecewise Linear Membership Function:** This function can capture more intricate and nonuniform membership patterns since it is composed of straight-line segments.

The utilization of fuzzy logic in decision-making, control systems, and other domains where imprecise information must be taken into account is made easier by these membership functions, which aid in the quantification of linguistic terms and fuzzy ideas. The particulars of the problem domain and the linguistic context influence the choice of the membership function.

2.2.2 LINGUISTIC VARIABLES AND FUZZY OPERATIONS

2.2.2.1 Linguistic Variables

A key idea in fuzzy logic is linguistic variables, which offer a way to express qualitative or nonnumeric words within an official mathematical framework. Linguistic variables are different from standard variables in that they use phrases that are conveyed in normal language, like "high," "medium," or "low." These words are necessary to convey subjective and ambiguous information in a way that makes sense to people. A linguistic variable is made up of three main components:

1. **Universe of Discourse:** The linguistic variable's range of values is defined by the discourse universe. For example, if the variable of language is "temperature," then all conceivable temperature values may be included in the universe of conversation.

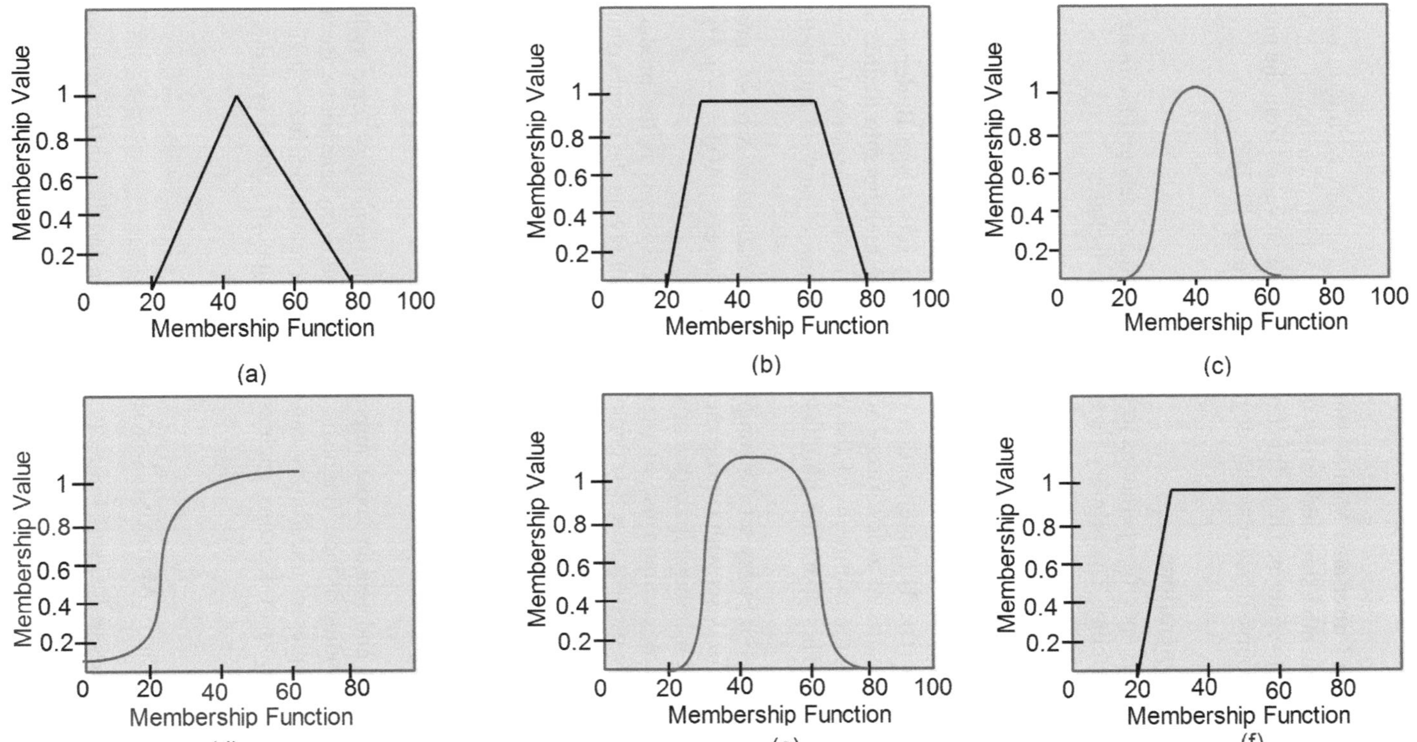

FIGURE 2.3 Different classes of parameterized membership functions: (a) triangular; (b) trapezoidal; (c) Gaussian; (d) sigmoidal (S-shaped); (e) bell; (f) piecewise linear.

2. **Terms:** Terms are the linguistic labels or categories associated with the variable. In the temperature example, terms could include "hot," "warm," and "cold."

3. **Membership Functions:** Membership functions associate each term with a degree of membership, indicating the extent to which a particular value belongs to a given term. These functions map numeric values from the universe of discourse to membership degrees in the range [0, 1].

2.2.2.2 Fuzzy Operations

Fuzzy logic employs specific operations to manipulate fuzzy sets and facilitate reasoning under uncertainty. The primary fuzzy operations include:

1. **Union (∪):** The union of two fuzzy sets A and B, denoted as A ∪ B, results in a new fuzzy set where the membership degree at each point is the maximum of the corresponding membership degrees in A and B. This operation captures the notion of "or" in linguistic terms.

2. **Intersection (∩):** The intersection of two fuzzy sets A and B, denoted as A ∩ B, results in a new fuzzy set where the membership degree at each point is the minimum of the corresponding membership degrees in A and B. This operation imitates the notation of "and" in linguistic terms.

3. **Complement (~):** Fuzzy set A's complement, represented as A or ~A, produces a new fuzzy set with flipped membership degrees, $\mu A(x) = 1 - \mu A(x)$ if $\mu A(x)$ is the membership degree of x in A.

4. **Product (×):** A common use in fuzzy inference systems is the product operation. The product A × B is computed by multiplying the membership degrees pointwise for two fuzzy integers, A and B. The evaluation of fuzzy rules requires this operation.

5. **Dilation and Contraction:** These operations change how wide apart a fuzzy collection is; contraction narrows the fuzzy set, whereas dilation widens it. These operations are useful for modifying fuzzy sets to satisfy particular requirements.

To properly model and reason about imprecise information, fuzzy logic systems require an understanding of these linguistic variables and fuzzy operations. Because of its capacity to represent uncertainty and human-like reasoning, fuzzy logic is a useful tool in a variety of domains, such as decision support, control systems, and artificial intelligence.

2.2.3 FUZZY LOGIC IN MODELING UNCERTAINTY

Fuzzy logic's capacity to deal with imprecise information and the inherent ambiguity of real-world circumstances makes it useful for modeling uncertainty in a variety of domains (Kumar & Schuhmacher, 2005). Among the noteworthy applications are shown in Figure 2.4.

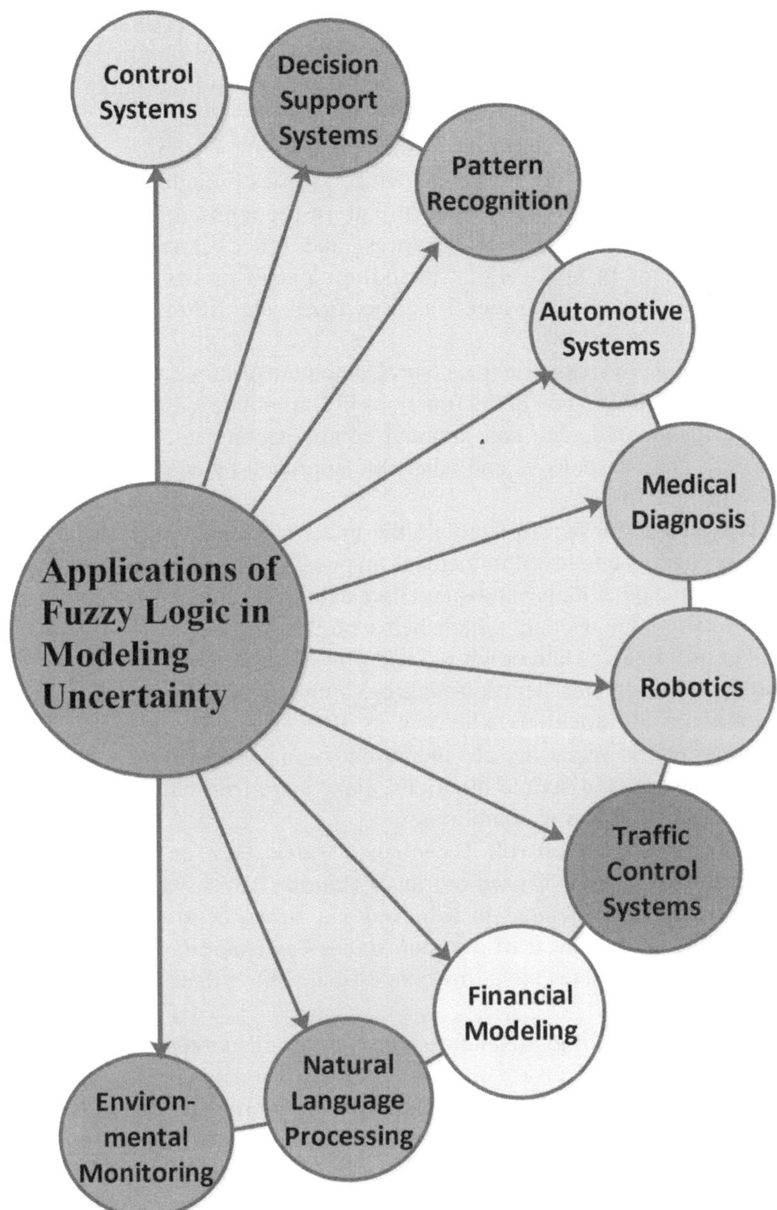

FIGURE 2.4 Applications of fuzzy logic in modeling uncertainty.

- **Autocontrol Systems:** In autocontrol systems where exact mathematical models are hard to formulate or contain uncertainties, FLCs are widely used. Systems with variable operating circumstances, such as automated vehicle systems, electric motor speed control, and HVAC system temperature control, are ideal for fuzzy controllers.

- **Decision Support:** To simulate and assess intricate decision-making processes involving multiple criteria, fuzzy logic is employed in decision support systems. This is especially helpful when handling qualitative data, preferences, and subjective evaluations. Applications include financial decision-making, project management, and strategic planning.
- **Pattern Recognition:** To manage uncertainties brought on by changes in input data, fuzzy logic is used in pattern recognition systems. It makes pattern detection in voice, pictures, and other signals more versatile and reliable. In applications including speech processing, handwriting recognition, and facial identification, fuzzy algorithms help to increase accuracy.
- **Automobile Systems:** Engine control, automatic gearbox, and antilock brake systems are only a few of the functions in automobile systems that use fuzzy logic. Compared with conventional control techniques, fuzzy controllers offer a more responsive and adaptable approach by adjusting to changing driving conditions.
- **Medical Diagnosis:** To simulate the uncertainty present in medical data, fuzzy logic is employed in medical diagnosis systems. It makes it possible to combine quantitative and qualitative data from a range of diagnostic tests with expert knowledge, which helps produce more precise and nuanced diagnoses in areas like prognosis and illness classification.
- **Robotics:** Complex decision-making, ambiguous settings, and imprecise inputs are all situations where fuzzy logic is used in robotics. Robots' autonomy and versatility are increased by fuzzy controllers, which allow them to navigate dynamic situations, make judgments based on sensor data, and adapt to changing conditions.
- **Transportation Control:** To enhance traffic flow and optimize signal timings, fuzzy logic is used in transportation control systems. More effective traffic management is achieved via fuzzy controllers, which take into account current traffic circumstances and modify signal timings in response to variables including congestion, vehicle density, and pedestrian activity.
- **Financial Modeling:** To manage uncertainties in market conditions, investment risks, and decision-making, financial models employ fuzzy logic. When accurate modeling is difficult, fuzzy algorithms help with portfolio optimization, risk assessment, and financial market trend prediction.
- **Natural Language Processing:** Tasks involving natural language processing, like sentiment analysis and language comprehension, make use of fuzzy logic; it makes it possible for systems to understand and react to linguistic inputs in a way that is more human-like by helping to capture the ambiguity and context-dependent quality of human language.
- **Environmental Monitoring:** To evaluate and forecast environmental characteristics, fuzzy logic is employed in environmental monitoring systems. It supports decision-making related to pollution control and environmental management, as well as the handling of imperfect sensor data and information integration from several sources.

These many uses demonstrate how flexible fuzzy logic is for handling uncertainty, which makes it an invaluable tool for simulating intricate real-world systems where more traditional methods would not be sufficient.

2.3 FUZZY INFERENCE SYSTEMS

The computational models known as fuzzy inference systems (FISs) use fuzzy logic to simulate human-like reasoning in uncertain and imprecise scenarios (Rizvi et al., 2020). Fuzzification, rule base, inference engine, rule evaluation, aggregation, and defuzzification are among their essential parts. Fuzzification is the process of converting crisp input values into fuzzy sets so that the system can handle linguistically imprecise data. The fuzzy input and output variable associations are defined by IF–THEN rules, which make up the rule base. The system's reaction is decided by the inference engine, which assesses these rules in light of the input data. Rule outputs are combined via aggregation to capture the overall behavior of the system. Finally, defuzzification transforms the aggregated fuzzy output into a crisp result. FIS is widely applied in control systems, decision support, pattern recognition, and various domains where modeling uncertainty is crucial for effective decision-making.

2.3.1 UNDERSTANDING FUZZY INFERENCE (MAMDANI AND SUGENO MODELS)

FISs are models that leverage fuzzy logic to make decisions and draw conclusions under uncertainty. Two popular types of FIS are the Mamdani model and the Sugeno model, each with distinct characteristics:

Mamdani Fuzzy Inference

1. **Fuzzification:** Crisp input values are transformed into fuzzy sets using membership functions.
2. **Rule Base:** IF–THEN rules define the relationship between fuzzy input sets and fuzzy output sets. These rules express linguistic rules in the form "IF input is A AND input is B THEN output is C."
3. **Inference Engine:** The inference engine evaluates the degree of membership for each rule's antecedent (IF part). Typically, the minimum operator is used for AND operations, and the maximum operator is used for OR operations.
4. **Aggregation:** The activated rule outputs are aggregated using the maximum operator, creating a combined fuzzy output set.
5. **Defuzzification:** The aggregated fuzzy output set is converted into a crisp output value using methods such as centroid or mean of max.

Sugeno Fuzzy Inference

1. **Fuzzification:** Similar to the Mamdani model, crisp input values are fuzzified into fuzzy sets.
2. **Rule Base:** IF–THEN rules in the Sugeno model express relationships in the form "IF input is A AND input is B THEN output is $f(x, y) = P + Q \times input1 + R \times input2 + \ldots$" where P, Q, R, etc., are constants.

3. **Inference Engine:** The inference engine calculates a crisp output value for each rule rather than a fuzzy set.
4. **Aggregation:** The calculated output values are aggregated using a weighted average or another aggregation method.
5. **Defuzzification:** The final output is obtained directly from the aggregation step, as it is already in crisp form.

Comparison

- **Nature of Output:**
  - Mamdani produces fuzzy sets as output.
  - Sugeno produces crisp, numeric values as output.
- **Interpretability:**
  - Mamdani's outputs are linguistically interpretable fuzzy sets.
  - Sugeno outputs are numeric and can lack linguistic interpretability.
- **Computational Efficiency:**
  - Sugeno models often have simpler computations, making them computationally more efficient than Mamdani models.
- **Applications:**
  - Mamdani models are preferred when linguistic interpretability is crucial and suitable for applications like control systems.
  - Sugeno models are often chosen for applications where precise numeric outputs are more important, such as in economic modeling or system optimization.

Understanding the distinctions between the Mamdani and Sugeno models allows practitioners to select the appropriate FIS for a given application based on the nature of the problem and the desired form of the output.

2.3.2 Components of Fuzzy Inference Systems

FISs consist of several integral components as shown in Figure 2.5 that collectively enable the modeling of complex systems and the extraction of meaningful information from imprecise inputs (Guillaume & Charnomordic, 2012).

The first key component is fuzzification, where crisp input values are transformed into fuzzy sets using membership functions. This step allows for representing linguistic variables and captures the uncertainty inherent in real-world data. The second component, rule base, articulates the knowledge and relationships between fuzzy input variables and output variables through a set of IF–THEN rules. These rules capture the reasoning behind the decisions made in response to the ambiguous input conditions; they are combined by the inference engine, which frequently uses techniques like the max–min composition to produce a fuzzy output set that depicts the system's reaction to the supplied inputs. Defuzzification, the fourth component, helps decision-makers take action by transforming the fuzzy output set into a crisp output value. Every component of the FIS is essential to its operation since it offers a methodical and organized framework for managing uncertainty and reaching well-informed conclusions in a variety of applications, such as pattern recognition and control systems.

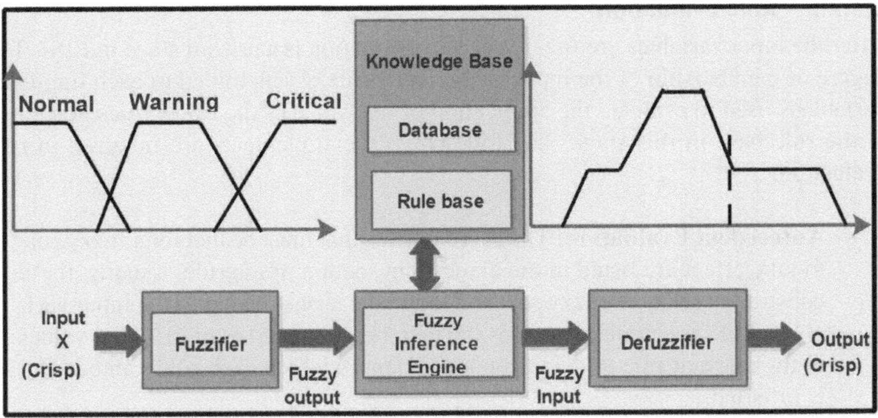

FIGURE 2.5 A fuzzy inference system.

2.3.2.1 Fuzzification

A critical stage in the FIS is fuzzification, which converts crisp (precise) input values into fuzzy sets. This procedure makes it possible for the FIS to handle ambiguous, qualitative, or linguistic data, which is crucial for simulating real-world systems that are rife with uncertainty.

As part of the fuzzification process, membership degrees are assigned to each input value concerning established fuzzy sets. Membership functions describing the extent to which input belongs to a specific linguistic phrase characterize these fuzzy collections. By defining the form and properties of the fuzzy sets, the membership functions offer a means of capturing the imprecision present in human language.

Fuzzification uses a variety of membership functions, each appropriate for a particular set of circumstances. Triangular, trapezoidal, Gaussian, and sigmoidal membership functions are common types. The selection of the membership function is contingent upon the characteristics of the input variable and the lexicon linked to it.

Fuzzification produces a set of membership degrees representing the extent to which each input value belongs to the fuzzy sets connected to the linguistic variables, and these fuzzy sets serve as the foundation for rule evaluation and subsequent inference within the FIS, now representing the imprecise input data. Consider, for practical purposes, the following scenario: the basic input variable is temperature and the linguistic phrases that correspond to it are low, medium, and high. Furnishing fuzzy sets such as "temperature is low," "temperature is medium," and "temperature is high" would include figuring out how closely each of these linguistic phrases corresponds to the recorded temperature. Fuzzification allows FISs to represent intricate, real-world systems by assiduously obtaining and interpreting ambiguous and imprecise data in a manner consistent with human thought processes. It prepares the groundwork for later FIS stages including defuzzification, inference, and rule evaluation.

2.3.2.2 Rule Evaluation

After the input variables are fuzzified, rule evaluation is a crucial stage in FISs. The degree of membership of the input values in the fuzzy sets linked to each linguistic variable is used to evaluate the application or relevance of the fuzzy rules generated in the rule base in this stage. The following crucial elements are involved in rule evaluation:

- **Antecedent Evaluation:** The prerequisites that must be met for a fuzzy rule to take effect are listed in the antecedent section of the rule. Usually, these constraints relate to fuzzy sets and language terms related to the input variables. Rule assessment assesses the membership degrees of the input values in the relevant rule to determine the extent to which each rule's antecedent is satisfied.
- **Rule Activation:** The calculated degrees of satisfaction for the antecedents determine the activation level of each rule: The higher the degree of satisfaction, the more the rule is activated. This activation level quantifies the strength of each rule's contribution to the overall inference process.
- **Degree of Activation:** The degree of activation reflects the combined influence of multiple rules. This is especially important when multiple rules are applicable due to overlapping fuzzy sets or linguistic terms in the antecedent parts. The minimum operator (for AND conditions) or maximum operator (for OR conditions) is often used to aggregate the degree of activation across rules.
- **Rule Weighting (optional):** Some FIS implementations include rule weighting, where individual rules are assigned different weights to reflect their importance or significance. Rule weights can influence the overall impact of a rule on the final output.

The outcome of rule evaluation is a set of activated rules with associated degrees of activation. These activated rules contribute to the subsequent steps in the FIS, such as the inference engine and aggregation, influencing the determination of the final output or decision.

In summary, rule evaluation plays a crucial role in fuzzy inference by determining the relevance and strength of individual rules based on the current input conditions. It allows the FIS to adaptively and contextually assess the applicability of rules in response to varying input values and linguistic terms, capturing the inherent flexibility and adaptability of fuzzy logic in handling complex, uncertain systems.

2.3.2.3 Defuzzification

Defuzzification is the process in a FIS where the fuzzy output, represented by fuzzy sets with associated membership degrees, is converted into a crisp, actionable result (Madanda et al., 2023). In other words, defuzzification transforms the imprecise and fuzzy output into a single, well-defined numeric value that can be used for decision-making or control actions. Defuzzification involves the following steps:

- **Aggregate Fuzzy Output Sets:** Before defuzzification, the individual fuzzy output sets corresponding to each activated rule need to be aggregated. This aggregation is typically done using fuzzy logic operators, such as the maximum operator for OR conditions or the sum operator.
- **Identify the Crisp Output Value:** The aggregated fuzzy output set, representing the overall response of the FIS, needs to be distilled into a single crisp value. Different defuzzification methods achieve this in various ways.
- **Defuzzification:** There are several common defuzzification methods:
  - **Centroid:** The centroid method calculates the center of mass or gravity of the aggregated fuzzy output set. It identifies the "center" of the fuzzy output and uses that point as the crisp output.
  - **Mean of Maximum:** The mean of maximum method identifies the maximum value of the aggregated fuzzy output set and calculates the mean of the values at which the membership function reaches its maximum. This mean is then considered the crisp output.
  - **Bisector Method:** The bisector method determines the point at which the area under the aggregated fuzzy output set is equally divided. The value at this point is taken as the crisp output.
  - **First of Maximum:** The first of maximum method selects the first value at which the aggregated fuzzy output set reaches its maximum. This value is then considered as the crisp output.
  - **Weighted Average Method (for Sugeno FIS):** In Sugeno-type FIS, where the output is a function of the input values, a weighted average of the output values from different rules is calculated based on their degrees of activation.
- **Generate Crisp Output:** The selected method yields a single numeric value that represents the crisp output of the FIS. This output is now a concrete result that can be utilized in subsequent processes, such as control actions or decision-making.

Defuzzification is a crucial step in the FIS process as it provides a clear, actionable output based on the fuzzy logic reasoning performed earlier in the system. The particular needs of the application and the intended interpretation of the fuzzy output are often determining factors when selecting a defuzzification technique.

2.3.3 FIS Applications in Engineering

FISs have many applications in engineering because they offer a flexible and effective way to understand and handle situations where uncertainty and imprecision are widespread (Ahmed et al., 2017). Figure 2.6 shows a few examples of applications from various technical disciplines as follows:

- **Control Systems:** In control systems, FLCs are frequently employed when it is difficult to develop accurate mathematical models for a process. Applications include motor speed control, level control in chemical processes, and temperature control in HVAC (heating, ventilation, and air conditioning) systems.

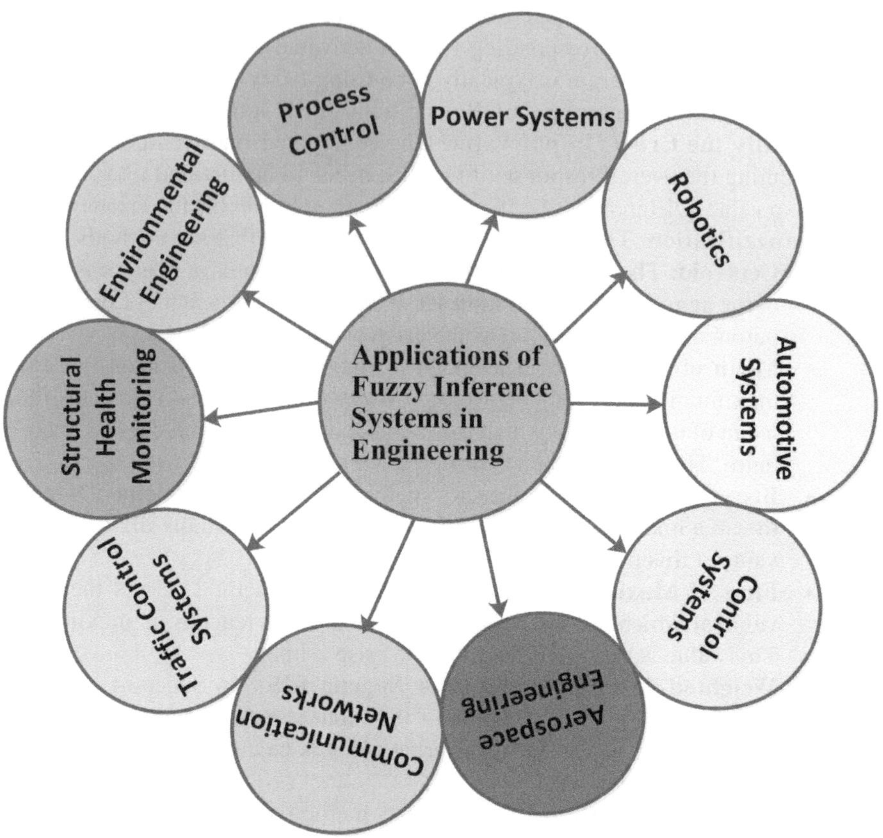

FIGURE 2.6 Some of the major applications of fuzzy inference systems in engineering.

- **Automotive Systems:** Engine control, automatic transmissions, antilock brake systems, and other automotive systems all make use of fuzzy logic. Because FLCs can adapt to changing road conditions, they improve vehicle performance, safety, and fuel efficiency.
- **Robotics:** In robotics, fuzzy logic is used to plan paths, avoid obstacles, and make decisions in changing contexts. By enabling real-time decision-making based on sensor inputs, FIS enhances the adaptability and autonomy of robots.
- **Power Systems:** To estimate demand, manage energy, and detect faults, fuzzy logic is employed in power systems. Fuzzy inference takes into account variables including generation capacity, system stability, and demand to assist power grids in operating as efficiently as possible.
- **Process Control:** Fuzzy logic is used for control and optimization in a variety of industrial processes to increase productivity and sustain intended performance under a variety of circumstances in settings like manufacturing, wastewater treatment, and chemical plants.

- **Environmental Engineering:** For tasks like pollution control, water treatment, and air quality monitoring, fuzzy logic is used in environmental engineering. Fuzzy inference assists with decision-making for sustainable practices, which can model intricate interactions between environmental elements.
- **Structural Health Monitoring:** To evaluate the state of bridges, buildings, and other infrastructure, fuzzy logic is employed in structural health monitoring. To identify abnormalities, evaluate structural integrity, and forecast maintenance requirements, fuzzy inference can interpret sensor data.
- **Traffic Control Systems:** Intelligent transportation systems use fuzzy logic to optimize routes, manage traffic, and control traffic signals. By adjusting to shifting traffic circumstances, fuzzy inference can lessen congestion and enhance overall traffic flow.
- **Communication Networks:** In the field of telecommunications, fuzzy logic is employed for fault detection, quality of service management, and dynamic bandwidth allocation. Fuzzy inference can enhance the efficiency of communication systems and maximize network resources.
- **Aerospace Engineering:** Autopilot systems, guiding, navigation, and control are among the functions in aerospace systems where fuzzy logic is utilized. Fuzzy inference can adjust to changing conditions and manage uncertainty in-flight situations.

These examples show the flexibility of fuzzy inference for handling challenging engineering problems in various fields. For engineers looking for reliable and flexible solutions in practical applications, fuzzy logic is an invaluable tool due to its capacity to represent and manage uncertainty.

2.4 FUZZY MCDM

Fuzzy multicriteria decision-making (MCDM) is a sophisticated approach that combines fuzzy logic with decision-making strategies to tackle intricate issues with multiple criteria (Kaya et al., 2019). With this method, fuzzy sets are used to define criteria and alternatives, enabling decision-makers to communicate vague and subjective information using language.

Decision-makers can genuinely evaluate the links between criteria and alternatives by using fuzzy pairwise comparisons and fuzzy weights, which are used to express the ambiguity in the importance of various criteria. Because the input data is fuzzy, fuzzy aggregation combine these individual evaluations to get overall fuzzy ratings for each option. Fuzzy MCDM approaches, including fuzzy TOPSIS (technique for order of preference by similarity to ideal solution) and fuzzy analytic hierarchy, expand on conventional methods to address imprecise language expressions and judgments. The method is particularly useful in real-world situations when there is a lot of uncertainty, inconsistency, and qualitative evaluations. Many industries, such as banking, environmental management, project selection, and resource allocation, have successfully used fuzzy multicriteria cost-benefit analysis. This technique gives decision-makers an effective tool for negotiating challenging decision-making scenarios.

2.4.1 THE CONCEPT OF MCDM

MCDM was developed to handle circumstances in which there are multiple, frequently opposing, criteria or elements to consider. Traditionally, choices are usually made without explicitly taking into account the interdependencies between a group of criteria. On the other hand, MCDM offers a structured methodology for assessing and ranking options based on a variety of criteria, acknowledging the complexity of decision problems. The following are important ideas related to MCDM:

- **Multiple Criteria:** MCDM addresses decision-making situations in which decision-makers must take into account multiple factors. These standards can take multiple forms, encompassing quantitative and qualitative elements, expenses, advantages, dangers, and additional pertinent characteristics.
- **Decision Alternatives:** MCDM entails assessing and contrasting a range of possibilities or decision alternatives. These options stand for various plans of action, approaches, initiatives, or fixes for a certain issue.
- **Criteria Weights:** Not all MCDM criteria have the same weight. Weights are assigned by decision-makers to factors to indicate their relative importance. The significance of every criterion in the decision-making process is measured by these weights.
- **Pairwise Comparisons:** Decision-makers frequently have to state which criteria they believe to be more important than others in the decision-making process. When comparing two sets of criteria side by side, pairwise comparisons are used to determine which is preferred or of more value.
- **Ranking and Scoring:** MCDM offers a system for ranking or scoring for the alternatives based on the allocated weights and criteria evaluations. The aim is to find the most preferred or ideal option according to the decision-maker's objectives.
- **Decision Matrix:** An essential MCDM tool, a decision matrix arranges and displays data on criteria assessments, weights, and alternative ranks. It presents the choice problem in an organized manner.
- **Decision Support Systems:** To facilitate the evaluation and analysis of alternatives according to a range of criteria, MCDM commonly employs decision support systems, including computational models and software tools. These tools support decision-makers in navigating difficult decisions.
- **Sensitivity Analysis:** By using MCDM, one can assess how changes to the values or weights of criteria affect the final rankings. This aids in decision-makers' comprehension of how resilient their choices are in various situations.
- **Subjectivity and Stakeholder Involvement:** MCDM promotes stakeholder involvement while acknowledging the subjectivity that is a part of decision-making. The preferences, values, and viewpoints of the stakeholders are vital in determining the criteria, weights, and general decision-making process.

MCDM is used in many different domains, including engineering, project management, business, finance, and environmental management. It gives decision-makers a clear, methodical framework for managing intricate decision-making situations involving multiple, frequently incompatible criteria.

2.4.2 TRADITIONAL MCDM METHODS AND LIMITATIONS

Decision-makers have long relied on traditional MCDM techniques to help them assess and choose options based on a variety of criteria (Taherdoost & Madanchian, 2023). These techniques do have certain restrictions, though. Here are a few popular classical MCDM techniques along with their disadvantages:

- **Weighted Sum Model:** This model calculates the total score for each alternative by multiplying the weights by the normalized scores. Each criterion is given a weight that indicates its significance. The subjectivity of weight assignment, possible bias, and the presumption of linear connections among criteria are some of the shortcomings.
- **Weighted Product Model:** The normalized scores for each criterion increased to their corresponding weight and are multiplied to determine the total score for each choice in the Weighted Product Model. Sensitivity to extreme values, the presumption of criteria independence, and difficulties determining acceptable weights are some of the limitations.
- **Simple Additive Weighting:** The third method is called simple additive weighting, which combines criteria by multiplying the normalized score of each criterion by its weight and adding the results. Sensitivity to weight fluctuations and the presumption that criteria are equally significant within a given weight category are two limitations.
- **Analytic Hierarchy Process:** Analytic hierarchy creates a structured hierarchical framework for decision-making by comparing criteria in pairs. The process reduces the possibilities for inconsistency, the difficulty of achieving consistent judgments in paired comparisons, and other limitations in handling a large number of criteria.
- **ELECTRE (Elimination et Choix Traduisant la Realité):** This group of outranking techniques evaluates substitutes according to how well or poorly they meet predetermined standards. Sensitivity to shifts in preference thresholds, challenges with handling imprecise information, and the possibility of rank reversal are some of the limitations.
- **TOPSIS:** TOPSIS arranges options according to how close they are to the positive ideal solution and how far they are from the negative ideal solution. Sensitivity to the normalization method selected, reliance on predetermined ideal solutions, and susceptibility to extreme values are some of the limitations.
- **Compromise Programming:** This technique looks for solutions that strike a balance between opposing requirements. Limitations include the impact of decision-makers' differing tastes and the challenge of identifying a suitable compromise option.
- **Limitations across Traditional MCDM Methods:**
  - **Subjectivity:** The dependability of the results might be impacted by the subjectivity introduced by traditional approaches, which frequently rely

on subjective inputs such as pairwise comparisons, preference values, and criteria weights.

- **Inconsistency:** Unreliable rankings and decision results can result from inconsistent judgments or weights.
- **Sensitivity to Parameter Changes:** Conventional techniques, such as weights, thresholds, or normalization, may be sensitive to modifications in parameter values.
- **Handling Uncertainty:** These techniques can make it difficult to deal with ambiguous information and uncertainties when making decisions.

Although decision science has historically relied on classic MCDM methods, resolving these constraints necessitates careful thought and, in certain situations, the use of more sophisticated approaches, including fuzzy MCDM or contemporary decision-making strategies.

2.4.3   Introduction to Fuzzy MCDM

Fuzzy MCDM improves on conventional techniques by using fuzzy logic to address the uncertainties and imprecise information present in decision-making procedures (Abdullah, L., 2013). When criteria, preferences, and data are not well defined, fuzzy MCDM offers a more practical and adaptable framework for assessing and prioritizing choices based on multiple factors. The goal of this method is to use fuzzy set theory to capture and model the inherent subjectivity and ambiguity in decision-making. Fuzzy MCDM is useful in many domains where decision-makers must navigate complicated and ambiguous choice landscapes, such as finance, engineering, environmental management, and decision support systems. This method allows for more robust and nuanced choice outcomes in the presence of imprecise information by utilizing fuzzy logic to describe decision-making scenarios in a more realistic and nuanced manner.

2.4.4   Addressing Imprecise Criteria Through Fuzzy MCDM Approaches

The issues presented by imprecise criteria in decision-making processes are explicitly addressed by fuzzy MCDM techniques. Due to the subjectivity, ambiguity, and uncertainty present in real-world decision-making situations, imprecise criteria frequently result. Fuzzy MCDM offers a methodical framework for efficiently modeling and managing imprecise criteria (Mohamed et al., 2023). The following are the main ways that fuzzy MCDM techniques deal with imprecise criteria:

- **Fuzzy Sets and Linguistic Terms:** A key idea in fuzzy MCDM is fuzzy sets, which give decision-makers a way to articulate and represent fuzzy criteria in language. Fuzzy membership functions that represent the degrees to which items belong to particular language phrases (such as "high," "medium," and "low") are used to construct criteria rather than strict categories. This makes it possible to convey vague information in a more sophisticated manner.

- **Fuzzy Weighting:** By enabling decision-makers to give fuzzy weights to criteria, fuzzy MCDM takes into account imprecise preferences. Using fuzzy numbers, decision-makers can convey the ambiguity or uncertainty surrounding the weights of certain criteria instead of presenting clear-cut weights. This eliminates the need for exact quantitative evaluations and reflects the ambiguous nature of their preferences.
- **Fuzzy Pairwise Comparison:** To address imprecise judgments, fuzzy MCDM expands on the conventional pairwise comparison. Using ambiguous language phrases, decision-makers can convey the degree of preference or dominance of one criterion or option over another. This makes it possible to depict erroneous and subjective judgments more realistically.
- **Fuzzy Aggregation:** To integrate fuzzy assessments, fuzzy logic operators including the minimum, maximum, and average operators are used in the aggregation step. These operators make sure that the aggregation process reflects the uncertainty associated with imprecise criteria while taking into account the imprecise nature of the incoming data. Fuzzy sets that reflect the overall evaluation of options based on multiple imprecise criteria are produced by aggregation.
- **Fuzzy Inference Systems:** To address imprecise reasoning and decision-making, fuzzy inference is incorporated into fuzzy MCDM models. Decision-makers can model and reason under uncertainty, imprecision, and subjectivity using fuzzy inference, which uses fuzzy rules to infer conclusions from fuzzy input information.
- **Fuzzy TOPSIS, Fuzzy AHP, and Other Fuzzy Variants:** To deal with imprecise criteria, fuzzy MCDM comprises modifications of conventional MCDM techniques. More reliable decision outcomes are produced by fuzzy versions of techniques like fuzzy TOPSIS and fuzzy analytic hierarchy, which take into account imprecision in criteria evaluations and preferences.
- **Sensitivity Analysis:** With fuzzy MCDM, sensitivity analysis can be performed to evaluate how modifications to linguistic terms, criteria weights, or other parameters affect the results of the final decision. This helps decision-makers comprehend how resilient their choices are to various outcomes and degrees of imprecision.

Fuzzy MCDM offers a comprehensive and efficient framework for handling imprecise criteria in decision-making by using fuzzy logic concepts and combining language phrases, fuzzy weights, and fuzzy operators. Decision-makers can now navigate complex and uncertain decision-making situations with greater assurance.

2.5 FUZZY GOAL PROGRAMMING

To deal with uncertainty and imprecision in decision-making, fuzzy goal programming (FGP) incorporates fuzzy logic into conventional goal programming. FGP uses fuzzy sets and linguistic concepts to express goals and constraints (Rivaz et al., 2020). This enables decision-makers to convey their goals more flexibly and realistically. Fuzzy objectives allow for imperfect boundaries on decision variables, whereas fuzzy restrictions allow for subjectivity and ambiguity in the objective specification. To reach a compromise solution, the strategy seeks to strike a balance between conflicting and ambiguous objectives.

Fuzzy numbers represent an attempt to limit departures from fuzzy aims. FGP models are solved using optimization strategies that account for the fuzzy nature of the problem. This provides decision-makers with a set of compromise solutions and permits sensitivity analysis. Many domains, including project management, finance, production planning, and resource allocation, employ fuzzy goal programming to help decision-makers deal with imprecise data, conflicting objectives, and the need for flexible goal formulation.

2.5.1 Understanding Fuzzy Goal Programming

FGP stands as an advanced extension of conventional goal programming, integrating fuzzy logic to effectively manage imprecision and uncertainty inherent in multi-objective decision-making scenarios (Tiwari et al., 1987). Figure 2.7 shows the different steps followed by FGP to make decisions in various scenarios.

FGP introduces the application of fuzzy sets and linguistic terms, such as "low," "medium," and "high," to express objectives as fuzzy goals, accommodating the inherent vagueness in goal specifications. Fuzzy constraints, characterized by fuzzy sets and membership functions, are employed to model imprecise limitations on decision variables. FGP seeks an optimal compromise solution that minimizes deviations from fuzzy goals, measured using fuzzy numbers while adhering to fuzzy constraints.

FGP iteratively traverses the terrain of fuzzy objectives and constraints using optimization techniques to produce a set of compromise solutions. Sensitivity analysis gives decision-makers important information for sound decision-making by enabling them to assess how modifications to ambiguous goals and constraints affect the best possible outcome. Applications of FGP may be found in many different fields, including finance, production planning, and resource allocation. It excels in situations where there is a lack of accurate information, competing goals, and a requirement for flexible goal specification. This methodology proves to be a powerful instrument, providing decision-makers with a sophisticated and flexible way to successfully traverse difficult and unpredictable choice environments.

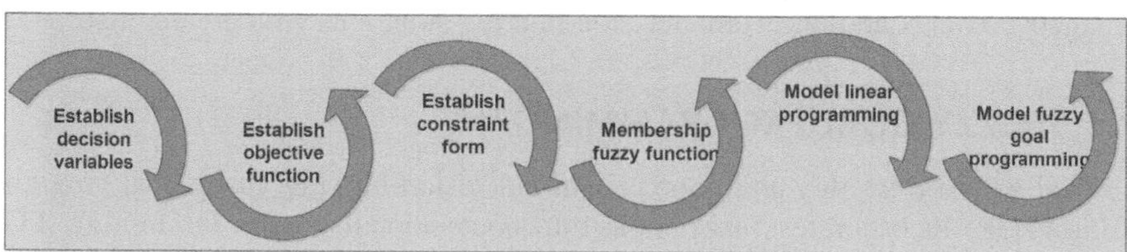

FIGURE 2.7   Steps followed by fuzzy goal programming.

2.5.2 Integrating FGP with MCO

Integrating MCO with FGP creates a powerful approach to decision-making, particularly in intricate and uncertain scenarios (Komsiyah et al., 2018). This integration combines the flexibility of FGP in handling unclear goals and limitations with the rigorous evaluation of many MCO criteria. The primary components of this integration are as follows:

- **Fuzzy Representation of Objectives and Constraints:** FGP enables fuzzy goal and fuzzy constraint representations. This considers the uncertainty and imprecision resulting from the decisions and limitations made by decision-makers.
- **Incorporation of Multiple Criteria:** Multicriteria optimization involves evaluating multiple criteria at the same time when each reflects a different aspect of the decision-making problem. Finding a compromise approach that best balances these conflicting goals—which may have different objectives— is the goal.
- **Fuzzy Aggregation of Criteria:** FGP and MCO can be joined by mixing the different criteria in an unclear manner. Fuzzy logic operators are utilized to aggregate the fuzzy assessments of every criterion, including the minimum, maximum, and average operators, allowing for thoroughly evaluating the options is using the integrated fuzzy criteria.
- **Joint Optimization:** Taking into account the trade-offs and synergies between the various criteria, integrating aims to cooperatively optimize the fuzzy goals and constraints. This enables decision-makers to investigate options that accomplish a balanced performance across a variety of metrics in addition to satisfying vague goals.
- **Solving Complex Decision Landscapes:** FGP combined with MCO is especially helpful when decisions must be based on uncertainty regarding the relative relevance of criteria as well as diverse objectives. The integration makes it easier to explore the choice space holistically, taking into account the effects of multiple criteria as well as vague goals.
- **Adaptive Decision Support:** Decision-makers can benefit from adaptive decision assistance by integrating FGP and MCO. It makes it possible to investigate compromise alternatives that optimize across a variety of parameters and are in line with hazy goals. Sensitivity analysis can be used to determine how resilient the integrated solution is to changing circumstances.
- **Applications in Real-World Scenarios:** This integrated approach is useful in a variety of real-world settings where decision-makers must weigh several, frequently contradictory criteria and deal with imprecise information, such as project management, resource allocation, and strategic planning.

The amalgamation of FGP and MCO capitalizes on the merits of both approaches, providing an all-encompassing framework for decision-making that tackles the intricacies of practical issues. When decision-makers are searching for solutions that satisfy imprecise objectives while simultaneously optimizing performance across a range of metrics in a unified and flexible manner, this integration is particularly beneficial.

2.5.3   HANDLING CONFLICTING OBJECTIVES USING FGP

Managing conflicting goals is a common issue in decision-making processes. When goals are imprecise and ambiguous, fuzzy goal programming provides a dependable framework for negotiating trade-offs between competing aims (Zare et al., 2021). Fuzzy goal programming allow decision-makers to represent conflicting goals by capturing the ambiguity and imprecision in their specifications. To specify each competing goal, fuzzy sets, and linguistic terms allow for a more flexible representation of the choice environment using the following processes:

- **Deviation Variables and Fuzzy Numbers:** FGP adds deviation variables to gauge the extent to which each goal is achieved or not. These variances are measured using fuzzy numbers, which take into consideration the imprecision in defining objective performance. The fuzzy numbers' range of possible deviations captures the uncertainty that exists during the decision-making process.

- **Objective Hierarchy and Priority Weights:** FGP has an objective hierarchy that prioritizes conflicting goals based on their relative relevance. Each goal is given a priority weight by decision-makers indicating its relative importance. FGP seeks a compromise by balancing competing goals based on their priorities, guided by the objective hierarchy.

- **Optimization:** Taking into account the priority weights given to competing objectives, FGP seeks to minimize deviations from the fuzzy goals. The process of optimization entails weighing the inherent trade-offs and determining a compromise solution that strikes a balance between competing objectives.

- **Trade-Off Analysis:** FGP gives decision-makers information about the compromises that come with having competing goals. Decision-makers can comprehend the effects of giving one aim priority over another and make well-informed judgments based on the trade-offs by looking at the compromise options produced by the model.

- **Sensitivity Analysis:** FGP facilitates evaluating the effects of modifications to fuzzy goals and priority weights on the best compromise solution. Decision-makers can make changes depending on changing conditions or preferences by understanding the resilience of the solution with the use of this study.

- **Adaptive Decision-Making:** By enabling decision-makers to repeatedly hone their priorities and objectives, FGP promotes adaptive decision-making. Decision-makers can review the FGP model again to investigate new compromise alternatives that fit their changing goals if preferences or outside circumstances change.

- **Applications in Real-World Scenarios:** FGP has been effectively implemented in multiple fields in which competing goals are frequent, such as strategy planning, project management, and resource allocation. The approach is a useful tool in complex decision-making because it can handle competing goals while taking imprecision into consideration landscapes.

In conclusion, fuzzy goal programming, which combines fuzzy representations, prioritization, and optimization approaches, offers a methodical and adaptable way to handle conflicting objectives. It helps decision-makers balance competing objectives while considering the trade-offs and inherent uncertainties seen in real-world decision-making situations.

2.5.4 Applications of FGP in Engineering

There are several uses for FGP in engineering contexts where decision-makers have to make difficult choices with conflicting objectives and incomplete information (Malik et al., 2023). Here are some notable examples of fuzzy goal programming in engineering:

- **Project Management:** By taking into account fuzzy goals about project completion time, cost, and resource utilization, FGP is employed in project management to optimize project scheduling, resource allocation, and task assignment. FGP can be used to balance conflicting objectives, such as reducing expenses while prolonging the project's timeline.
- **Supply Chain Optimization:** In engineering supply chain management, FGP helps optimize choices about inventory control, manufacturing scheduling, and distribution. Conflicting goals like minimizing costs, maximizing customer satisfaction, and reducing lead times are common in supply chain scenarios, and FGP provides a framework for achieving a balanced solution.
- **Process Optimization:** FGP is used in chemical and manufacturing engineering for process optimization. It helps in balancing conflicting objectives related to production rates, energy consumption, and environmental impact. Fuzzy goals can be set to account for uncertainties in raw material properties or equipment performance.
- **Energy Management:** FGP is employed in energy systems engineering to optimize energy production, distribution, and consumption. Conflicting goals, such as maximizing energy efficiency while minimizing costs or environmental impact, can be addressed using FGP to find compromise solutions under uncertainty.
- **Water Resources Management:** In civil and environmental engineering, FGP assists in water resources management by addressing conflicting goals related to water allocation, reservoir operation, and environmental conservation. Fuzzy goals may include uncertainties in water demand projections and climate variability.
- **Structural Design and Optimization:** FGP is utilized in structural engineering for optimizing the design of structures, considering conflicting goals such as minimizing material costs while ensuring safety and structural

integrity. Fuzzy goals can capture imprecise information related to material properties or loading conditions.

- **Traffic Management:** FGP is applied in transportation engineering to optimize traffic flow, signal timing, and route planning. Conflicting goals, such as minimizing travel time while reducing fuel consumption or emissions, can be addressed using FGP in the presence of uncertainties in traffic patterns.
- **Quality Control in Manufacturing:** FGP is used for quality control in manufacturing engineering, where conflicting goals involve minimizing defects and production costs. Fuzzy goals can represent imprecise information related to variations in material properties or manufacturing processes.
- **Optimal Design in Aerospace Engineering:** In aerospace engineering, FGP assists in optimal design processes considering conflicting goals related to weight reduction, structural integrity, and cost efficiency. Fuzzy goals can account for uncertainties in material properties and loading conditions.
- **Robotic Systems Design:** FGP is applied in the design and optimization of robotic systems, addressing conflicting goals such as maximizing operational efficiency while minimizing energy consumption or production costs. Fuzzy goals can capture uncertainties in task specifications or environmental conditions.

In these engineering scenarios, FGP proves valuable for decision-makers by providing a systematic and flexible approach to handling imprecise information, uncertainties, and conflicting objectives, ultimately contributing to more robust and adaptive solutions.

2.6 FUZZY TOPSIS

Fuzzy TOPSIS is an advanced decision-making method that extends the traditional TOPSIS approach by incorporating fuzzy logic to handle uncertainty and imprecision. This method begins by representing criteria and alternatives using fuzzy sets and linguistic terms, allowing decision-makers to express the degree of membership with greater nuance. A fuzzy decision matrix is then normalized to ensure comparable and equally weighted criteria. The most and least acceptable results across all criteria are represented by fuzzy ideal and anti-ideal solutions. The relative proximity of each alternative to the optimal solution is determined by fuzzy TOPSIS using similarity measurements, such as the Euclidean distance modified for fuzzy sets. The final ranking, which shows the alternative that is most favored, is produced using these closeness indicators. When evaluating criteria involves inherent ambiguity and imprecision, this method provides decision-makers with a flexible and reliable instrument for making judgments based on several criteria.

2.6.1 INTRODUCTION TO FUZZY TOPSIS

Fuzzy logic is a complex technique of decision-making that builds upon the TOPSIS method to manage imprecisions and ambiguities in multicriteria decision scenarios

(Palczewski & Salabun, 2019). Fuzzy TOPSIS is based on fuzzy sets and language expressions' capacity to represent and analyze imprecise data. Unlike traditional methods of decision-making, which typically rely on discrete numerical values, fuzzy TOPSIS enables decision-makers to convey their subjective assessments using language, such as "low," "medium," and "high." This allows for a more detailed and accurate description of their tastes.

Using fuzzy sets, the approach calculates the similarity between each alternative and the ideal and anti-ideal solutions. This determines the options' relative proximity, which determines the preferred order. Fuzzy TOPSIS delivers a robust and adaptable framework for ranking and making decisions about preferences in composite choice settings. It functions especially well in decision-making scenarios when criteria evaluations are inevitably inaccurate or ambiguous.

2.6.2 Steps Involved in Fuzzy TOPSIS

Fuzzy TOPSIS goes through several steps in MCDM before arriving at a preference order or choice among a set of alternatives. Figure 2.8 shows the main steps that make up fuzzy TOPSIS, and we discuss them in more detail as well.

- **Fuzzy Decision Matrix:** Make a fuzzy decision matrix first, with each row denoting an option and each column denoting a criterion. To capture imprecision and uncertainty, the degree of membership of each alternative about each criterion is expressed using fuzzy sets and linguistic expressions.
- **Normalize the Fuzzy Decision Matrix:** To make sure that every criterion is equally important and on a comparable scale, normalize the fuzzy decision matrix. The fuzzy TOPSIS algorithm's subsequent computations depend on this phase. Linear transformation and vector normalization are two popular techniques.
- **Fuzzy Ideal and Anti-Ideal Solutions:** Compile the best and worst results for each criterion to define the fuzzy ideal and anti-ideal solutions. To do this, each criterion's maximum and lowest values must be determined. Fuzzy sets representing the ideal and anti-ideal solutions must then be created.
- **Similarity Measures:** Using the proper similarity measures, determine how similar each alternative is to the fuzzy ideal and anti-ideal solutions. The Euclidean distance, the Minkowski distance, or other distance metrics modified to handle fuzzy sets are examples of common measures. The degree to which each option resembles the ideal and anti-ideal solutions is measured by the similarity metrics.
- **Relative Closeness to the Ideal Solution:** Assess the degree to which each option is similar to the ideal and anti-ideal solutions. Usually, this is accomplished by using a criterion for decision-making that takes into account both the distance from the anti-ideal solution and the proximity to the ideal answer. The relative closeness measure and the closeness coefficient are popular techniques.

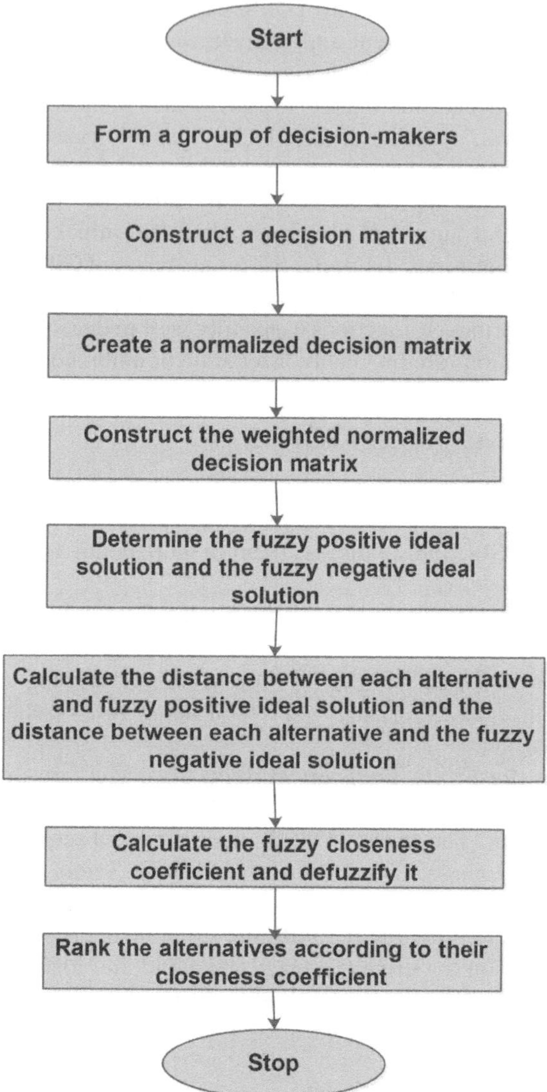

FIGURE 2.8 Steps in fuzzy TOPSIS.

- **Ranking Alternatives:** Sort the options according to how close they are to each other. In the decision-making process, the option with the highest closeness value is regarded as the most favored or ideal option, reflecting its priority.
- **Sensitivity Analysis:** Perform sensitivity analysis to assess the robustness of the rankings to changes in fuzzy membership values or weights assigned to criteria. This step helps decision-makers understand the stability of the fuzzy TOPSIS results under different scenarios.

Fuzzy TOPSIS provides decision-makers with a systematic and comprehensive approach to handling imprecise information and conflicting criteria in multicriteria decision-making situations. It allows for a nuanced representation of preferences and effectively balances the trade-offs among different criteria, making it a valuable tool in complex decision landscapes.

2.6.3 Fuzzy TOPSIS in Engineering Decision-Making

Fuzzy TOPSIS has found diverse applications in engineering decision-making where multiple criteria need to be considered, and imprecise information is prevalent (Irvanizam, I., 2018). Here are some notable applications of fuzzy TOPSIS in engineering.

- **Supplier Selection:** In supply chain and manufacturing engineering, fuzzy TOPSIS is employed to evaluate and rank potential suppliers based on criteria such as cost, quality, reliability, and delivery time. The method accommodates uncertainties in supplier performance and provides a systematic approach to supplier selection.
- **Material Selection:** Fuzzy TOPSIS aids in selecting materials for engineering projects by considering criteria such as mechanical properties, cost, and environmental impact. The imprecise nature of material properties and the need to balance conflicting criteria make fuzzy TOPSIS well suited for material selection processes.
- **Project Risk Management:** In civil engineering and construction, fuzzy TOPSIS assists in project risk management by evaluating and ranking alternative risk mitigation strategies. Criteria may include cost, time, and the effectiveness of risk reduction measures, and fuzzy TOPSIS helps in decision-making under uncertainty.
- **Environmental Impact Assessment:** Fuzzy TOPSIS is applied to assess the environmental impact of engineering projects. Criteria related to air and water quality, habitat preservation, and energy consumption are considered, and fuzzy TOPSIS facilitates the ranking of alternatives with imprecise environmental consequences.
- **Energy System Planning:** Fuzzy TOPSIS is utilized for planning and selecting optimal energy systems. Criteria may include cost, efficiency, environmental impact, and reliability. Fuzzy TOPSIS helps decision-makers navigate the complexities of energy system planning under uncertain and imprecise conditions.
- **Optimal Design of Structures:** Fuzzy TOPSIS assists structural engineers in optimizing designs by considering criteria such as cost, safety, and environmental impact. The method helps balance conflicting design objectives and considers uncertainties in material properties and design parameters.
- **Water Resources Management:** In civil and environmental engineering, fuzzy TOPSIS is applied to evaluating and ranking alternative strategies for water allocation, reservoir operation, and conservation efforts, considering conflicting criteria and uncertainties in hydrological data.

- **Manufacturing Process Optimization:** Fuzzy TOPSIS is employed in manufacturing engineering for optimizing production processes. Criteria may include production cost, efficiency, and product quality, and fuzzy TOPSIS aids decision-makers in selecting the most favorable process configuration under uncertain manufacturing conditions.
- **Transportation System Planning:** Fuzzy TOPSIS is applied in transportation engineering for planning and selecting transportation systems. Criteria may include cost, environmental impact, and safety. Fuzzy TOPSIS assists in ranking alternatives, considering uncertainties in traffic patterns and infrastructure conditions.
- **Renewable Energy Project Selection:** Fuzzy TOPSIS is used in evaluating and selecting renewable energy projects by considering criteria such as energy output, cost-effectiveness, and environmental impact. The method assists in ranking alternative projects and supports decision-making in the renewable energy sector.

In these engineering applications, fuzzy TOPSIS provides a systematic and effective approach to handling imprecise information, uncertainties, and conflicting criteria, ultimately aiding decision-makers in selecting optimal solutions in complex decision landscapes.

2.6.4 CASE STUDIES DEMONSTRATING THE ADVANTAGES OF FUZZY TOPSIS

In the dynamic landscape of business strategy, the utilization of case studies proves instrumental in gaining nuanced insights and shaping informed decision-making. Here, we offer two hypothetical scenarios to illustrate the advantages of fuzzy TOPSIS in diverse engineering decision-making contexts (Sun & Lin, 2009). Next, we describe a real-world case study in depth.

2.6.4.1 Supplier Selection in Manufacturing

In this example, a manufacturing company needs to select a supplier for a critical component used in its production process. The supplier's financial stability, cost, quality, and delivery time are just a few of the many factors that go into the decision. However, the information that is accessible is vague, with vendors offering hazy approximations and subjective judgments. The company can use fuzzy sets and linguistic phrases to represent the imprecision in supplier evaluations thanks to fuzzy TOPSIS. Fuzzy similarity metrics account for supplier performance uncertainty, making it possible for the decision-makers to evaluate the providers according to how near they are to the ideal source while taking into account all of the benefits and drawbacks.

2.6.4.2 Optimal Design of Sustainable Structures

Creating a sustainable building that balances construction costs, environmental impact, and structural safety is the job of an engineering business. The design team must balance competing goals, and because site conditions and material qualities vary, the data provided for environmental assessments is imprecise: The design team can use fuzzy TOPSIS to represent the imprecision in environmental requirements and to compare and normalize contradictory criteria, such as reducing construction costs while maintaining environmental impact. The approach computes the relative proximity of various designs to the optimal sustainable structure while accounting for uncertainty in material qualities.

These fictitious case studies show the benefits of fuzzy TOPSIS in managing imprecise information, accounting for uncertainties, and offering decision-makers a methodical and adaptable approach to MCDM. The approach is a useful tool for handling the complexity of real-world decision landscapes because it may combine fuzzy sets, linguistic concepts, and subjective assessments.

2.7   A REAL-WORLD CASE STUDY

In this extended case study, we delve into a specific instance where the application of case study methodology significantly contributed to the refinement of business strategies, providing detailed details and a comprehensive analysis.

Context: A leading multinational corporation facing challenges in a fiercely competitive market sought to enhance its market positioning and overall performance. The decision-makers recognized the need for a strategic overhaul but were confronted with the complexities of devising an effective plan in the ever-evolving business environment.

Case Study Details: We focused on a critical period in the corporation's history, examining the challenges faced, the strategic decisions made, and the subsequent outcomes. We scrutinized facets including market trends, competitor analyses, internal organizational dynamics, and customer feedback.

- **Challenges Identified:**
  - The analysis unveiled challenges related to shifting consumer preferences, emerging technologies, and intensified competition.
  - Internal inefficiencies, communication gaps, and outdated processes were also identified as hindrances to the company's success.
- **Strategic Decision-Making:**
  - Through an extensive examination of the case, strategic decisions such as diversification, digital transformation, and organizational restructuring emerged.
  - We detailed the rationale behind each decision, considering both internal and external factors.
- **Implementation and Results:**
  - The company implemented the proposed strategies, with a focus on agility, innovation, and customer-centricity.
  - Performance metrics, market share, and financial indicators were meticulously tracked to assess the impact of the strategic changes.

Comprehensive Analysis

- **Successes and Failures:**
  - We were able to highlight successful aspects of the implemented strategies, such as increased market share and improved customer satisfaction.
  - Failures, including resistance to change and initial operational disruptions, were analyzed to extract valuable lessons.

- **Adaptability and Learning:**
  - The corporation's ability to adapt to market dynamics and learn from both successes and failures was a key aspect of the analysis.
  - Insights into the importance of a learning culture and agile decision-making processes were drawn from the case study.
- **Long-Term Implications:**
  - The long-term implications of the implemented strategies were assessed, considering sustainability, scalability, and future market trends.
  - Lessons learned from this case study informed the company's ongoing strategic planning and decision-making processes.

Specific Criteria Considered in Fuzzy System Case Studies

1. Precision and Uncertainty:
- Criteria might include the level of precision required in a given engineering task and the uncertainties involved. Fuzzy systems excel in situations where uncertainties and imprecisions are inherent.

2. Adaptability to Dynamic Environments:
- Evaluation of how well the optimization technique adapts to dynamic and changing conditions. Fuzzy systems are known for their adaptability, making them suitable for scenarios with varying parameters.

3. Complexity of Relationships:
- Examination of the complexity in the relationships between variables. Fuzzy systems are favored when dealing with nonlinear relationships that might be challenging for traditional optimization techniques.

4. Handling Multivariable Systems:
- Assessment of the optimization technique's capacity to handle systems with multiple variables and interconnected parameters. Fuzzy systems are effective in managing complex relationships in multivariable systems.

Decision-Making Processes Employed

1. Problem Formulation:
- Decision-makers articulate the goals, constraints, and variables involved in the system.

2. Model Development:
- Developing the fuzzy system model involves defining linguistic variables, membership functions, and fuzzy rules. This step is crucial for accurately representing the system's behavior.

3. Rule Base Construction:
- Construction of a rule base that captures the decision-making logic of the fuzzy system. This step involves incorporating expert knowledge and domain-specific insights.

4. Inference and Defuzzification:
- Application of fuzzy rules to input data and then defuzzification to convert fuzzy outputs into crisp values. These steps determine the system's decision.

5. Validation and Testing:
- Rigorous validation and testing of the fuzzy system model to ensure its accuracy and reliability in real-world scenarios.

Comparative Analysis of Results

1. Performance Metrics:
- Performance metrics between fuzzy system approaches and traditional optimization techniques may include efficiency, accuracy, and computational speed.

2. Robustness:
- Fuzzy systems are often valued for their robustness in the face of uncertainties.

3. Resource Utilization:
- Considering computational resources and time helps in understanding the efficiency and practicality of each approach.

4. Sensitivity Analysis:
- Determining how changes in input parameters affect the performance of each optimization technique.

5. Real-World Applicability:
- Comparing the applicability of fuzzy systems and traditional optimization techniques in real-world engineering scenarios considering the adaptability of each approach to practical challenges.

This detailed case study not only provided a retrospective view of the challenges faced by the corporation but also served as a blueprint for businesses navigating similar complexities. The analysis offered actionable insights, emphasizing the importance of adaptability, continuous learning, and strategic agility in a rapidly changing business landscape. Through the lens of this case study, readers gain a holistic understanding of the intricacies involved in refining business strategies, making vital links between theoretical concepts and practical applications. It's important to note that the actual details of the criteria and analysis would depend on the specific engineering domain, the nature of the optimization problem, and the goals of the study.

2.7.1 Fuzzy Systems in Engineering Domains

Fuzzy systems have found diverse applications across various engineering domains, leveraging their ability to model and manage uncertainties, imprecisions, and complex relationships (Gupta, 2021; Vyas et al., 2022). Some notable applications include:

- **Control Systems:** Fuzzy logic control systems are extensively used in engineering for applications like temperature control, speed regulation, and process optimization. Fuzzy controllers provide a more intuitive and flexible approach, particularly in systems where precise mathematical modeling is challenging due to uncertainties.
- **Robotics and Automation:** Fuzzy systems play a crucial role in robotics and automation for tasks such as path planning, object recognition, and robotic decision-making. Fuzzy logic helps robots adapt to dynamic environments and make context-aware decisions based on imprecise sensor data.
- **Process Optimization:** In chemical engineering and manufacturing, fuzzy systems are applied to optimize complex processes. Fuzzy models can

capture the nonlinear relationships and uncertainties in variables, improving efficiency, yield, and quality in industrial processes.

- **Power Systems and Energy Management:** Fuzzy systems are utilized in power systems for load forecasting, energy distribution, and fault detection. They contribute to smart grid technologies by handling uncertainties in renewable energy sources, demand-side management, and energy storage systems.
- **Civil Engineering and Structural Health Monitoring:** Fuzzy logic is applied in civil engineering for structural health monitoring and assessment. Fuzzy systems assist in evaluating the condition of structures based on imprecise sensor data, contributing to early detection of potential issues and improving safety.
- **Environmental Engineering:** Fuzzy systems are employed in environmental engineering for tasks such as air and water quality assessment, pollutant dispersion modeling, and environmental impact analysis. Fuzzy models help handle uncertainties inherent in environmental data and regulatory standards.
- **Traffic Control and Transportation Systems:** Fuzzy logic is used in traffic signal control, route planning, and transportation management systems. Fuzzy systems can adapt to changing traffic conditions and provide effective solutions in scenarios where precise modeling of traffic patterns is challenging.
- **Communication Networks:** Fuzzy systems optimizes tasks like dynamic bandwidth allocation, load balancing, and quality of service management. Fuzzy models aid in handling uncertainties and variations in network traffic.
- **Aerospace Engineering:** Fuzzy systems contribute to aerospace engineering for tasks such as flight control, navigation, and fault diagnosis. They provide adaptive control solutions that can handle variations in aircraft dynamics and environmental conditions.
- **Biomedical Engineering:** Fuzzy logic is applied in biomedical engineering for medical diagnosis, image processing, and decision support systems. Fuzzy systems support dealing with the in-built imprecision and subjectivity in medical data.

These examples demonstrate how flexible fuzzy systems are in tackling the problems presented by ambiguities and imprecisions in a range of engineering fields. Fuzzy systems improve the flexibility, resilience, and effectiveness of engineering solutions in complex and dynamic situations by offering a framework for modeling and reasoning with ambiguous or incomplete data.

2.7.2 OUTCOMES AND BENEFITS OF FUZZY SYSTEMS

Fuzzy system applications have shown numerous significant benefits and positive effects across a variety of fields in engineering, demonstrating their resilience to uncertainty, which is essential in dynamic environments. Fuzzy systems provide flexible frameworks for modeling complicated problems in decision-making contexts, allowing for better informed and effective decision-making processes. Table 2.1

highlights the outcomes and advantages of fuzzy system applications in engineering. Control engineers gain from improved control techniques because fuzzy logic controllers provide a more human-like and intuitive approach, which is especially useful in dynamic environments. By addressing nonlinear interactions and uncertainty, fuzzy systems help optimize complicated processes in chemical engineering and manufacturing, ultimately enhancing productivity and resource consumption. Fuzzy logic in robotics provides enhanced robustness to systems by accepting incorrect sensor data and generating context-aware decisions.

In conclusion, the outcomes and benefits derived from the application of fuzzy systems in engineering are substantial and multifaceted. From improved decision-making and adaptability to uncertainty to enhanced control systems and optimized processes, fuzzy systems contribute significantly to the efficiency, reliability, and sustainability of engineering solutions across various domains.

2.7.3 ADVANTAGES OF FUZZY SYSTEMS OVER TRADITIONAL OPTIMIZATION SYSTEMS

Fuzzy systems, especially fuzzy logic-based approaches like fuzzy goal programming and fuzzy TOPSIS, are designed to address the challenges of uncertainty and imprecision in multicriteria engineering optimization. These fuzzy approaches have a number of advantages over traditional optimization methods for tackling these challenges:

1. **Handling Uncertainty and Imprecision**
   - **Fuzzy Sets and Membership Functions:**
     - Fuzzy logic allows for the representation of uncertainty and imprecision through fuzzy sets and membership functions. Variables are described not as crisp values but as fuzzy sets with degrees of membership.
   - **Linguistic Variables:**
     - Fuzzy systems uses linguistic variables, qualitative terms (e.g., high, medium, low) that describe the imprecise nature of certain parameters. This allows for a more human-like representation of preferences and criteria.

2. **Fuzzy Goal Programming**
   - Fuzzy goal programming extends traditional goal programming by allowing decision-makers to express goals in fuzzy terms, considering the trade-offs and compromises between conflicting objectives. It considers the imprecision in setting and achieving goals, providing a more realistic optimization framework.
     - Fuzzy goal programming also identifies solutions that are optimal under uncertainty considering the fuzziness associated with decision criteria. This is particularly beneficial in situations where precise numeric values are hard to obtain.

3. **Fuzzy TOPSIS**
   - **Ranking under Uncertainty:**
     - Fuzzy TOPSIS is used for ranking alternatives when the decision criteria are imprecise. It measures the distance between each

TABLE 2.1

Features, Outcomes, and Advantages of Fuzzy System Applications in Engineering

Fuzzy System Feature	Outcomes	Advantages
Adaptability to Uncertainty	Engineering solutions demonstrate the ability to adjust to dynamic and unpredictable settings.	Fuzzy systems excel in handling imprecision and uncertainty, providing flexibility crucial in real-world situations where accurate modeling is challenging.
Improved Decision-Making	Fuzzy systems offer a flexible framework for modeling and analyzing complex situations.	Decision-makers benefit from incorporating imprecise information, linguistic words, and subjective assessments, leading to more informed and effective decision-making processes.
Enhanced Control Systems	Fuzzy systems provide control engineers with improved, resilient control techniques.	Fuzzy logic controllers enhance system performance in dynamic contexts by offering a more human-like and intuitive approach to control.
Optimization of Processes	Fuzzy systems contribute to making complex processes in chemical engineering and manufacturing more efficient.	Fuzzy models increase productivity, improve quality, and optimize resource usage by incorporating nonlinear relationships and addressing uncertainty.
Increased Robustness in Robotic	Fuzzy systems improve the resilience of robotic systems in automation.	Fuzzy logic-equipped robots perform better by accepting inaccurate sensor input, adapting to changes, and making context-aware decisions, enhancing overall system robustness.
Smart Grid and Energy Management	Fuzzy systems play a vital role in effective energy management within smart grid technology.	By managing risks associated with renewable energy sources, aiding in demand forecasting, and optimizing energy distribution, fuzzy systems contribute to creating dependable and sustainable power systems.
Structural Health Monitoring	Fuzzy systems assist in assessing the state of structures using inaccurate sensor data in civil engineering.	Fuzzy systems enhance infrastructure longevity and safety by facilitating the early discovery and resolution of potential problems.
Efficient Traffic Control	Fuzzy logic in traffic control systems results in fewer traffic jams and more effective traffic flow.	Fuzzy systems provide dynamic solutions that adapt to changing traffic conditions, improving the overall efficiency of transportation networks.
Environmental Impact Assessment	Fuzzy systems assist in environmental engineering by assessing and mitigating the environmental impact of engineering projects.	Handling uncertainties in environmental data and regulatory standards, fuzzy systems facilitate more accurate impact assessments, promoting environmentally sustainable engineering practices.
Enhanced Biomedical Diagnostics	Fuzzy systems contribute to medical diagnostics in biomedical engineering by handling imprecision and subjectivity in medical data.	Decision support systems based on fuzzy logic aid health care professionals in making more accurate diagnoses, thereby improving overall patient care and treatment outcomes.

alternative and the ideal solutions, accounting for uncertainties, and considers the positive ideal solution and the negative ideal solution. It offers a more nuanced ranking in the presence of uncertainty and provides a more robust approach in situations where the exact values of criteria are uncertain.

4. Advantages over Traditional Optimization Methods
- **Flexibility and Adaptability:**
  - Fuzzy logic-based approaches are more flexible and adaptable to dynamic and uncertain environments. They can handle changes in criteria, preferences, and data with greater ease than traditional methods.
  - **Expressiveness of Uncertainty:**
  - Linguistic variables and fuzzy sets provide a way to capture and articulate the imprecision inherent in real-world decision-making.
  - **Realistic Decision-Making:**
  - Fuzzy systems acknowledge that decision-making is inherently uncertain and imprecise. By incorporating these aspects, they provide more realistic and practical solutions that align with the complexities of the real world.
  - **Handling Qualitative Information:**
  - Fuzzy logic-based approaches can effectively handle qualitative information and expert knowledge, which can be challenging for traditional optimization methods that rely on precise numerical data.

In short, fuzzy logic-based approaches such as fuzzy goal programming and fuzzy TOPSIS offer a more exact and realistic approach to multicriteria engineering optimization by explicitly addressing uncertainties and imprecisions. Their flexibility, adaptability, and ability to handle qualitative information make them advantageous over traditional optimization methods in complex and dynamic decision-making scenarios.

2.8   FUTURE TRENDS AND CONCLUSION

Looking ahead, the future of fuzzy systems in engineering holds promising trends that reflect a continued evolution toward more intelligent, adaptive, and explainable solutions. The integration of fuzzy systems with artificial intelligence (AI), particularly machine learning and deep learning, is anticipated to enhance learning capabilities and foster the development of more sophisticated models. The demand for explainable AI aligns well with the transparent and interpretable nature of fuzzy logic, positioning it as a key player in developing AI applications where understanding decision processes is crucial. Additionally, the convergence of fuzzy systems with edge computing and the Internet of Things (IoT) is poised to bring real-time decision-making capabilities to the forefront of engineering applications. Hybrid systems combining fuzzy logic with other computational intelligence techniques and the nascent field of quantum fuzzy systems underscore the versatility and adaptability of fuzzy systems in addressing complex engineering challenges. As technology continues to advance, the collaborative synergy between fuzzy logic and emerging trends positions it as a resilient and essential tool for navigating uncertainties in diverse engineering domains.

2.8.1 Emerging Trends in Fuzzy Systems for Multicriteria Optimization

Emerging trends in fuzzy systems for MCO reflect the ongoing evolution of these methodologies to address complex decision-making challenges. Several key trends are shaping the future landscape:

- **Integration with Metaheuristic Algorithms:** Fuzzy systems are increasingly being integrated with metaheuristic algorithms, such as genetic algorithms, particle swarm optimization, and simulated annealing. This integration aims to enhance the global search capabilities of fuzzy optimization systems, allowing them to navigate complex solution spaces more effectively.
- **Explainable Fuzzy Systems:** With the growing importance of explainable AI, there is a trend toward developing fuzzy systems that provide clear and interpretable insights into decision processes. Explainable fuzzy systems enhance transparency, enabling decision-makers to understand and trust the optimization outcomes, especially in critical applications.
- **Hybridization with Deep Learning:** There is growing interest in combining deep learning methods with fuzzy systems. The goal of this hybridization is to improve the learning and adaptability elements of fuzzy systems by utilizing deep learning's pattern recognition and feature extraction capabilities, especially in large-scale, high-dimensional optimization issues.
- **Dynamic and Adaptive Fuzzy Systems:** With time, fuzzy systems that can dynamically adjust to shifting conditions and standards can produce more reliable answers by self-adjusting and re-optimizing in response to changes in the decision landscape.
- **Blockchain-Based Fuzzy Systems:** The combination of fuzzy systems and blockchain technology is a developing trend in the context of decentralized decision-making and safe data exchange. This guarantees traceability, transparency, and data integrity in MCO processes, especially in situations where decision-making is distributed and cooperative.
- **Experiments with Quantum Fuzzy Logic:** Quantum fuzzy systems provide a new paradigm for MCO by utilizing concepts from quantum computing to handle uncertainty and state superposition.
- **Real-time and Edge-Based Optimization:** Fuzzy systems are being modified to enable optimization procedures to take place nearer to data sources in edge computing and real-time decision-making scenarios. Applications needing quick reactions, such as those in edge analytics, IoT devices, and smart systems, should pay particular attention to this trend.
- **Meta-Optimization and Auto-Tuning:** To automatically adjust their structures and tune their parameters, fuzzy systems are progressively integrating meta-optimization approaches. This tendency makes it easier to create fuzzy systems that can optimize themselves based on the features of the current optimization challenge.

Various growing themes are shaping the area as we investigate the current landscape of fuzzy systems for MCO. The incorporation of metaheuristic algorithms, such as genetic algorithms and particle swarm optimization, improves global search capabilities for effectively traversing complex solution spaces (Table 2.2).

Together, these new developments push fuzzy systems for MCO forward, improving their adaptability, efficiency, and suitability for a variety of challenging decision-making situations. Fuzzy logic's compatibility with these new developments is expected to be crucial in influencing the development of the upcoming generation of intelligent optimization tools as technology advances.

2.8.2 KEY INSIGHTS

To summarize, researchers have made many important discoveries during the investigation of fuzzy systems for MCO in engineering. Due to their capacity to represent imprecision and uncertainty, fuzzy systems have become essential for handling the complexity involved in decision-making processes. Because real-world engineering scenarios are complex and involve balancing conflicting aims and preferences, multicriteria optimization is highly valuable.

With its roots in managing ambiguity, fuzzy logic has become a potent instrument by providing decision-makers with adaptable frameworks for articulating preferences through the use of language variables and fuzzy operations. Fuzzy logic foundations such as fuzzy sets, membership functions, and linguistic variables serve as the foundation for understanding and modeling fuzzy data. Fuzzy inference systems, exemplified by Mamdani and Sugeno models, serve as the core engines

TABLE 2.2

Emerging Trends in Fuzzy Systems for MCO

Trend	Example
Integration with Metaheuristic Algorithms	Genetic algorithms enhance global search in fuzzy optimization.
Explainable Fuzzy Systems	Fuzzy systems can provide clear insights into decision processes.
Hybridization with Deep Learning	Integrating deep learning enhances pattern recognition in fuzzy optimization.
Dynamic and Adaptive Fuzzy Systems	Fuzzy systems can dynamically adapt to evolving decision landscapes.
Blockchain-based Fuzzy Systems	Blockchain can ensure traceability and transparency in fuzzy optimization.
Quantum Fuzzy Logic	Quantum fuzzy logic can handle uncertainty in optimization problems.
Real-Time, Edge-Based Optimization	Fuzzy systems offer real-time optimization in edge computing and IoT applications.
Meta-Optimization and Auto-Tuning	Meta-optimization can automatically adjust fuzzy system parameters.

driving decision-making processes, employing fuzzification, rule evaluation, and defuzzification to provide nuanced and context-aware outcomes.

The application of fuzzy systems extends beyond traditional domains, finding utility in multicriteria decision-making and fuzzy goal programming, where the emphasis is on accommodating imprecise criteria and handling conflicting objectives. Case studies exemplifying the advantages of fuzzy TOPSIS underscore its effectiveness in diverse engineering scenarios, including supplier selection, material choice, and project risk management. Looking forward, the future trends in fuzzy systems point toward enhanced integration with artificial intelligence, the development of explainable models, and the exploration of quantum fuzzy logic, among other innovations. As fuzzy systems continue to evolve, their applications in engineering are set to become more adaptive, intelligent, and aligned with the demands of an ever-changing technological landscape.

2.8.3 CONTRIBUTIONS AND FUTURE PROSPECTS IN THE ENGINEERING DOMAIN

The contributions of fuzzy systems in the engineering domain have been substantial, offering innovative solutions to complex problems characterized by uncertainty, imprecision, and conflicting criteria. Some significant contributions and upcoming prospects comprise:

- **Handling Uncertainty:** In engineering, fuzzy systems have significantly improved the way uncertainty is handled in decision-making. In situations where conventional approaches are insufficient, they offer a strong framework for modeling and reasoning with imprecise information, enabling more precise and flexible solutions.
- **Multicriteria Optimization:** The development of MCO approaches has been greatly aided by fuzzy systems. Fuzzy logic has given engineers a powerful tool for traversing trade-offs and creating optimal solutions in a variety of domains, including manufacturing, transportation, and energy systems, by accommodating conflicting aims and preferences.
- **Adaptive Control Systems:** The integration of fuzzy logic in control systems has enhanced their adaptability to dynamic and uncertain environments. Fuzzy control systems excel in applications such as robotics, process control, and autonomous vehicles, where real-time decision-making is critical.
- **Transparent Decision Models:** Fuzzy systems contribute to transparent and interpretable decision models. This interpretability is crucial in applications where understanding the decision process is as important as the outcome, such as in medical diagnosis, financial decision-making, and structural health monitoring.
- **Advancements in Artificial Intelligence:** Fuzzy systems have influenced the evolution of artificial intelligence, particularly in explainable AI. The transparent nature of fuzzy logic models aligns with the growing demand for AI systems that can provide understandable explanations for their decisions, contributing to the ethical deployment of AI in engineering applications.

- **Quantum Fuzzy Logic:** The exploration of quantum fuzzy logic represents a frontier in handling complex optimization problems. Prospects involve further research into quantum-inspired computing, potentially opening new avenues for solving engineering problems with unprecedented efficiency and accuracy.
- **Real-Time and Edge Computing:** Fuzzy systems are increasingly finding applications in real-time decision-making and edge computing scenarios. This trend is likely to grow, especially with the rise of IoT and the need for efficient, decentralized decision processes in engineering applications.
- **Hybrid Systems and Integration with Emerging Technologies:** The integration of fuzzy systems with emerging technologies, such as metaheuristic algorithms, deep learning, and blockchain, holds promise for creating hybrid systems that can address complex and large-scale engineering challenges more effectively.

Looking to the future, prospects in the engineering domain involve the continual refinement and integration of fuzzy systems with cutting-edge technologies. The ongoing development of explainable models, exploration of quantum-inspired computing, and application of fuzzy logic in emerging fields like edge computing and blockchain are indicative of the dynamic role fuzzy systems are poised to play in shaping the future of engineering solutions. As technology continues to advance, the adaptability and versatility of fuzzy systems position them as integral components in the engineer's toolkit for tackling the intricacies of real-world decision landscapes.

REFERENCES

Abdullah, L. (2013). Fuzzy multi-criteria decision making and its applications: A brief review of category. Procedia—Social and Behavioral Sciences, 97, 131–136.

Ahmed, A. A., Masrur, A., & Shah, S. M. A. (2017). Application of Adaptive Neuro-Fuzzy Inference System (ANFIS) to estimate the Biochemical Oxygen Demand (BOD) of Surma River. Journal of King Saud University-Engineering Sciences, 29(3), 237–243.

Akram, M., Wasim, F., Alcantud, J. C. R., & Al-kenani, A. N. (2021). Multi-criteria optimization technique with complex pythagorean fuzzy N-soft information. International Journal of Computational Intelligence Systems, 14(167).

Diaz-Curbelo, A., Andrade, R. A. E., & Municio, A. M. G. (2020). The role of fuzzy logic to dealing with epistemic uncertainty in supply chain risk assessment: Review standpoints. International Journal of Fuzzy Systems, 22(9).

Guillaume, V., & Charnomordic, B. (2012). Fuzzy inference systems: An integrated modeling environment for collaboration between expert knowledge and data using FisPro. Expert Systems with Applications, 39(10), 8744–8755.

Gupta, A. K. (2021). Fuzzy logic and their application in different areas of engineering science and research: A survey. International Journal of Scientific Research in Science and Technology, 8(2).

Irvanizam, I. (2018). Application of the fuzzy TOPSIS multi-attribute decision-making method to determine scholarship recipients. Journal of Physics: Conference Series, 978.

Kaya, I., Colak, M., & Terzi, F. (2019). A comprehensive review of fuzzy multi-criteria decision-making methodologies for energy policy making. Energy Strategy Reviews, 24, 207–228.

Komsiyah, S., Meiliana, M., & Centika, H. E. (2018). A fuzzy goal programming model for production planning in furniture company. Procedia Computer Science, 135, 544–552.

Kose, U. (2012). Fundamentals of fuzzy logic with an easy-to-use, interactive fuzzy control application. International Journal of Modern Engineering Research (IJMER), 2(3), 1198–1203.

Kumar, V., & Schuhmacher, M. (2005). Fuzzy uncertainty analysis in system modeling. Computer Aided Chemical Engineering, 20, 391–396.

Leitmann, G. (1996). On the fundamentals of fuzzy set theory. Journal of Mathematical Analysis and Applications, 201(0285), 786–826.

Madanda, V. C., Sengani, F., & Mulenga, F. (2023). Applications of fuzzy theory-based approaches in tunnelling geomechanics: A state-of-the-art review. Mining, Metallurgy & Exploration, 40, 819–837.

Malik, Z. A., Kumar, R., Pathak, G., Roy, H., & Malik, M. A. (2023). Application of fuzzy goal programming approach in the real-life problem of agriculture sector. Brazilian Journal of Operations and Production Management, 20(1).

Mohamed, A. A. A., Mohamed, S. A., & Zino, M. (2023). Application of fuzzy multicriteria decision-making model in selecting pandemic hospital site. Future Business Journal, 9(14).

Palczewski, K., & Salabun, W. (2019). The fuzzy TOPSIS applications in the last decade. Procedia Computer Science, 159, 2294–2303.

Rivaz, S., Nasseri, S. H., & Ziaseraji, M. (2020). A fuzzy goal programming approach to multiobjective transportation problems. Fuzzy Information and Engineering, 12(2).

Rizvi, S., Mitchell, J., Razaque, A., Rizvi, M. R., & Williams, I. (2020). A Fuzzy Inference System (FIS) to evaluate the security readiness of cloud service providers. Journal of Cloud Computing, 9(42).

Sun, C. C., & Lin, G. T. R. (2009). Using fuzzy TOPSIS method for evaluating the competitive advantages of shopping websites. Expert Systems with Applications, 36(9), 11764–11771.

Taherdoost, H., & Madanchian, M. (2023). Multi-Criteria Decision Making (MCDM) methods and concepts. Encyclopedia, 3(1), 77–87.

Tiwari, R. N., Dharmar, S., & Rao, J. R. (1987). Fuzzy goal programming—An additive model. Fuzzy Sets and Systems, 24(1), 27–34.

Vyas, S., Gupta, S., Bhargava, D., & Boddu, R. (2022). Fuzzy logic system implementation on the performance parameters of health data management frameworks. Advances in Feature Transformation based Medical Decision Support Systems for Health Informatics.

Xu, C. (2022). An improved fuzzy multi-criteria algorithm for optimizing Concentrated Solar Power (CSP) hybridized systems based on Pythagorean fuzzy set. Cleaner Engineering and Technology, 6.

Zare, H., Saraji, M. K., Tavana, M., Streimikiene, D., & Cavallaro, F. (2021). An integrated fuzzy goal programming—Theory of constraints model for production planning and optimization. Sustainability, 13(22).

3 Optimizing Ti-6Al-4V Milling under MQL Conditions Using SVR, NSGA-II, and TOPSIS

*Van-Canh Nguyen, Dung Hoang Tien,
Thuy-Duong Nguyen, Van Hung Pham,
Ba Nghien Nguyen, Duy Trinh Nguyen,
and Van Que Nguyen*

3.1 INTRODUCTION

Machining processes, including milling, drilling, boring, and turning, hold significant importance across various industries (Abellan-Nebot and Romero Subirón, 2010; Okokpujie et al., 2019). During these operations, materials undergo a transformative process wherein they are separated into chips, contributing approximately 5% to the GDP of developed nations (Lindgren et al., 2017). The elimination of undesired components leads to the plastic deformation of the workpiece, converting a substantial portion of machining energy into thermal energy (Balogun et al., 2016; Kundrak et al., 2011). Recognizing the pivotal role of these procedures across diverse sectors, particularly in the aerospace industry—characterized by high expertise, effective resource management, and continual development—becomes imperative to address escalating demands (Soni et al., 2023). Machining titanium alloys in general often encounters many challenges due to the mechanical and physical characteristics of this material (Klocke et al., 2015; Ravi Teja et al., 2018).

Ti-6Al-4V is the most commonly used titanium alloy, accounting for over 50% of total global titanium production. This α-β phase alloy is extensively utilized in the aviation and medical industries, which comprise around 80% of Ti-6Al-4V usage (Salvo et al., 2017; Wang et al., 2023). Titanium alloys constitute up to 14% of the structural composition of the Airbus A350-900 XWB, 15% of the Boeing 787, and 25% of GE CF6 aircraft engines (Shokrani et al., 2019). Titanium and its alloys are also utilized in the automotive, shipbuilding, and chemical industries due to advantageous properties including good corrosion resistance and high biocompatibility, etc. (Lin et al., 2021; Peters et al., 2003).

DOI: 10.1201/9781032635170-3

Titanium demonstrates superior properties such as high heat and hardness resistance, corrosion resistance, and fracture toughness over alternative materials. It has the highest strength-to-weight ratio among structural materials along with excellent machinability and biocompatibility. However, these same characteristics also make machining titanium alloys difficult and challenging (Pramanik, 2014). The main difficulties when machining titanium alloys come from their high chemical reactivity at elevated temperatures and heat concentration at the machining area due to low thermal conductivity (Gupta et al., 2021; Saini et al., 2016).

Titanium alloys have relatively low thermal conductivity at only 15% that of steel. Consequently, about 80% of heat produced during titanium alloy machining concentrates on the cutting tool, a far larger proportion than the roughly 50% of heat absorbed by tools when machining steel. This causes the phenomenon of high temperatures concentrating in the area between the cutting tool and workpiece, causing rapid tool wear (Ezugwu, 2005). In addition, titanium alloy, with its superior elasticity to that of other metals (Pramanik, 2014), tends to rebound more during machining due to its low elastic modulus. This rebounding tendency can cause the workpiece to disengage from the cutting tool, leading to friction rather than cutting on the workpiece surface. This friction increases heat and wear at the cutting zone, adversely affecting surface finish and dimensional accuracy. This characteristic of titanium can be analyzed by comparing its vibrational behavior during machining with that of other alloys and materials.

In machining, the majority of expended energy plasticly deforms and frictionally shears the workpiece to form chips, largely converting this energy into heat; this significantly elevates cutting zone temperatures. With a thermal conductivity of just 6.7 W/m $^{\circ}$K, compared with 45 W/m $^{\circ}$K in carbon steel, titanium alloys dissipate much less heat during machining. Thus, titanium alloys reach higher cutting temperatures than standard steel materials (Materials and Company, n.d.). Indeed, up to 50% of the cutting heat can be transferred to the cutting tool depending on the material of the tool (Gao et al., 2016). However, in the case of machining titanium alloys, this heat transfer can reach nearly 80%, significantly increasing the temperature of the cutting tool.

The increased heat transfer to the cutting tool is a primary reason that machining titanium alloys result in lower tool life than with other challenging alloys. To reduce cutting heat in titanium alloy machines and general metal cutting, cooling fluids are commonly used in both industrial production and experimental studies. These fluids are usually water, oil, or other diluted solvents that work to rapidly dissipate heat and lower friction at the tool-workpiece interface (Serope Kalpakjian, 2009). This not only cools the cutting area but also cleans the machined surface, reducing cutting force and power consumption and enhancing surface quality, thereby significantly extending tool life. However, using flood cooling has drawbacks, including high maintenance and post-processing costs, accounting for 8%–16% of total machining costs.

Studies have highlighted environmental and health risks associated with metal cutting oils. The permissible exposure limits set by OSHA (Ghuge and Palande, 2022) and NIOSH for metalworking fluid mists are 5 mg/m^3 and 0.5 mg/m^3, respectively.

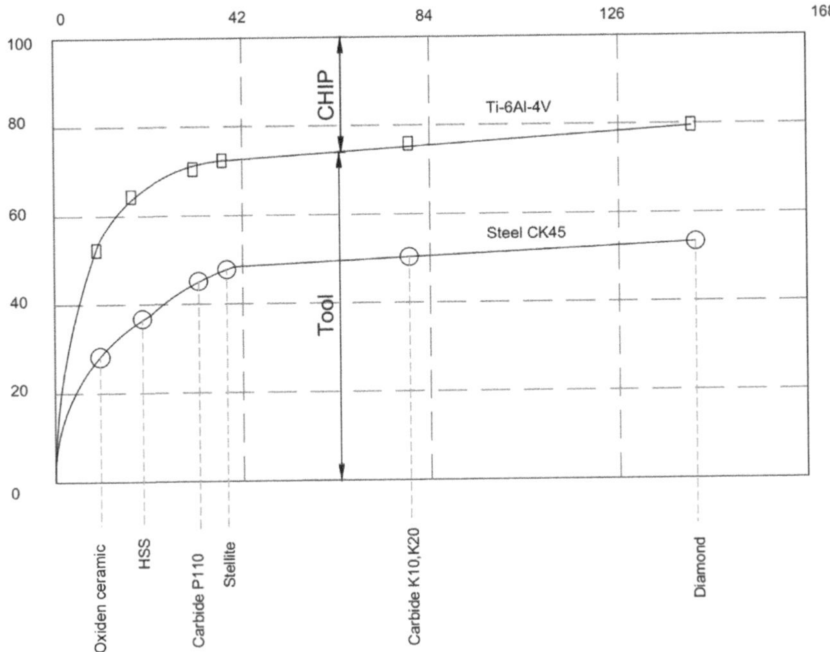

FIGURE 3.1 Distribution of thermal load when machining titanium and steel.

However, traditional cooling and lubrication methods in automotive manufacturing facilities often exceed these limits, potentially leading to cancer and mutations, mainly due to polycyclic aromatic hydrocarbons in mineral oils (Apostoli et al., 1993).

Improper disposal of cooling fluids can lead to environmental pollution and product quality issues such as surface oxidation. Excessive use of these fluids results in waste and increased production costs. Additionally, improper application can reduce the precision of the finished product. Due to these drawbacks, traditional flood cooling methods are gradually being replaced by alternative lubrication and cooling techniques (Stephenson and Agapiou, 2018).

Dry machining, an alternative solution, eliminates almost all cutting fluids, but its application is limited due to poor surface quality, rapid tool wear, and high airborne particle concentration. To overcome these challenges, minimum quantity lubrication (MQL) has been researched and widely implemented, especially in Germany and Japan from 1996 to 2000, and is becoming increasingly popular in other developing countries including India, China, and Vietnam. MQL uses a minimal amount of lubricant, typically between 5 and 50 ml/h, compared with up to 50,000ml/h in traditional flood lubrication. Often utilizing biodegradable vegetable oils, MQL reduces environmental and occupational health risks, as well as maintenance and processing costs, while ensuring the required quality.

Artificial neural networks (ANNs) can model intricate input–output relationships by detecting patterns in data without requiring explicit programming (Kishore et al., 2022; Wang et al., 2022). Drawing inspiration from biological neural functions like those of the brain, the concept of ANNs originated from the pioneering work by McCulloch and Walter Pits (Cooper, 2005; Sankar et al., 2019). In machining, ANNs along with genetic algorithms (GAs) have emerged as notable options, demonstrating the ability to optimize cutting parameters in complex environments. Their increasing popularity is attributed to their adaptability to nonlinear behaviors (M, 2006).

In a milling experiment on P20 steel, Escamilla et al. used ANN to predict MRR and R_a (M, 2006), and Maheshwera used regression models and ANNs to predict R_a in the hard turning of AISI 52100 steel; they showed the efficacy of utilizing ANNs for these predictive applications (Maheshwera Reddy Paturi et al., 2018). Similarly, Boga and Koroglu (2021) leveraged ANNs and a GA to predict minimum surface roughness and identify ideal cutting parameters. The authors delved deeper into GAs, aiming to minimize production time in multi-axis CNC machining while ensuring surface roughness. However, the performance of GAs heavily depends on the initial configuration, thus requiring substantial computational resources. This dependency can make GAs less practical for smaller operations or under limited computational constraints.

For these reasons, multicriteria decision-making (MCDM) is often an optimal solution. Common MCDM techniques utilized include MOORA, VIKOR, simple additive weighting, and TOPSIS (technique for order of preference by similarity to ideal solution); for instance, Que et al. (2023) successfully applied TOPSIS to evaluate various machining strategies based on different criteria. Their research highlights TOPSIS's superiority in offering a systematic approach to consider and rank multiple alternative options, integrating various objectives.

Each method—AI, GA, and MCDM—brings unique strengths to machining optimization but also faces distinct challenges. The predictive accuracy of AI, GA's robust optimization abilities, and MCDM's structured decision framework suggest an appealing case for integration. Such a combined approach could leverage each method's strengths while minimizing their limitations, for example, AI for precise prediction and modeling, GA for finding optimal solutions in complex scenarios, and MCDM for effective evaluation and selection. This integrated approach could address broader challenges in machining optimization, offering a more comprehensive and effective solution.

In this chapter, we combine support vector regression (SVR), the nondominated sorting genetic algorithm II (NSGA-II), and TOPSIS with entropy weight to optimize the multi-objective milling of titanium alloy Ti-6Al-4V under minimum quantity lubrication (MQL). The objectives are minimizing surface roughness (R_a) and cutting force (F_c) and maximizing material removal rate (MRR). These goals are crucial for improving the efficiency and sustainability of the machining process, especially with challenging materials like titanium alloys. SVR models predicted the impacts of various machining parameters on R_a, F_c, and MRR, then NSGA-II optimized these parameters. Finally, the entropy-weighted TOPSIS provided a structured decision framework for selecting the most effective machining strategy.

3.2 EXPERIMENT PROCEDURE

3.2.1 Design of Experiments

In this study, we utilized the Taguchi orthogonal array L_{27} experimental design to decrease the number of trials. We analyzed five 3-level machining and lubrication parameters.

3.2.2 Experiment Setup

We prepared experimental workpieces measuring $35 \times 35 \times 100$ mm. All faces of the specimens were ground to ensure high accuracy in the positioning process. Experimental runs utilized a 16 mm diameter silent tool holder and 390R-070202M-MM 1040 inserts (Sandvik), selected for suitability in face milling optimization based on experimental Ti-6Al-4V alloy data.

Cutting and lubrication parameters aligned with cutting tool manufacturer recommendations, expert input, and previous findings. Table 3.1 presents the experimental variant levels. In the experiment, we employed the DMU50 CNC device shown in Figure 3.2, operating with an MQL setup. We affixed the milling jig to a force sensor, shown in Figure 3.2a, which is part of the experimental setup for machining. The cutting force F_c was derived from F_x, F_y, and F_z forces measured by a Kistler dynamometer 9139AA, and we computed MRR using Equation (3.1). For each test piece, we measured Ra using a Mitutoyo Surftest JS-201, following the ISO 1997 standard, as depicted in Figure 3.2c and following the formula $R_a = f1(P,Q,V_c,f_z,a_p)$, $Fc = f2(P,Q,V_c,f_z,a_p)$ with the data in Table 3.1. The outcomes derived from the experimental runs were methodically computed and systematically recorded in Table 3.2.

3.3 THEORETICAL BACKGROUND

For the study for this chapter, we combined SVR, NSGA II, and TOPSIS for the multi-objective optimization of milling Ti-6Al-4V alloy under MQL conditions. The

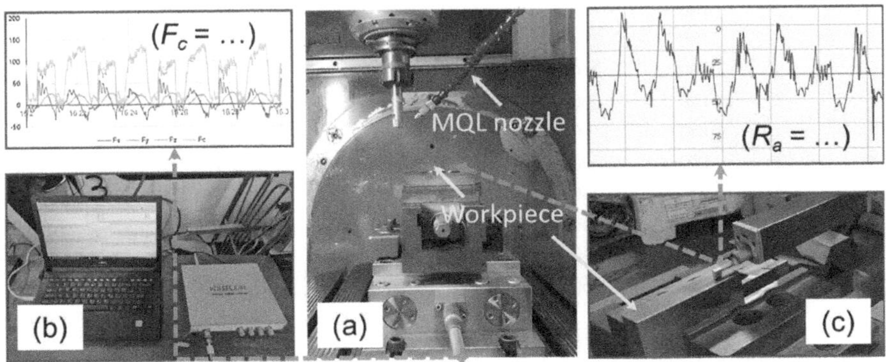

FIGURE 3.2 The experimental setup: (a) surface milling, (b) measuring cutting forces, and (c) measuring surface roughness.

TABLE 3.1

The Cutting and Lubrication Parameters for the Ti-6Al-4V Alloy

Variant	Unit	Level 0	Level 1	Level 2
P	bar	1	3	5
Q	ml/h	50	100	150
a_p	mm	0.1	0.5	0.9
V_c	m/min	60	150	240
f_z	mm/tooth	0.02	0.06	0.10

TABLE 3.2

The L_{27} Experimental Array and Measured Results

Run	P bar	Q ml/h	V_c m/min	f_z mm/tooth	a_p mm	R_a μm	F_c N	MRR mm3/min
1	1	50	60	0.02	0.1	0.157	85.863	54.60
2	1	50	60	0.02	0.5	0.170	156.055	272.98
3	1	50	60	0.02	0.9	0.164	265.870	491.36
4	1	100	150	0.06	0.1	0.234	142.638	409.46
5	1	100	150	0.06	0.5	0.207	264.657	2047.32
6	1	100	150	0.06	0.9	0.239	453.540	3685.17
7	1	150	240	0.1	0.1	0.481	116.257	1091.90
8	1	150	240	0.1	0.5	0.545	317.218	5459.51
9	1	150	240	0.1	0.9	0.653	628.896	9827.12
10	3	50	150	0.1	0.1	0.350	118.840	682.44
11	3	50	150	0.1	0.5	0.416	141.163	3412.19
12	3	50	150	0.1	0.9	0.476	211.536	6141.95
13	3	100	240	0.02	0.1	0.184	149.629	218.38
14	3	100	240	0.02	0.5	0.197	237.397	1091.90
15	3	100	240	0.02	0.9	0.276	391.824	1965.42
16	3	150	60	0.06	0.1	0.212	126.610	163.79
17	3	150	60	0.06	0.5	0.220	243.248	818.93
18	3	150	60	0.06	0.9	0.204	379.741	1474.07
19	5	50	240	0.06	0.1	0.247	92.917	655.14
20	5	50	240	0.06	0.5	0.388	100.872	3275.71
21	5	50	240	0.06	0.9	0.763	136.224	5896.27
22	5	100	60	0.1	0.1	0.650	159.149	272.98
23	5	100	60	0.1	0.5	0.695	267.386	1364.88
24	5	100	60	0.1	0.9	0.663	454.039	2456.78
25	5	150	150	0.02	0.1	0.168	101.847	136.49
26	5	150	150	0.02	0.5	0.173	173.709	682.44
27	5	150	150	0.02	0.9	0.179	266.895	1228.39

SVR constructed the regression models that form the foundation for the optimization challenge using NSGA II. The feasible optimal milling parameters (Pareto Set), determined by NSGA II, were then prioritized through TOPSIS to identify the best parameter set.

3.3.1 Support Vector Regression

Support vector machine (SVM) stands out as a leading supervised learning method extensively applied in machine learning for classification purposes. Its primary aim is to identify the optimal hyperplane, which might be linear or nonlinear, that distinctly segregates data into two separate classes. This objective is accomplished by strategically placing the hyperplane to maximize the margin, or the space, between the hyperplane and the closest data point from each class. Figures 3.3 and 3.4 illustrate the configuration of a hyperplane along with its associated margin. Given a hyperplane equation $w.x + b = 0$, the core objective of SVM is to calculate w and b to ensure the largest margin between the two classes.

The SVM algorithm, well-known for its proficiency in classification tasks, extends its capabilities to regression problems via its variant known as support vector regression. SVR aims to construct a function that precisely represents the relationship between input features and a continuous outcome variable, focusing particularly on reducing the discrepancies in predictions. While SVR is grounded in the same underlying principles as SVM, it is specifically adapted to cater to regression contexts.

At the heart of SVR is a unique loss function introduced by Vapnik (Cortes and Vapnik, 1995; Vapnik, 2000, 1998). This loss function is characterized by its tolerance for errors within a specified epsilon range for points that are not close to the true value. In essence, the function assigns zero error to all training set points that fall within this range of epsilon.

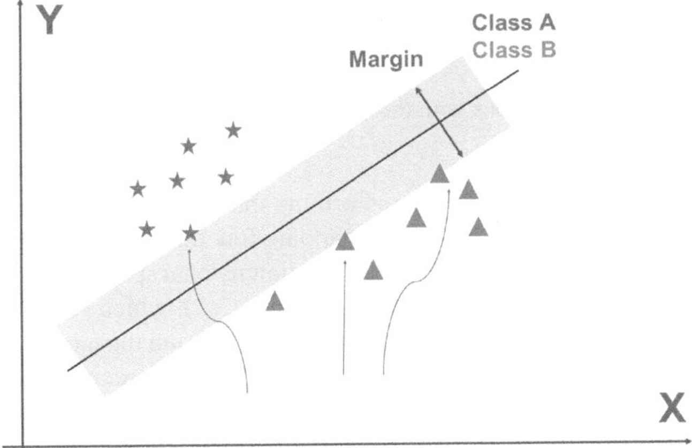

FIGURE 3.3 The hyperplane and the margin.

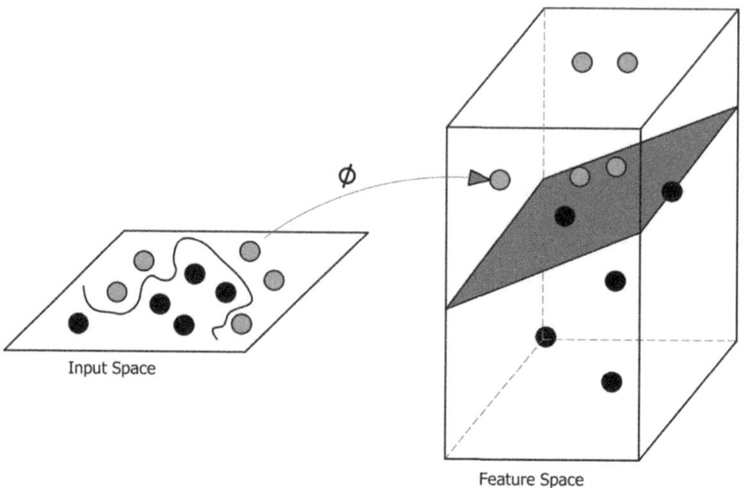

FIGURE 3.4   The role of the kernel.

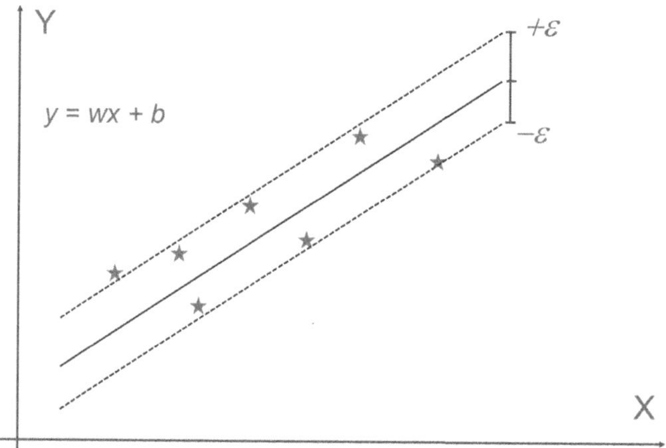

FIGURE 3.5   Linear regressions with the range of epsilon.

In SVR, the initial step involves transforming the input variable x into a m-dimensional feature space through a nonlinear mapping function. After this transformation, a linear model is developed within this higher-dimensional space. The construction of this linear model (Figure 3.5) adheres to Equation (3.1), which succinctly describes the association between the transformed input variables and the target variable within the enlarged feature space. This approach allows SVR to accommodate nonlinear relationships (Figure 3.6) in the data by leveraging the power of linear models in a higher-dimensional context:

$$f(x,w) = \sum_{i=1}^{m} w_i \cdot g_i(x) + b \tag{3.1}$$

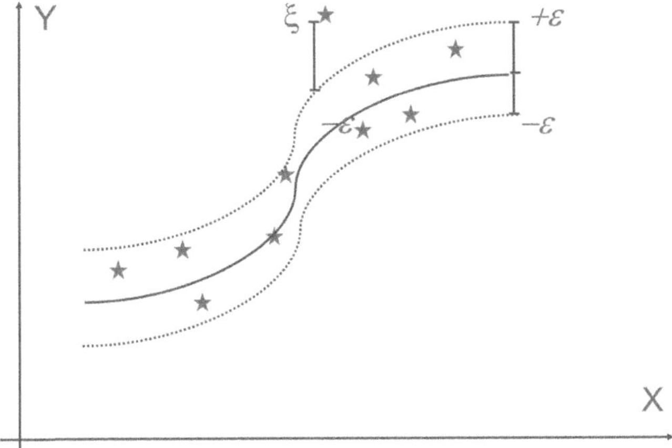

FIGURE 3.6 Nonlinear regression with the range of epsilon.

where $g_i(x) i = 1, 2, \ldots m$ is the set that comprises the nonlinear mapping functions.

The accuracy of the estimate is evaluated by loss function $L(y, f(x,w))$. SVR uses a loss function called epsilon (Equation (3.2)), an insensitive loss function proposed by Vapnik (Cortes and Vapnik, 1995; Vapnik, 2000, 1998):

$$L = \begin{cases} 0 \ \textit{if} \ |y - f(x,w)| \le \varepsilon \\ |y - f(x,w)| \ \textit{otherwise} \end{cases} \tag{3.2}$$

That is, SVM performs linear regression in multidimensional feature space using function L and minimizing $\|W\|^2$ to decrease the complexity of the model. This problem can be solved by introducing slug variables ζ_i and ζ_i^* with i = 1, 2, . . . n to measure the deviation of the training samples which lies outside of the epsilon range. Therefore, SVR is minimized by the function Equation (3.3) with the constraints of Equation (3.4):

$$min \frac{1}{2}\|w\|^2 + C \cdot \sum_{i=1}^{n} \left(\zeta_i + \zeta_i^* \right) \tag{3.3}$$

$$\begin{cases} y_i - f(x_i, w) \le \varepsilon + \zeta_i^* \\ f(x_i, w) - y_i \le \varepsilon + \zeta_i \\ \zeta_i, \zeta_i^* \succ 0 \forall i = 1,...,n \end{cases} \tag{3.4}$$

Applying the duality theorem for minimizing problems, we finally obtain the function f(x) (Equation (3.5)):

$$f(x) = \sum_{i=1}^{n} (\alpha_i - \alpha_i^*) \cdot K \cdot (x_i, x) + b \tag{3.5}$$

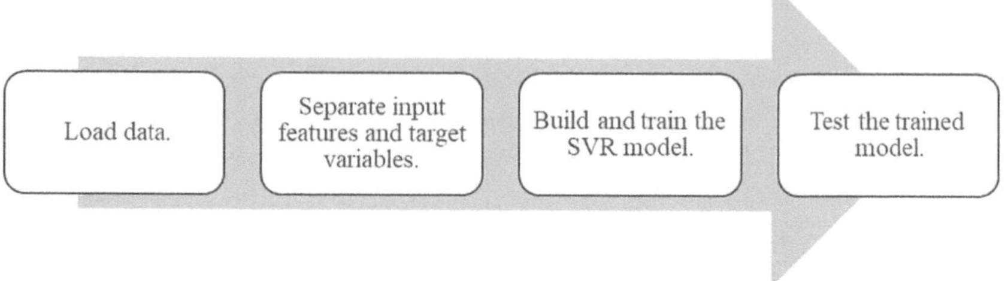

FIGURE 3.7   Implementing the SVR model using the Python sklearn library.

where nSV represents the count of support vectors, and the kernel function, denoted as κ, is defined as Equation (3.6):

$$K(x_t, x) = \sum_{j=1}^{m} g_j(x_i) \cdot g_j(x) \tag{3.6}$$

We ran the SVR model in Python using the sklearn library of the scikit-learn framework following the steps shown in Figure 3.7, and using the data in Table 3.2, which had been saved as a CSV file. See Appendix A for the Python code used to build the SVR model for R_a and F_c.

3.3.2   NSGA-II

In general, multi-objective optimization has several objective functions with inequality and equality constraints to optimize. Equation (3.7) defines an optimization problem:

$$\min f_m(x)\, m = 1, 2, 3, \ldots, M$$

$$s \cdot t \cdot g_j(x) < 0 \; j = 1, 2, 3, \ldots, J$$

$$h_k(x) = 0 \; k = 1, 2, 3, \ldots, K$$

$$x_i^L \leq x_i \leq x_i^U \; i = 1, 2, 3, \ldots, N \tag{3.7}$$

The objective of a multi-objective optimization problem is to identify a set of solutions that simultaneously satisfy two criteria: They must be free of any constraint violations, and they must excel in all their objective values. The ultimate result of this type of optimization is a collection of solutions that represent a compromise between various objectives. These solutions are known as Pareto optimal solutions.

Pareto optimal solutions are defined by their property of nondominance, meaning that it is not possible to improve any aspect of these solutions without worsening at least one other objective; enhancing one criterion would necessitate a compromise in another. The aggregate of these viable nondominated solutions

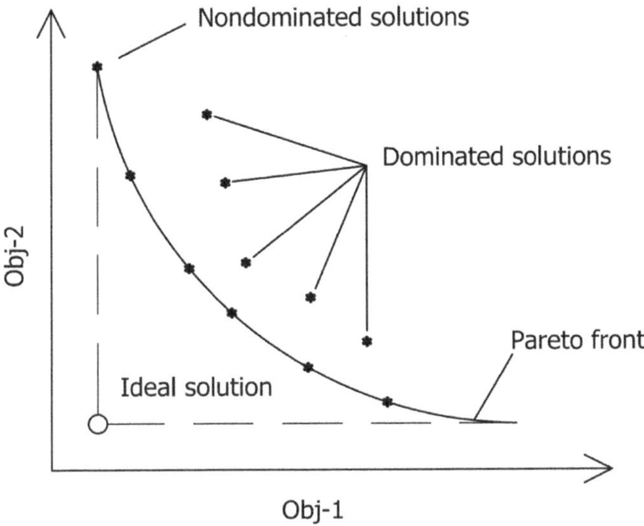

FIGURE 3.8 Pareto front for the two objectives

constitutes the Pareto optimal set. Similarly, the set of objective values reached by these solutions is known as the Pareto front (Gunantara, 2018). Figure 3.8 visually demonstrates this principle by depicting the Pareto front that arises when optimizing two objectives, labeled as Obj-1 and Obj-2. This graphical depiction aids in comprehending the trade-offs and concessions fundamental to multi-objective optimization.

Within the field of multi-objective optimization, NSGA-II marks a notable progression. This evolutionary algorithm utilizes a swift nondominated sorting technique, offering computational efficiency over older methodologies. NSGA-II incorporates a distinct selection mechanism that creates a mating pool by combining the parent and offspring populations. This mechanism prioritizes solutions using measures of both their performance (fitness) and the variety among solutions (spread). Simulation outcomes across a range of difficult test scenarios reveal that NSGA-II reliably surpasses competing approaches in two crucial aspects: It secures a wider distribution of solutions and exhibits enhanced convergence than the genuine Pareto optimal front.

NSGA-II retains the core framework of a genetic algorithm while incorporating particular adjustments in the mating and survival selection strategies. It initially selects individuals based on their 'fronts', a method that may entail splitting a front if the count of individuals surpasses a set survival threshold. Within these segmented fronts, the selection process is further honed using crowding distance, determined by the Manhattan Distance in the objective space, which plays a crucial role (Gunantara, 2018). Crowding distance promotes diversity among solutions by favoring regions that are less crowded; to preserve the boundaries of the solution set, the algorithm bestows an infinite crowding distance on the boundary points (first and last) in every generation, ensuring that the extremities are maintained.

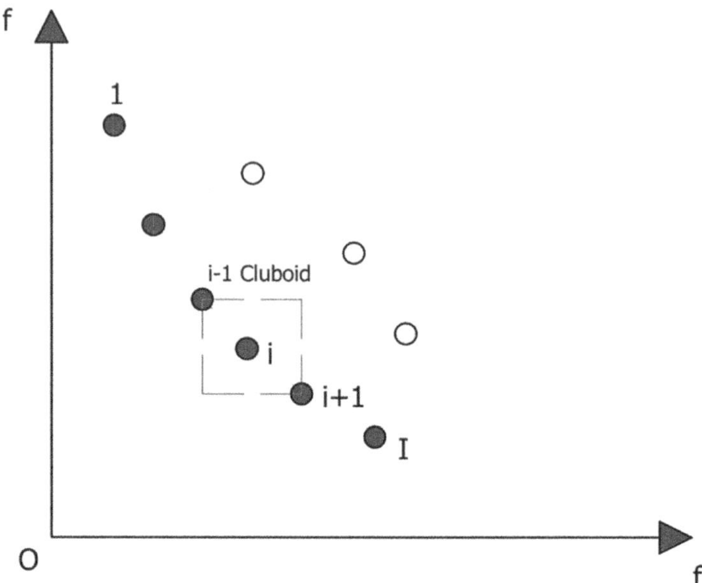

FIGURE 3.9 Calculating crowding distance.

Figure 3.9 visually represents the crowding-distance calculation, where disc points indicate solutions within the same nondominated front; the procedural flow of NSGA-II is depicted in Figure 3.10, providing a comprehensive understanding of the algorithm's mechanics.

NSGA-II is a simple and straightforward algorithm. First, all individuals in the previous population (P) and current population (Q) are included in R_t. Then, population is sorted according to nondomination algorithm described by pseudocode Appendix C.

In the context of NSGA-II, forming a new population P_{t+1} from a merged population is methodical and prioritizes the highest quality solutions. Initially, the top nondominated set, F_1, that contains the best solutions in the merged population is given the highest priority. All members of F_1 are included in P_{t+1} if the size of F_1 is less than or equal to the desired population size N. If there is still room in P_{t+1} after including all members of F_1, the algorithm proceeds to select members from subsequent nondominated fronts, based on their rank order. This means that solutions from F_2 are chosen next, followed by those from F_3, and so on, until the population P_{t+1} reaches its capacity or no more sets can be included.

If the cumulative count of solutions from F_1 to the last accommodated nondominated set Fl exceeds N, a selection must be made from the last set Fl. To do this, the solutions in Fl are sorted using the crowded-comparison operator, which ranks solutions in descending order based on their crowding distance. The algorithm then selects the best solutions from Fl to fill the remaining slots in P_{t+1} until the population count reaches exactly N. This approach ensures that P_{t+1} is composed of the best

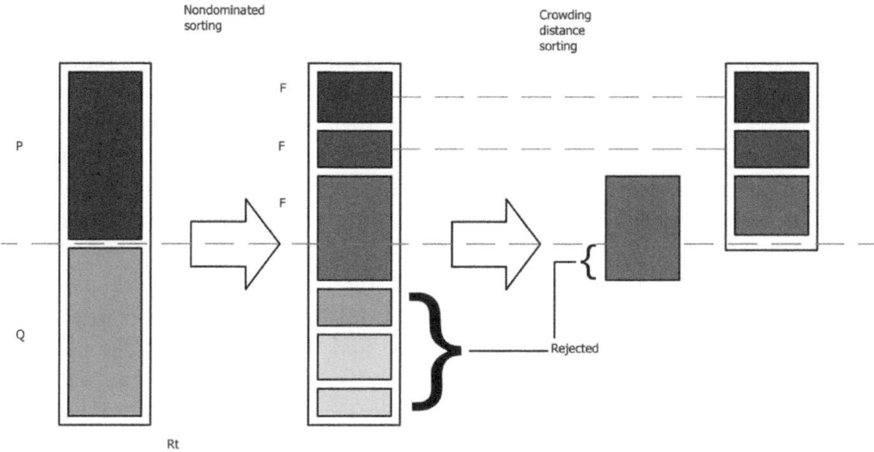

FIGURE 3.10 The NSGA-II procedure.

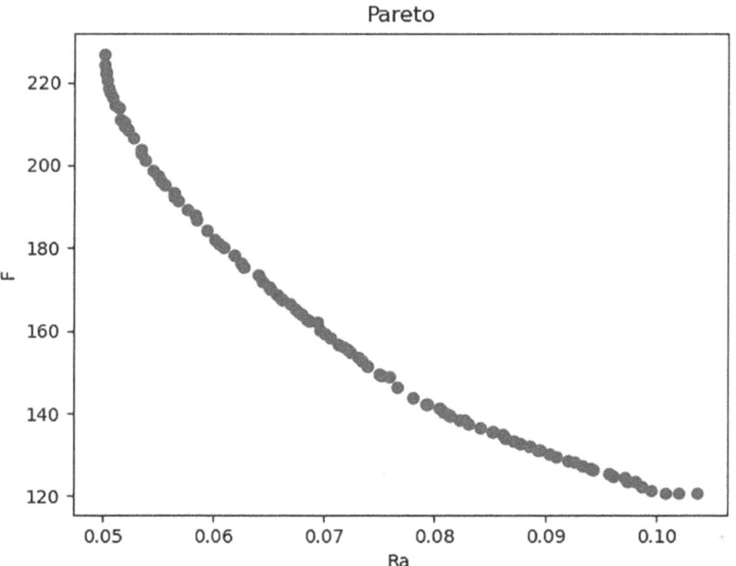

FIGURE 3.11 The nondominated front.

possible solutions from the merged population, balancing both the quality of solutions and the diversity represented in the population.

To implement SVR and multi-objective optimization, we used two Python libraries, sklearn and pymoo. We configured NSGA-II's hyperparameters with a population size of 50 and a total of 100 generations; Appendix B shows the Python code that combined the SVR model and NSGA-II to find the optimal R_a, and F_c. The nondominated solutions are shown in Figure 3.11.

3.3.3 TOPSIS

MCDM is utilized for decision-making based on multiple criteria, commonly used in multi-objective optimization scenarios. The core of MCDM lies in harmonizing diverse criteria when each criterion bears different significance, influencing the ultimate decision. This approach is centered around assessing and contrasting alternatives across several criteria to facilitate decision-making. Among the array of MCDM techniques, TOPSIS stands out as one of the most prevalent and extensively used methods.

TOPSIS, developed by Hwang and Yoon in 1981, is used to evaluate and select options within a multicriteria system. It assesses the preference level of options based on their distance from the optimal ideal value and the least ideal value. TOPSIS has the advantage of being straightforward to understand and implement, with high practical application. However, it is sensitive to extreme values and does not handle conflicting criteria well. One of the main drawbacks of TOPSIS is the determination of weights among criteria. To mitigate this, entropy weighting is often used in conjunction. Entropy-weighted TOPSIS is implemented through several steps (Zhu et al., 2020):

Step 1: Calculate the determinant of a matrix, with the elements of the matrix determined according to Equation (3.8).

$$
X = \begin{bmatrix}
x_{11} & \cdots & x_{1j} & x_{1n} \\
x_{21} & \cdots & x_{2j} & x_{2n} \\
\cdots & \cdots & \cdots & \cdots \\
x_{i1} & \cdots & x_{ij} & x_{in} \\
\cdots & \cdots & \cdots & \cdots \\
x_{m1} & \cdots & x_{mj} & x_{mn}
\end{bmatrix}
\tag{3.8}
$$

where x_{ij} is the value of criterion j in alternative i, n is the number of criteria, and m is the number of alternatives.

Step 2: Calculate the normalized matrix $R = (p_{ij})_{mxn}$, where p_{ij} is determined by Equation (3.9):

$$
p_{ij} = \frac{x_{ij}}{\sqrt{\sum_{i=1}^{n} x_{ij}^2}}, j \in [1,n]
\tag{3.9}
$$

Step 3: Entropy e_i of the ith criterion is defined as Equation (3.10):

$$
e_i = -\frac{1}{\ln(m)} \sum_{i=1}^{m} p_{ij} \ln\left(p_{ij}\right), i = 1\ldots m; j = 1\ldots n
\tag{3.10}
$$

where entropy e_i ranges between [0, 1]. A higher e_i indicates greater differentiation for the ith criterion, meaning that more information can be gleaned from that criterion.

Therefore, a higher weight should be assigned to that criterion. Subsequently, the weight w_i for the ith criterion is calculated using Equation (3.11):

$$w_i = \frac{1-e_i}{\sum_{i=1}^{m}(1-e_i)} \qquad (3.11)$$

Apply w_i corresponding to each optimization criterion in the subsequent steps.

Step 4: Calculate and construct the weighted normalized decision matrix, where wj is the weight of criterion j, calculated according to Equation (3.12):

$$Y = w_j \cdot p_{ij} \qquad (3.12)$$

Step 5: Identify the best and worst solutions using Equations (3.13) and (3.14):

$$A^+ = \left\{ y_1^+, y_2^+, \ldots, y_j^+, \ldots, y_n^+ \right\} \qquad (3.13)$$

$$A^- = \left\{ y_1^-, y_2^-, \ldots, y_j^-, \ldots, y_n^- \right\} \qquad (3.14)$$

where y_j^+ và y_j^- correspond to the best and worst alternatives, respectively, for criterion j.

Step 6: Calculate the positive ideal value S_i^+ and the negative ideal value S_i^- using Equations (3.15) and (3.16):

$$S_i^+ = \sqrt{\sum_{j=1}^{n}\left(y_{ij} - y_j^+\right)^2} \quad i = 1, 2, \ldots, m \qquad (3.15)$$

$$S_i^- = \sqrt{\sum_{i=j}^{n}\left(y_{ij} - y_j^-\right)^2} \quad i = 1, 2, \ldots, m \qquad (3.16)$$

Step 7: Determine the evaluation criteria and rank the solutions C_i^* using Equation (3.17):

$$C_i^* = \frac{S_i^-}{S_i^+ + S_i^-} \quad i = 1, 2, \ldots, m; \ 0 \leq C_i^* \leq 1 \qquad (3.17)$$

Step 8: Rank C_i^* to select the best alternative.

In the following section, we detail the synergistic combination of these methods to optimize the milling of the Ti-6Al-4V alloy. This integration represents a pivotal step in enhancing the efficiency and effectiveness of this complex machining operation.

3.4 SOLVING THE MULTI-OBJECTIVE OPTIMIZATION PROBLEM

Here, we explore combining SVR, NSGA-II, and TOPSIS to tackle the multi-objective optimization challenge. The objective is to pinpoint an optimal set of technological parameters that concurrently minimizes surface roughness (R_a) and cutting force (F_c) while maximizing the material removal rate (MRR).

3.4.1 Defining the Problem

The optimization problem is the following: Find x = [P, Q, V_c, f_z, a_p] to minimize {R_a, F_c}, and maximize *{MRR}* with the constraint in Equation (3.18):

$$\begin{cases} 1 \leq P \leq 5 \\ 50 \leq Q \leq 150 \\ 60 \leq V_c \leq 240 \\ 0.02 \leq f_z \leq 0.10 \\ 0.1 \leq a_p \leq 0.9 \end{cases} \tag{3.18}$$

where P (bar) is the MQL air pressure, Q (ml/h) is the MQL oil flow rate, V_c (m/min) is the cutting speed, f_z (mm/tooth) is the feed per tooth, and a_p (mm) is the depth of the cut.

3.4.2 Integrating SVR and NSGA-II with Entropy-Weighted TOPSIS

As we have noted, for this study, we employed a hybrid approach of combining SVR and NSGA-II with entropy-weighted TOPSIS to determine the ideal technological parameters for milling Ti-6Al-4V alloy with minimal lubrication. We utilized SVR to establish a regression model linking R_a (Figure 3.12); F_c (Figure 3.13); and cutting parameters P, Q, V_c, f_z, and a_p. This model formed the foundation for multi-objective optimization via NSGA-II.

The outcomes of multi-objective optimization obtained through NSGA-II culminate in a Pareto set, which comprises a collection of feasible optimal values. Then we evaluated this Pareto set using entropy-weighted TOPSIS to ascertain the optimal values. Table 3.3 summarizes the results of the optimization process using the combined SVR–NSGA-II–TOPSIS algorithm. We then used the predicted technological parameters to verify the reliability of the optimization methodology.

3.4.3 Validated Experiments

We conducted the verification experiments under the same conditions as the initial experimental study. The experimental results are measured and summarized in Table 3.4, showing the differences between the measured and predicted R_a, F_c, and MRR corresponding to the three different experiments. We calculated the prediction errors for Ra, Fc, and MRR based on the differences between the predicted and measured values.

Experimental confirmation result No. 22 shows that the prediction error for MRR was very small, reflecting that MRR was calculated by formula and depended on the technological parameters of the machining process. Similarly, MRR was negligible

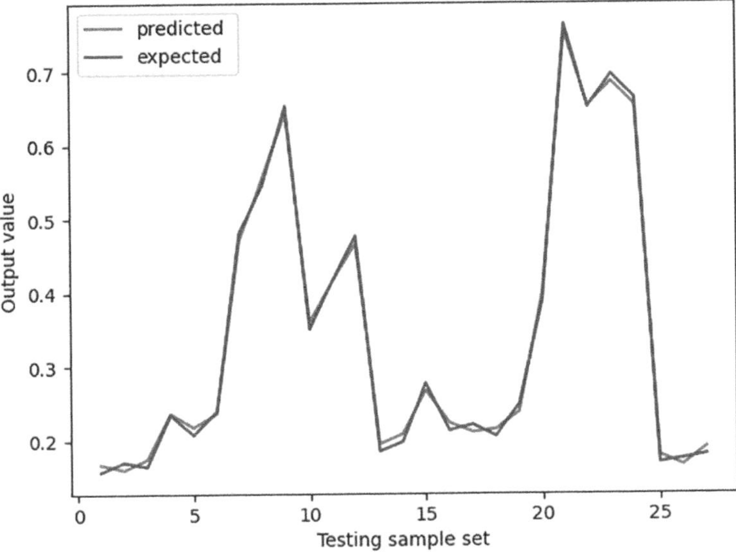

FIGURE 3.12 SVR predicted vs expected R_a.

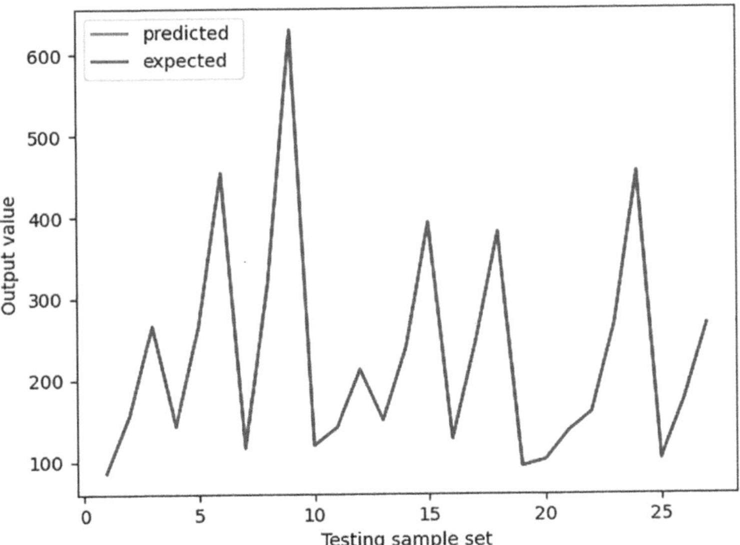

FIGURE 3.13 SVR predicted vs expected F_c.

for experiments 16 and 14, 0.03% and 0.04%, respectively. The prediction errors for the optimized R_a and F_c ranged from 1.87% to 4.93%, demonstrating the reliability of the research methodology. Future researchers or practitioners can combine SVR and NSGA-II as MCDM techniques in similar studies or for practical machining processes.

TABLE 3.3

Optimization Results from Combining SVR, NSGA-II, and TOPSIS

No	P bar	Q ml/h	V_c m/min	f_z mm/tooth	a_p mm	R_a µm	F_c N	MRR mm³/min	Rank
22	3.8	124.5	249.92	0.09	0.80	0.670	775.993	8306.0679	1
16	4.5	103.1	258.76	0.06	0.84	0.670	775.993	6123.8484	2
14	4.3	107.4	280.02	0.06	0.80	0.670	775.993	5949.2332	3

TABLE 3.4

Predicted and Measured R_a, F_c, and MRR

	Predicted			Measured			Errors		
	R_a	F_c	MRR	R_a	F_c	MRR	R_a	F_c	MRR
No	µm	N	mm³/min	µm	N	mm³/min	%	%	%
22	0.670	775.99	8306.07	0.6427	830.49	8309.07	4.14%	7.02%	0.04%
16	0.6705	775.99	6123.85	0.6894	802.33	6125.85	2.82%	3.39%	0.03%
14	0.6705	775.99	5949.23	0.6859	814.28	5951.23	2.30%	4.93%	0.03%
			Minimum Errors				**2.30%**	**1.87%**	**0.03%**
			Maximum Errors				**4.14%**	**4.93%**	**0.04%**

3.5   CONCLUSION

In the experimental research outlined in this chapter, we successfully demonstrated the efficacy of combining SVR with NSGA-II and TOPSIS in the multi-objective optimization of finishing milling for Ti-6Al-4V alloy under MQL conditions. The mechanical properties of the alloy were integral to the optimization process, providing a basis for accurate prediction and optimization of the technological parameters. Confirmation tests validated the reliability and precision of the hybrid method, with prediction errors not exceeding 5% for surface roughness Ra and cutting force Fc, and a mere 0.04% material removal rate. These findings not only contribute valuable insights for the machining of Ti-6Al-4V alloy but also serve as a reference point for advancing research in various machining processes. Combining SVR, NSGA-II, and entropy-weighted TOPSIS has the potential to extend beyond cutting and machining, offering a robust tool for multi-objective optimization in computational research.

REFERENCES

Abellan-Nebot JV, Romero Subirón F. A review of machining monitoring systems based on artificial intelligence process models. Int J Adv Manuf Technol 2010;47:237–257. https://doi.org/10.1007/s00170-009-2191-8.

Apostoli P, Crippa M, Fracasso ME, Cottica D, Alessio L. Increases in polycyclic aromatic hydrocarbon content and mutagenicity in a cutting fluid as a consequence of its use. Int Arch Occup Environ Health 1993;64:473–477. https://doi.org/10.1007/BF00381094.

Balogun VA, Edem IF, Adekunle AA, Mativenga PT. Specific energy based evaluation of machining efficiency. J Clean Prod 2016;116:187–197. https://doi.org/10.1016/j.jclepro.2015.12.106.

Boga C, Koroglu T. Proper estimation of surface roughness using hybrid intelligence based on artificial neural network and genetic algorithm. J Manuf Process 2021;70:560–569. https://doi.org/10.1016/j.jmapro.2021.08.062.

Cooper SJ. Donald O. Hebb's synapse and learning rule: A history and commentary. Neurosci Biobehav Rev 2005;28:851–784. https://doi.org/10.1016/j.neubiorev.2004.09.009.

Cortes C, Vapnik V. Support-vector networks. Mach Learn 1995;20:273–297. https://doi.org/10.1007/BF00994018.

Davim P. Machining of Titanium Alloys. Springer; 2014.

D'addona DM, Teti R. Genetic algorithm-based optimization of cutting parameters in turning processes. Procedia Cirp. 2013 Jan 1;7:323–8. https://doi.org/10.1016/j.procir.2013.05.055

Ezugwu EO. Key improvements in the machining of difficult-to-cut aerospace superalloys. Int J Mach Tools Manuf 2005;45:1353–1367. https://doi.org/10.1016/j.ijmachtools.2005.02.003.

Gao Y, Wang G, Liu B. Chip formation characteristics in the machining of titanium alloys: A review. Int J Mach Mach Mater 2016;18:155. https://doi.org/10.1504/IJMMM.2016.075467.

Ghuge NC, Palande DD. The emergence of MQL with vegetable oil as a green manufacturing technique: A Review. SAMRIDDHI A J Phys Sci Eng Technol 2022;14:66–71. https://doi.org/10.18090/samriddhi.v14i01.11.

Gunantara N. A review of multi-objective optimization: Methods and its applications. Cogent Eng 2018;5:1–16. https://doi.org/10.1080/23311916.2018.1502242.

Gupta MK, Song Q, Liu Z, Sarikaya M, Mia M, Jamil M, et al. Tribological performance based machinability investigations in cryogenic cooling assisted turning of α-β titanium Alloy. Tribol Int 2021;160:107032. https://doi.org/10.1016/j.triboint.2021.107032.

Kishore K, Sinha MK, Singh A, Archana, Gupta MK, Korkmaz ME. A comprehensive review on the grinding process: Advancements, applications and challenges. Proc Inst Mech Eng Part C J Mech Eng Sci 2022;236:10923–10952. https://doi.org/10.1177/09544062221110782.

Klocke F, Soo SL, Karpuschewski B, Webster JA, Novovic D, Elfizy A, et al. Abrasive machining of advanced aerospace alloys and composites. CIRP Ann 2015;64:581–604. https://doi.org/10.1016/j.cirp.2015.05.004.

Kundrak J, Mamalis AG, Gyani K, Bana V. Surface layer microhardness changes with high-speed turning of hardened steels. Int J Adv Manuf Technol 2011;53:105–112. https://doi.org/10.1007/s00170-010-2840-y.

Lin Z, Song K, Yu X. A review on wire and arc additive manufacturing of titanium alloy. J Manuf Process 2021;70:24–45. https://doi.org/10.1016/j.jmapro.2021.08.018.

Lindgren L-E, Hao Q, Wedberg D. Improved and simplified dislocation density based plasticity model for AISI 316L. Mech Mater 2017;108:68–76. https://doi.org/10.1016/j.mechmat.2017.03.007.

Maheshwera Reddy Paturi U, Devarasetti H, Kumar Reddy Narala S. Application of regression and artificial neural network analysis in modelling of surface roughness in hard turning Of AISI 52100 steel. Mater Today Proc 2018;5:4766–4777. https://doi.org/10.1016/j.matpr.2017.12.050.

Okokpujie IP, Bolu CA, Ohunakin OS, Akinlabi ET, Adelekan DS. A review of recent application of machining techniques, based on the phenomena of CNC machining operations. Procedia Manuf 2019;35:1054–1060. https://doi.org/10.1016/j.promfg.2019.06.056.

Peters M, Kumpfert J, Ward CH, Leyens C. Titanium alloys for aerospace applications. Adv Eng Mater 2003;5:419–427. https://doi.org/10.1002/adem.200310095.

Pramanik A. Problems and solutions in machining of titanium alloys. Int J Adv Manuf Technol 2014;70:919–928. https://doi.org/10.1007/s00170-013-5326-x.

Que N Van, Dung HT, Thien N Van, Dong P Van, Canh N Van. Multiple response prediction and optimization in thin-walled milling of 6061 aluminum alloy. Eng Technol Appl Sci Res 2023;13:10447–10452.

Ravi Teja S, Swapna Sri NC, Chaitanya GP, Vishnu Teja C, Bhavana, Tewari A, et al. Machinability studies of aerospace steel using hot end milling. Mater Today Proc 2018;5:27051–27057. https://doi.org/10.1016/j.matpr.2018.09.009.

Saini A, Pabla BS, Dhami SS. Developments in cutting tool technology in improving machinability of Ti6Al4V alloy: A review. Proc Inst Mech Eng Part B J Eng Manuf 2016;230:1977–1989.

Salvo C, Aguilar C, Cardoso-Gil R, Medina A, Bejar L, Mangalaraja R V. Study on the microstructural evolution of Ti-Nb based alloy obtained by high-energy ball milling. J Alloys Compd 2017;720:254–263. https://doi.org/10.1016/j.jallcom.2017.05.262.

Sankar MR, Saxena S, Banik SR, Iqbal IM, Nath R, Bora LJ, et al. Experimental study and artificial neural network modeling of machining with minimum quantity cutting fluid. Mater Today Proc 2019;18:4921–4931. https://doi.org/10.1016/j.matpr.2019.07.484.

Serope Kalpakjian SRS. Manufacturing Engineering and Technology. 6th ed. Prentice Hall; 2009.

Shokrani A, Al-Samarrai I, Newman ST. Hybrid cryogenic MQL for improving tool life in machining of Ti-6Al-4V titanium alloy. J Manuf Process 2019;43:229–243. https://doi.org/10.1016/j.jmapro.2019.05.006.

Soni R, Verma R, Kumar Garg R, Sharma V. A critical review of recent advances in the aerospace materials. Mater Today Proc 2023. https://doi.org/10.1016/j.matpr.2023.08.108.

Stephenson DA, Agapiou JS. Metal cutting theory and practice. Met Cut Theory Pract 2018. https://doi.org/10.1201/9781315373119.

Vapnik V. The Nature of Statistical Learning Theory. 2nd ed. Springer-Verlag New York; 2000. https://doi.org/10.1007/978-1-4757-3264-1.

Vapnik VN. Statistical Learning Theory. Wiley; 1998.

Wang M, Hu J, Zhu J, Zhang K, Kovalchuk D, Yang Y, et al. Microstructure and mechanical properties of Ti-6Al-4V cruciform structure fabricated by coaxial electron beam wire-feed additive manufacturing. J Alloys Compd 2023;960:170943. https://doi.org/10.1016/j.jallcom.2023.170943.

Wang Z, Li H, Yu T. Study on surface integrity and surface roughness model of titanium alloy TC21 milling considering tool vibration. Appl Sci 2022;12. https://doi.org/10.3390/app12084041.

Zhu Y, Tian D, Yan F. Effectiveness of entropy weight method in decision-making. Math Probl Eng 2020;2020:3564835. https://doi.org/10.1155/2020/3564835.

Appendix A: Regression Models for R_a and F_c

```python
import numpy as np
from sklearn.svm import SVR
import matplotlib.pyplot as plt
import pandas
# step 1: Load data from a csv file
df = pandas.read_csv('C:/research/data.csv')
# step 2: Separate input features and target
  variables
ra = np.array(df['Ra'])
f = np.array(df['F'])
x = df.drop('Ra',axis=1)
x = x.drop('F',axis=1)
X1 = np.array(x)
# step 3: Create and train the SVR model
svr1 = SVR(kernel='rbf', C=1e1, gamma=0.1,
  epsilon=.01)
svr2 = SVR(kernel='rbf', C=1e1, gamma=0.1,
  epsilon=.01)
svr1.fit(X1,ra)
svr2.fit(X1,f)
# step 4: Test the trained model
rap = svr1.predict(X1)
fp = svr2.predict(X1)
# step 5: Visulization the results
t = np.arange(1., len(rap)+1, 1.0)
plt.figure(1)
plt.subplot(2,1,1)
plt.plot(t,ra,'r',label="prediction")
plt.subplot(2,1,2)
plt.plot(t,rap,'b',label="expected")
plt.title('SVR prediction output vs expected output
  for Ra')
plt.xlabel('Testing sample set')
plt.ylabel('Output value')
plt.legend(loc="upper left")
plt.show()
```

Appendix B: Combining SVR-NSGA II

```python
import numpy as np
from sklearn.svm import SVR
import matplotlib.pyplot as plt
import pandas
from pymoo.model.problem import Problem
from pymoo.algorithms.nsga2 import NSGA2
from pymoo.optimize import minimize
from pymoo.visualization.scatter import Scatter
from platypus import NSGAII, Problem, Real
# step 1: Load data from a csv file
df = pandas.read_csv('C:/research/data.csv')
# step 2: Separate input features and target
  variables
ra = np.array(df['Ra'])
f = np.array(df['F'])
x = df.drop('Ra',axis=1)
x = x.drop('F',axis=1)
X1 = np.array(x)
# step 3: Create and train the SVR model
svr1 = SVR(kernel='rbf', C=1e1, gamma=0.1,
  epsilon=.01)
svr2 = SVR(kernel='rbf', C=1e1, gamma=0.1,
  epsilon=.01)
svr1.fit(X1,ra)
svr2.fit(X1,f)
# step 4: Test the trained model
rap = svr1.predict(X1)
fp = svr2.predict(X1)
class MyProblem(Problem):
  def __init__(self):
  super().__init__(n_var=5,
  n_obj=2,
  n_constr=0,
  xl=np.array([1.0, 50.0, 0.1,60,0.02]),
  xu=np.array([6.0, 150.0, 0.9,240,0.1]),
  elementwise_evaluation=True)
```

```python
    def _evaluate(self, x, out, *args, **kwargs):
    f1 = svr1.predict([x])
    f2 = svr2.predict([x])
    g1 = -svr1.predict([x])
    g2 = -svr2.predict([x])
    out["F"] = np.column_stack([f1, f2])
    out["G"] = np.column_stack([g1,g2])


 problem = MyProblem()


 algorithm = NSGA2(pop_size=50, return_least_
infeasible=False,eliminate_duplicates=True)
 res = minimize(problem,
  algorithm,
  ('n_gen', 100),
  seed=1,
  verbose=True)
plt.scatter([s.objectives[0] for s in res],
  [s.objectives[1] for s in res)
plt.xlim([0, 1.1])
plt.ylim([0, 1.1])
plt.title('Pareto')
plt.xlabel("Ra")
plt.ylabel("F")
plt.show()
```

Appendix C: Pseudocode and Explanation of NSGA-II

<u>fast-non-dominated-sort(P)</u>
for each $p \in P$
   $S_p = \phi$
   $n_p = 0$
   for each $q \in P$
      if $(p \prec q)$ then   If p dominates q
         $S_p = S_p \cup \{q\}$ Add q to the set of solutions dominated by p
      else if $(q \prec p)$ then
         $n_p = n_p + 1$ Increment the domination counter of p
   if $n_p = 0$ then p belongs to the first front
      $p_{\text{rank}} = 1$
      $\mathcal{F}_1 = \mathcal{F}_1 \cup \{p\}$
$i = 1$ Initialize the front counter
while $\mathcal{F}_i \neq t\ \phi$
   $Q = \phi$ Used to store the members of the next front
   for each $p \in \mathcal{F}_i$
      for each $q \in S_p$
         $n_q = n_q - 1$
         if $n_q = 0$ then q belongs to the next front
             $q_{\text{rank}} = i + 1$
             $Q = Q \cup \{q\}$
   $i = i + 1$
   $\mathcal{F}_i = Q$

Appendix D: Pareto Front Data Summary for SVR-NSGA II-TOPSIS Analysis in Multi-Objective Machining Optimization

No.	P	Q	V_c	f_z	a_p	R_a	F_c	MRR	S_i^+	S_i^-	C_i^+	Ranking
	bar	ml/h	m/min	mm/tooth	mm	µm	N	mm³/min				
22	3.8	124.5	249.92	0.09	0.80	0.6705	775.9935	8306.0679	0.0000	60.0040	1.0000	1
16	4.5	103.1	258.76	0.06	0.84	0.6705	775.9935	6123.8484	16.7037	43.3003	0.7216	2
14	4.3	107.4	280.02	0.06	0.80	0.6705	775.9935	5949.2332	17.9983	42.0057	0.7000	3
10	4.1	101.9	230.81	0.07	0.72	0.6704	775.9337	4971.2239	25.2491	34.7549	0.5792	4
19	2.3	94.2	234.23	0.06	0.70	0.6700	775.4553	4251.5518	30.6069	29.3971	0.4899	5
9	3.5	60.2	178.41	0.07	0.70	0.6705	775.9935	4173.1127	31.1884	28.8156	0.4802	6
18	3.4	57.8	220.71	0.08	0.46	0.6705	775.9935	3602.8198	35.4016	24.6024	0.4100	7
7	4.0	136.1	278.46	0.06	0.44	0.6705	775.9935	3030.1122	39.6476	20.3564	0.3393	8
12	5.8	120.8	297.29	0.04	0.47	0.6705	775.9935	2639.8967	42.5554	17.4486	0.2908	9
1	3.5	73.3	227.38	0.05	0.48	0.6705	775.9935	2576.9691	43.0071	16.9969	0.2833	10
11	2.3	110.8	144.27	0.05	0.70	0.6705	775.9934	2432.1981	44.0952	15.9087	0.2651	11
8	4.3	61.3	154.64	0.08	0.41	0.6705	775.9933	2417.1109	44.1923	15.8117	0.2635	12
6	4.1	58.8	151.55	0.09	0.40	0.6704	775.8000	2388.8050	44.4170	15.5870	0.2598	13
20	4.3	125.7	148.61	0.06	0.60	0.6705	775.9935	2336.8349	44.8023	15.2017	0.2533	14
3	2.9	122.0	122.33	0.10	0.39	0.6705	775.9935	2096.3092	46.5632	13.4408	0.2240	15
21	5.4	59.1	238.92	0.06	0.30	0.6704	775.8294	2033.2020	47.0311	12.9729	0.2162	16
5	4.4	135.1	217.86	0.06	0.31	0.6705	775.9935	1822.8466	48.6129	11.3911	0.1898	17
17	3.6	92.4	158.54	0.02	0.84	0.6705	775.9931	1411.3812	51.6708	8.3331	0.1389	18
24	2.1	143.3	188.53	0.10	0.15	0.6705	775.9935	1243.1059	52.9258	7.0782	0.1180	19
4	5.7	98.5	157.47	0.05	0.26	0.6703	775.8040	974.6922	54.8861	5.1178	0.0853	20
23	4.2	88.0	144.94	0.07	0.16	0.6705	775.9935	752.5325	56.5628	3.4411	0.0573	21
15	4.9	90.6	152.80	0.03	0.32	0.6705	775.9859	673.0498	57.1225	2.8815	0.0480	22
13	2.6	149.0	163.20	0.04	0.11	0.6705	775.9935	291.0797	59.9766	0.0274	0.0005	23
2	6.0	104.7	150.60	0.02	0.19	0.6666	770.6797	284.3820	60.0040	0.0000	0.0000	24

4 3D Printing Parameters for Optimum Tensile Strength Using the Taguchi-Based Response Surface Method

Trieu Khoa Nguyen, Nguyen Thy Ton That, and Binh-Duong Nguyen*

4.1 INTRODUCTION

Today, 3D printing is a prominent manufacturing methodology characterized by remarkable advancements in fabricating intricate geometries that pose challenges for conventional manufacturing techniques. Its applications span a wide spectrum, encompassing diverse sectors including manufacturing and education. It can be classified into liquid-based, solid-based, or powder-based printing, with many different variations.

Among the array of 3D printing technologies, solid-based fused deposition modeling (FDM) distinguishes itself as the most prevalent approach for producing plastic parts. This prevalence owes itself to FDM's user-friendly operational characteristics and cost-effectiveness and the availability of a broad spectrum of compatible printing materials. Notably, the materials compatible with FDM range from pure plastics to varying proportions of metals or even wood. Invariably, the quality of a 3D-printed object is contingent not only upon its design and the selected material but also on printing parameters including layer thickness, infill density, printing speed, deposition temperature, chamber temperature, and printing angle. As a result, many researchers have undertaken the task of optimizing these printing parameters to cater to the distinct characteristics of various polymeric materials.

There has been much research on optimizing FDM 3D printing, for instance research to optimize the mechanical properties of polylactic acid (PLA) based on the parameters of build orientation, layer thickness, and feed rate [1]. In this study, because there were only three inputs, the authors used a full factorial setup; then they used (ANOVA) to analyze experimental data with Statgraphics Centurion XVII software. They also analyzed the impact of each individual factor on tensile strength

*Corresponding author: nguyenkhoatrieu@iuh.edu.vn

DOI: 10.1201/9781032635170-4

and found their own optimal 3D printing conditions. However, due to limitations of experimental setup and data analysis methods, optimal parameters for all three factors could not be given simultaneously. Next, a different research group used ANOVA to study the impacts of layer thickness and raster angle on the tensile strength of FDM 3D-printed samples for ABS (acrylonitrile butadiene styrene) and PLA [2].

In addition, V. Kovan et al. [3] studied the impacts of layer thickness and printing orientation angle on the mechanical properties of PLA 3D-printed specimens. Since there were only two parameters, they used a full factorial design and found the appropriate parameters for maximum tensile strength. Meanwhile, in addition to raster angle and layer thickness, M. Leite et al. [4] also researched infill density and extrusion temperature and used ANOVA to determine the optimal parameters for their printer; however, they did not calculate the contribution level, in other words, the significance of the parameters.

Also using ANOVA with FDM printing using PLA material, the authors in [5] used central composite design for experimental planning. With only three parameters, namely deposition angle, extruder temperature, and printing speed, they needed as many as 17 experiments; moreover, they could not conclude on the importance of the parameters. To overcome those disadvantages, others used the Taguchi method to improve the tensile properties of 3D-printed PLA samples [6]. These authors optimized infill density, nozzle temperature, honeycomb infill, and extrusion width and concluded that infill density plays a major role in influencing the tensile strength. This conclusion was readily predictable and lacked any exceptional characteristics. Also using the Taguchi method, researchers listed 14 process parameters but selected only three: infill density, printing speed, and layer thickness; therefore, their study was not comprehensive, but they also found that infill density was the most important parameter [7].

Meanwhile, because authors who only studied two parameters, printing temperature and printing speed, used full factorial planning and concluded that the tensile strength of PLA samples was proportional to both temperature and printing speed [8]; however, because they did not apply experimental planning, they could not evaluate the influence of these two parameters. Also studying only two parameters—printing temperature and printing speed—and using full factorial design, a more recent research group used both PLA and PETG plastic [9]; in line with previous research, they also concluded that the mechanical properties of PLA samples were proportional to printing temperature and speed. Also using full factorial design, others studied only the effect of infill design on tensile strength and showed the importance of infill type in optimizing tensile strength on FDM 3D printers [10].

The authors of [11] studied seven parameters to gain a more general view in optimizing the process; using the Taguchi method for PLA plastic samples, the authors concluded that infill density and infill pattern were the two most important parameters affecting tensile strength, 36% and 29.2%, respectively. They also determined that 3D-printed objects were highly orthotropic materials, so their mechanical properties were greatly influenced by the geometry (scalability effects). Hence, printing position factors can also have an important impact on tensile strength results. Also using the Taguchi method, researchers studied only four parameters—printing temperature, infill density, infill angle, and layer thickness—instead of seven to achieve maximum tensile strength and did find optimized 3D printing parameters for their equipment and materials [12].

4.2 PROBLEM STATEMENT

In the Industrial Revolution 4.0, 3D printing is an important component. Therefore, it is now critical to include 3D printing, or additive manufacturing, in education. As a typical technique, FDM is easy to use, easy to understand, and suitable for new learners, and it is often included at universities as well as in short-term STEM courses. While giving lessons, lecturers not only have to guide students to operate the machine but also explain the factors that affect the quality of printed products. Regarding printed product quality, there are two main criteria: tensile strength and surface quality [13]. While the surface finish of FDM-printed products is mainly affected by the thickness of the printed layer, the factors that affect tensile strength are much more complex. While initial thoughts suggest that printing temperature or printing speed are the most important parameters, this issue needs to be carefully studied.

Specific to our work, it is necessary to teach 3D printing as part of supporting students completing their university graduation theses for the Faculty of Mechanical Engineering (FME), Industrial University of Ho Chi Minh City (IUH); practical situations at the school have shown that tensile strength was the most important criterion. Normally, for a mechanical engineering thesis, the students design and machine a few typical parts made of metal, using 3D printing to print secondary parts to speed up the process of completing the thesis. The main goal of a graduation thesis is to help students apply the knowledge they have learned to manufacture a device, test it, and document the outcomes; the goal is not to mass-produce products. Therefore, the application of 3D printing to support graduation theses is appropriate and meets the training needs for Industry 4.0.

We consider that researching the parameters that affect the tensile strength of FDM 3D-printed products is necessary both for FME lecturers guiding students and for training facilities in general. For this study, we propose a two-step optimization process to help operators understand and select the optimal printing parameters. However, the process applies not only to FDM 3D printers but to additive manufacturing in general.

4.3 MATERIALS AND METHODS

4.3.1 THE FDM 3D PRINTER

For this work, we used a Cubicon Style® 3D FDM printer to fabricate specimens, as illustrated in Figure 4.1. The printer was characterized by its rectangular form with dimensions of 322 × 350 × 486 mm and a weight of 15 kg. The printer was equipped with three stepper motors for its operational control. Specifically, one of these stepper motors facilitated movement along the Z-axis through a lead screw mechanism, while the remaining two motors governed movement along the X-axis and Y-axis employing a belt mechanism. We used the default printer nozzle diameter of 0.4 mm. The printer could achieve a maximum printing speed of 150 mm/s. In terms of its spatial printing capabilities, the printer can produce polymeric objects with dimensions up to 150 × 150 × 150 mm in length × width × height, respectively.

FIGURE 4.1   The FDM-type Cubicon Style® 3D printer utilized in this study.

4.3.2 THE PLA FILAMENT

The material utilized here was cheap PLA filament [14], a biodegradable thermoplastic derived from renewable sources. This plastic can be composed of many additives such as cornstarch, sugarcane, cassava, and even potato starch. PLA printing filament is one of the most popular FDM 3D printing polymeric materials on the market. This type of plastic is low-cost and easy to print, and it comes in many vibrant colors.

PLA has many outstanding advantages over other commercial petrochemical plastic products such as ABS or polyvinyl alcohol. With the ability to decompose in the environment, PLA is widely used to produce daily utensils such as food packaging, trays, and cups. Therefore, it is suitable for use by students to familiarize themselves with additive manufacturing. Some PLA properties include that it is a thermoplastic and has a melting temperature between 190 °C and 220 °C; just like other printing plastics, it can also be used well for all types of FDM printers. PLA has good elasticity and comes in diverse colors for customers to freely choose.

Additionally, as a thermoplastic, PLA has the special property of being able to be softened many times under heat and become solid when cooled. Thermal impact changes only its physical but not its chemical properties, so it can be recycled many times, including the waste products generated during the 3D printing process. The

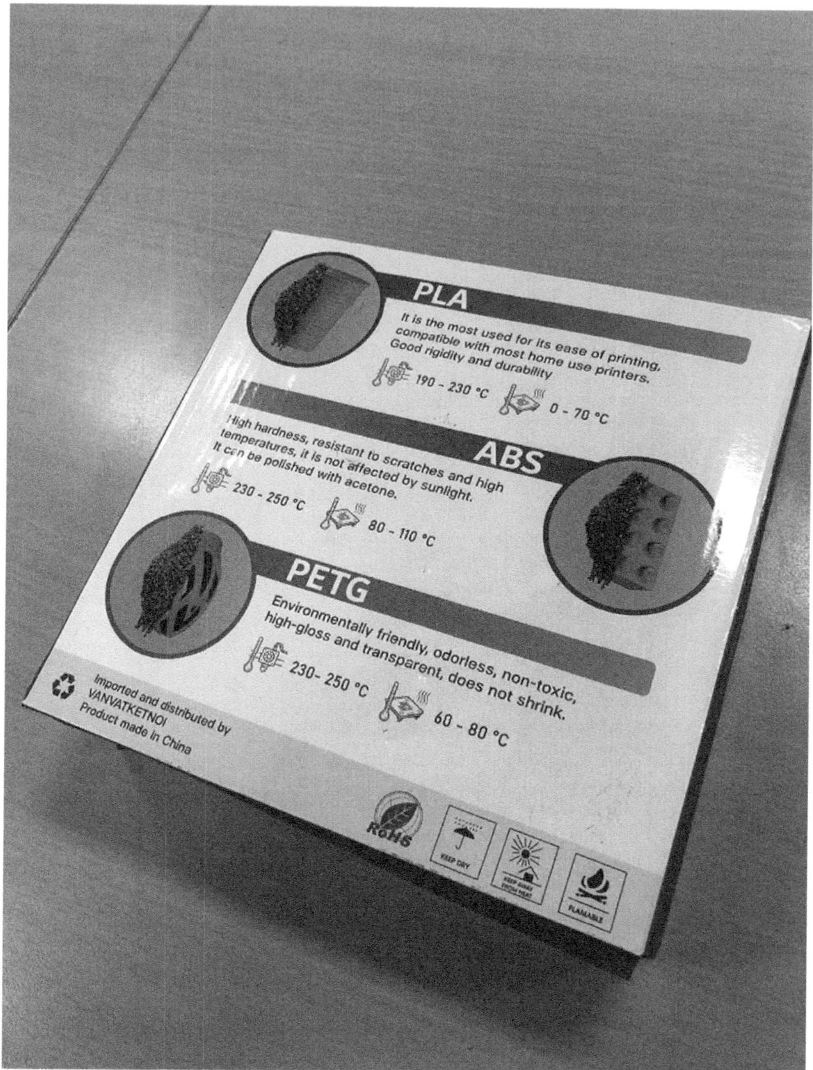

FIGURE 4.2   PLA filament from China used for the study.

material we selected for the present investigation was HUTI brand PLA filament as shown in Figure 4.2, manufactured in China; we selected it for its cost-effectiveness and widespread utilization. Detailed information regarding the filament's properties and recommended printer parameters can be found in Table 4.1.

4.3.3 THE TAGUCHI METHOD

The Taguchi method is a prevalent experimental design employed to systematically examine the influence of multiple parameters on both the mean outcomes

TABLE 4.1

The Recommended Parameters and Characteristics of PLA Filament

No	Performance	Parameters	Units
1	Printing temperature	190–230	0C
2	Printing bed temperature	0–70	0C
3	Diameter of filament	1.75	mm
4	Tolerance of diameter	±0.03	mm

and variances of a designated process [15]. This approach conventionally employs an orthogonal array to methodically explore the repercussions of distinct parameters while concurrently reducing the number of experimental trials. The Taguchi method is esteemed in engineering circles for its capacity to robustly evaluate the significance of all pertinent parameters that impact the output of a given process.

The Taguchi method entails measuring each parameter's impact using the signal-to-noise (S/N) ratio, the mean of iterated experiments divided by their corresponding variance. In the context of this investigation, the attribute under consideration was tensile strength, and we aimed maximize it following Equation (4.1) [16]:

$$\frac{S}{N} = -10 log_{10} \left[\frac{1}{n} \sum_{1}^{n} \frac{1}{y_i^2} \right] \qquad (4.1)$$

where y signifies the observed measurement and n signifies the total number of experimental trials conducted. Calculating the S/N ratio allows for calculating the individual influence of each parameter following Equation (4.2) [17]:

$$C_i = \frac{R_i}{\sum Ri} \times 100\% \qquad (4.2)$$

Here, Ri designates the range, that is, the difference between the minimum S/N ratio and the maximum S/N ratio corresponding to each parameter level. Calculating this range enables identifying the specific parameters that yield the highest S/N ratio. Consequently, this identification facilitates the prediction of optimal parameters that lead to the minimal deviation from the desired property in the final output.

4.3.4 Tensile Measurement

4.3.4.1 The Measurement Method

We measured tensile strength following ASTM D638 and using a Shimadzu Autograph AG-X Plus 20kN universal testing machine. The testing velocity was set at 5 mm/min. The pertinent specifications of this universal testing apparatus were a load capacity of 20 kN, a crosshead speed range spanning from 0.0005 to 1000 mm/min, a maximum return speed of 1200 mm/min, crosshead speed precision at ±0.1%, and a crosshead-table clearance measuring 850 mm [18].

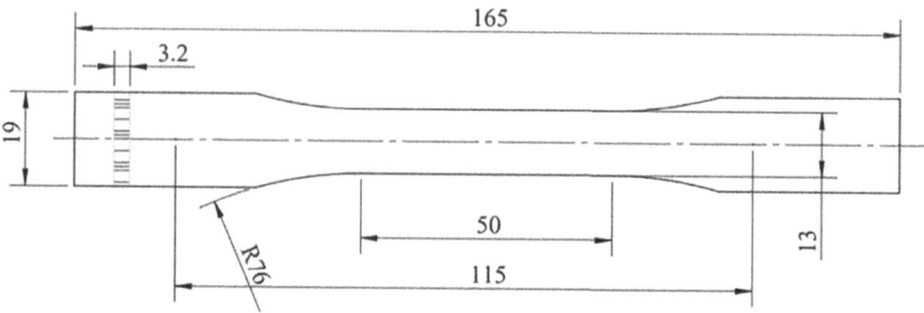

FIGURE 4.3   The dimensions of the test sample used according to ASTM D638.

4.3.4.2   The Dog-Bone Sample

We printed the sample for the tensile tests according to ASTM D638. The ASTM D638 test method is one of the most popular methods of measuring the tensile strength of reinforced and unreinforced plastics using a standard dumbbell or dog-bone-shaped sample under consistent temperature, humidity, and test speed [12]. In this case, the printed sample had a dog-bone shape with a length of 165 and a thickness of 3.2 mm, and the distance between grips was 115 mm.

4.4   RESULTS AND DISCUSSION

4.4.1   Process Parameter Selection

Numerous setting parameters are involved in the 3D printing process, encompassing both general practices and those specific to the Cubicon Style® printer. In the context of this research, a comprehensive examination of the literature and preliminary experiments guided the selection of seven key factors that we deemed exerted the most pronounced influence on tensile strength: infill style, printing density, layer thickness, nozzle speed, extrusion temperature, raster angle, and housing temperature; we held all other printing parameters at their default settings. The study parameters we employed in this research are detailed in Table 4.2.

In contrast to more intricate printers that afford a wide array of infill types, such as line, grid, cubic, concentric, tetrahedral, gyroid, zigzag, octet, or tri-hexagonal [19], the current printer, which is typically employed for rudimentary training or straightforward tasks, offers only two available infill types: line and concentric. The remaining six parameters were characterized by three levels each, and their central values adhered to the manufacturer's recommended settings for PLA filament.

4.4.2   Orthogonal Array Selection

Considering the number of selected parameters along with their respective levels, we employed the L18(2^1 3^6) orthogonal array (OA) delineated in Table 4.3. Following that experimental design, we printed a total of 18 samples and subjected them to tensile testing; Table 4.3 also provides both the computation index and the corresponding S/N ratio for each tested sample. As noted earlier, our ultimate aim was to maximize the tensile strength based on the highest S/N ratio.

TABLE 4.2
Process Parameters and Levels

No.	Symbol	Printing Factors	Printing Levels		
			1	2	3
1	A	Infill type (-)	Line	Concentric	
2	B	Printing density (%)	30.0	40.0	50.0
3	C	Layer thickness (mm)	0.1	0.15	0.2
4	D	Nozzle speed (mm/s)	80.0	100.0	120.0
5	E	Extrusion temperature (0 °C)	205.0	210.0	215.0
6	F	Raster angle (0°)	0	22.5	45.0
7	G	Housing temperature (0 °C)	35.0	45.0	55.0

TABLE 4.3
The L18 Orthogonal Array Used in the Current Study

No.	Printing Parameters							Max Stress (MPa)	MSD	S/N Ratio
	A	B	C	D	E	F	G			
1	Line	30	0.1	80	205	0	35	18.590	345.603	25.386
2	Line	30	0.15	100	210	22.5	45	23.119	534.470	27.279
3	Line	30	0.2	120	215	45	55	27.556	759.333	28.804
4	Line	40	0.1	80	210	22.5	55	20.021	400.824	26.030
5	Line	40	0.15	100	215	45	35	24.483	599.437	27.777
6	Line	40	0.2	120	205	0	45	28.029	785.647	28.952
7	Line	50	0.1	100	205	45	45	20.667	427.133	26.306
8	Line	50	0.15	120	210	0	55	25.661	658.492	28.186
9	**Line**	**50**	**0.2**	**80**	**215**	**22.5**	**35**	**30.943**	**957.444**	**29.811**
10	Con.	30	0.1	120	215	22.5	45	19.224	369.570	25.677
11	Con.	30	0.15	80	205	45	55	26.536	704.170	28.477
12	Con.	30	0.2	100	210	0	35	29.780	886.819	29.478
13	Con.	40	0.1	100	215	0	55	20.988	440.492	26.439
14	Con.	40	0.15	120	205	22.5	35	24.944	622.213	27.939
15	Con.	40	0.2	80	210	45	45	30.201	912.106	29.600
16	Con.	50	0.1	120	210	45	35	20.978	440.056	26.435
17	Con.	50	0.15	80	215	0	45	26.636	709.466	28.509
18	Con.	50	0.2	100	205	22.5	55	30.511	930.897	29.689
			Average					24.937		27.821

Notes: MSD, mean standard deviation; con., concentric infill type; S/N, signal to noise.

4.4.3 S/N Ratio and ANOVA Results

The S/N ratios we calculated are graphically depicted in Figure 4.4, elucidating the extent of influence exerted by the individual factors on the resultant outcomes [20]. Layer thickness emerged as the predominant influence on tensile

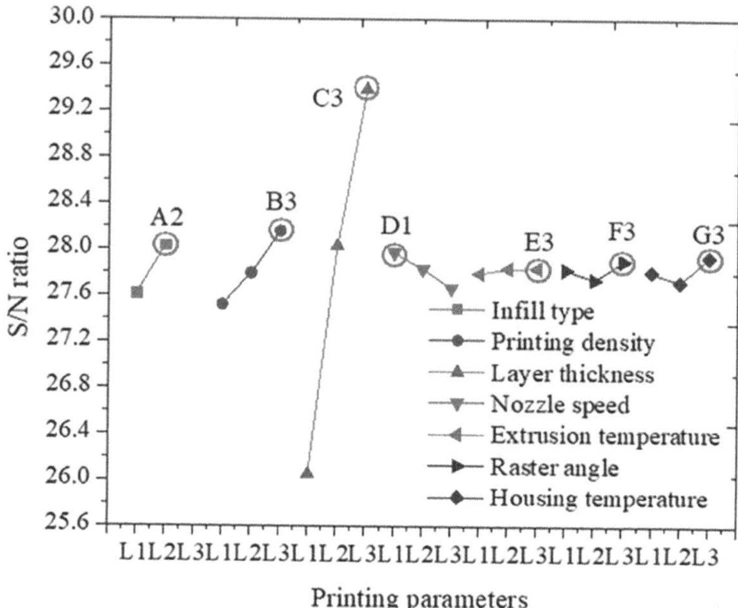

FIGURE 4.4   The S/N ratios of the tested samples.

strength, followed sequentially by printing density and infill type. We found these results surprising, but they do align with the findings of several prior studies on this subject. These observations underscored the distinctions between the present printer and its associated printing methodology when compared with previous investigations.

Specifically, the Cubicon Style® printer's interlayer fusion exhibited suboptimal performance, and the tensile strength was in turn notably influenced by the intrinsic properties of the extruded plastic filament. Specifically, larger extruded plastic filament diameters required greater force to break them, and we in turn observed a proportional increase in the tensile strength of the 3D-printed model. As depicted in Figure 4.5, the parameter contributions revealed that (C) layer thickness exerted the most substantial influence, accounting for 65.27%, followed by (B) printing density at 12.47%, and (A) infill type at 8.05%, all contributing to the overall tensile strength.

To validate the S/N ratio findings and undertake a more comprehensive investigation into the contributions of the pertinent factors, we conducted an ANOVA and present the results in Table 4.4. Although we observed minor numerical disparities, the ANOVA outcomes exhibited a substantial concurrence with the S/N ratio results; only layer thickness and printing density exhibited F values equal to or exceeding the critical $F_{(0.05, x, 4)}$ threshold, signifying their statistical significance. We subsequently validated the calculated results using the commercial software Minitab 18; the Minitab results corroborated the earlier calculations.

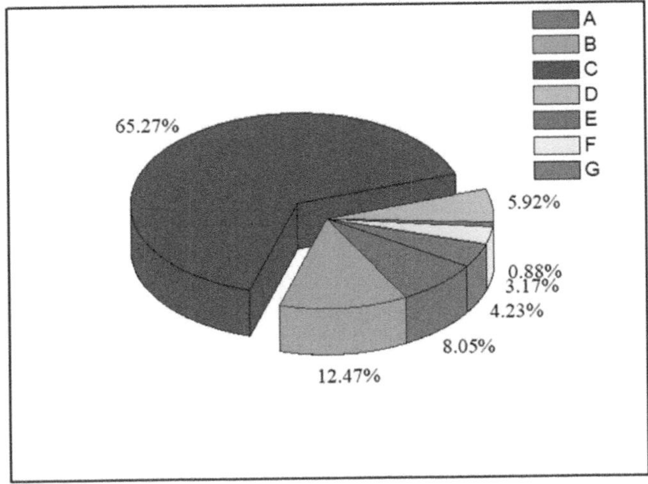

FIGURE 4.5   The contributions of each printing parameter.

Notes: (A) Infill style, (B) printing density, (C) layer thickness, (D) nozzle speed, (E) extrusion temperature, (F) raster angle, (G) housing temperature.

TABLE 4.4
ANOVA Results

	S	f	V	F	$F_{(0.05,x,4)}$	P%	Rank
A	6.394	1	6.394	9.274	7.709	2.20	3
B	9.574	2	4.787	6.943	6.944	3.29	2
C	267.277	2	133.639	193.846	6.944	91.92	1
D	3.559	2	1.780	2.582	6.944	1.22	4
E	0.030	2	0.015	0.022	6.944	0.01	7
F	0.231	2	0.115	0.167	6.944	0.08	6
G	0.963	2	0.482	0.699	6.944	0.33	5
Error	2.758	4	0.689				
Total	290.786	17					

4.4.4   THE FIRST VALIDATION EXPERIMENT

To corroborate the obtained optimal setting parameters (A2, B3, C3, D1, E3, F3, G3), we conducted a supplementary validation experiment. The tensile test results for this additional sample are presented in Figure 4.6, revealing a maximum of 32.748 MPa. Notably, this value exceeded the values obtained in the preceding 18 experimental trials, thus underscoring the improved outcome achieved through using the Taguchi method.

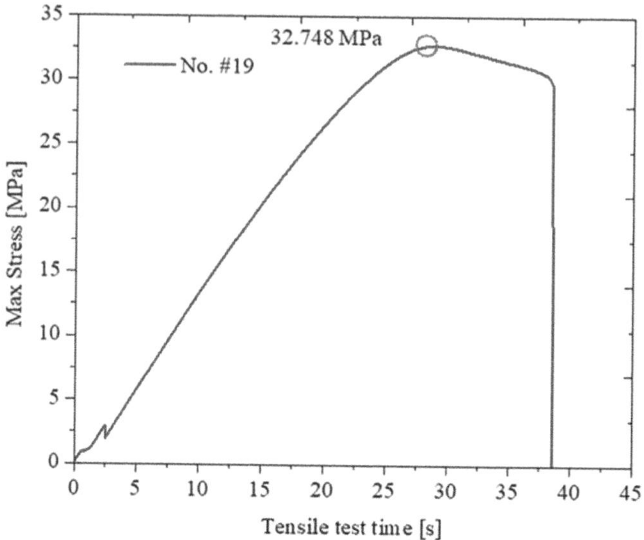

FIGURE 4.6 The tensile strength of the first validation sample.

4.4.5 DEVELOPING THE RESPONSE SURFACE MODEL

The Taguchi method represents a potent and efficient tool capable of quantifying the relative importance of each individual factor as well as identifying enhanced parameter combinations to maximize the tensile strength of printed samples. However, the Taguchi method confines the optimized values to specific levels among the available options for each factor. Consequently, we used response surface methodology (RSM) to further optimize the tensile strength of the PLA samples. In this investigative endeavor, we constructed a first-order response model based on the experimental outcomes presented in Table 4.3. Employing the least-squares method, we calculated all model coefficients and arrived at the simplified model in Equation 4.3:

$$MaxStress = 12.574 + 1.192 \times A + 0.882 \times B + 4.712 \times$$
$$C - 0.545 \times D + 0.046 \times E + 0.061 \times F + 0.130 \times G \qquad (4.3)$$

Within this model, we calculated that layer thickness held the most substantial influence, as indicated by its largest coefficient. This outcome aligns with the earlier findings derived from both the S/N ratio analysis using the Taguchi method and the ANOVA. Notably, the first-order model also underscores the greater importance of infill type than that of printing density in maximizing tensile strength. This observation was reinforced by the relatively modest error margin ($R^2 = 0.984$) in the regression analysis, suggesting the potential for further model refinement, even though there is limited scope for improvement.

4.4.6 THE SECOND VALIDATION EXPERIMENT

In the pursuit of enhancing the RSM model, the subsequent phase of the analysis incorporated second-order terms for all factors. Notably, due to the absence of statistically significant correlations between factors, the inclusion of second-order terms accounting for the interaction between the two factors was omitted. Consequently, the second-order response model took the following form:

$$\begin{aligned}
MaxStress = {}&-44.196 - 91267.786 \times A - 1.035E - 1 \times B + 146.976 \\
&\times C - 3.659E - 2 \times D + 5.818E - 1 \times E - 1.641E - 2 \times F - 3.798 \\
&\times G + 91268.978 \times A^2 + 2.397E - 3 \times B^2 - 175.744 \\
&\times C^2 + 4.681E - 5 \times D^2 - 1.363E - 3 \times E^2 + 4.254E - 4 \\
&\times F^2 + 4.364E - 3 \times G^2
\end{aligned} \tag{4.4}$$

It is worth noting that this model achieved an R^2 of 0.991, indicating a modest improvement over the first-order model. Furthermore, it exhibited a nonlinear regression with an acceptable level of error.

Utilizing this model, we could formulate a response surface to visually depict the influences of the factors under examination. Figure 4.7 shows a representative plot of the response surface of this study considering printing density and layer thickness. Within these plots, the highest points denote the optimal values associated with their respective factors.

We used the established surface response model to ascertain the optimal values for each factor and achieve the maximum tensile strength. As detailed in Table 4.5, the two methodologies yielded identical predictions for the optimal values of parameters

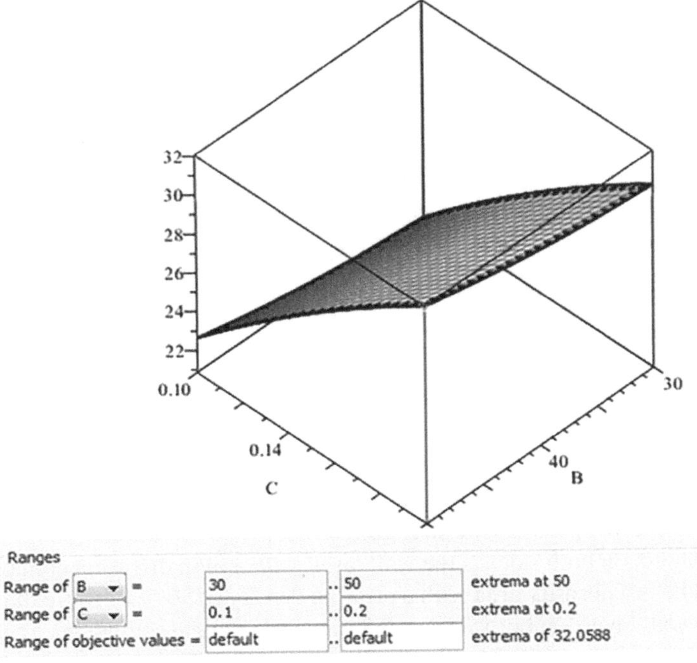

FIGURE 4.7 The response surface generated by B, printing density, and C, layer thickness.

TABLE 4.5
Validation Results

Symbol	Printing Factors	Taguchi	RSM
A	Infill type	Concentric	Concentric
B	Printing density	50.0	50.0
C	Layer thickness	0.2	0.2
D	Nozzle speed	80.0	80.0
E	Extrusion temperature	215.0	213.4
F	Raster angle	45.0	45.0
G	Housing temperature	55.0	55.0
	Prediction values	N/A	32.059
	Validation results	32.748	0.16200

A, B, C, D, F, and G, but in the case of parameter E (extrusion temperature), RSM yielded a more precise estimate than the S/N ratio.

Table 4.5 presents the optimal factor values obtained through both approaches. We also formulated and utilized a straightforward MATLAB® script to determine these optimal values based on the second-order response model, and the Maple® optimization module produced results that aligned with those derived from the MATLAB® script, thus validating the outcomes. This underscores the feasibility of conducting studies with limited resources by employing tools like MATLAB®, negating the necessity for complex algorithms or commercial software solutions.

It is noteworthy that when considering all seven factors for the construction of the second-order response model, established Design of Experiments techniques such as Box–Behnken design and central composite design typically necessitate 108 and, at times, up to 180 experimental trials, respectively. Consequently, this approach allowed for the efficient determination of optimal combinations to maximize tensile strength, obviating the need for supplementary screening of printing parameters. Figure 4.8 shows the findings for the second validation experiment.

4.4.7 Fracture Morphology Analysis

We took SEM imaging using a JEOL JSM-IT500 scanning electron microscope equipped with an Oxford ULTIM MAX SDD-EDS detector. The resolution in high vacuum mode was 3.0 nm (30 kV) × 15.0 nm (1.0 kV), and in low vacuum mode, it was 4.0 nm (30 kV BED). The direct magnification was from ×5 to ×300,000 (defined with a display size of 128 mm × 96 mm). Figure 4.9 presents the fracture morphology of sample 20; we took the image at the fracture surface of the sample after testing the sample's tensile strength.

In the photo, on both sides is the wall layer, with a high diffusion connection, and in the middle is a porosity area with a printing density of 50%. Because the thickness of the two wall layers on both sides is 0.8 mm each and the sample thickness is quite small, 3.2 mm, the porosity area has a small thickness. The porous area shows poor adhesion or diffusion between layers. The local deformation of the plastic fiber at the

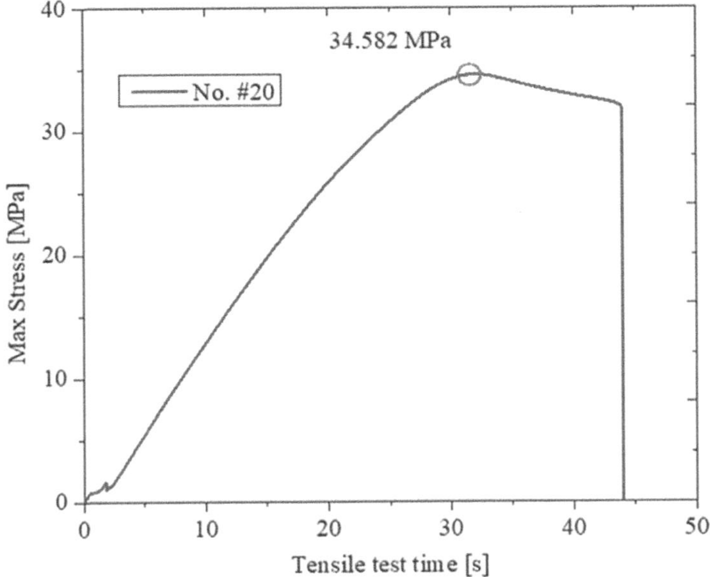

FIGURE 4.8 The tensile strength of the second validation sample.

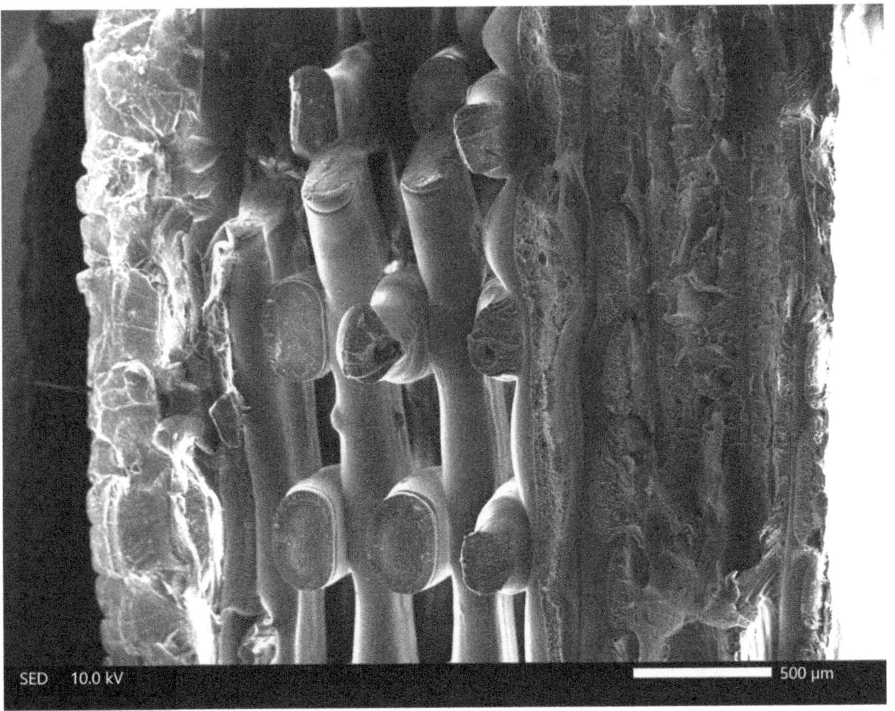

FIGURE 4.9 The SEM imaging of the second validation sample.

fracture site can also be observed in the SEM image. Therefore, the tensile strength of the sample depends on the extruded strand size. This again validates the ANOVA results that layer thickness is the most important factor affecting tensile strength.

4.5 CONCLUSION

In this chapter, we used Taguchi RSM to experimentally study the influence of seven 3D printing parameters on the tensile strength of FDM-printed samples of PLA filament to select the optimal parameters. For our work, we simply utilized a Taguchi orthogonal array in conjunction with ANOVA and RSM, and in a simple FDM 3D printing process using a Cubicon Style® machine, we studied infill style, printing density, layer thickness, nozzle speed, extrusion temperature, raster angle, and housing temperature.

We determined from the S/N ratio analyses and the ANOVAs based on the Taguchi method with an OA L18 (21 36) that layer thickness was the most important parameter for increasing tensile strength, followed by printing density (65.27% and 12.47%, respectively). This outcome was also confirmed by Minitab commercial software and an additional validation experiment. The results from this study corroborate that the method proposed here can be used to set 3D printer parameters for higher tensile strength when using conventional Taguchi methods or ANOVA. More importantly, these findings help both lecturers and students better understand the impact of setting parameters on the tensile strength results of printed samples.

REFERENCES

[1] Chacón, J. M., Caminero, M. A., García-Plaza, E., and Núñez, P. J.: Additive manufacturing of PLA structures using fused deposition modelling: Effect of process parameters on mechanical properties and their optimal selection. Materials & Design 124, 143–157 (2017).

[2] Abdullah, Z., Ting, H. Y., Ali, M. A. M., Fauadi, M. H. F. M., Kasim, M. S., Hambali, A., Ghazaly, M. M., and Handoko, F.: The effect of layer thickness and raster angle on tensile strength and flexural strength for Fused Deposition Modeling (FDM) parts. Journal of Advanced Manufacturing Technology 12(1(4)), 147–158 (2018).

[3] Kovan, V., Tezel, T., Çamurlu, H., and Topal, E.: Effect of printing parameters on mechanical properties of 3D printed plaCarbon fibre composites. materials science. Non-Equilibrium Phase Transformations 4(4), 126–128 (2018).

[4] Leite, M., Fernandes, J., Deus, A., Reis, L., and Vaz, M. F.: Study of the influence of 3D printing parameters on the mechanical properties of PLA. Proceedings of the 3rd International Conference on Progress in Additive Manufacturing (Pro-AM 2018), 547–552 (2018).

[5] Ouhsti, M., Haddadi, B., and Soufiane, B.: Effect of printing parameters on the mechanical properties of parts fabricated with open-source 3D printers in PLA by fused deposition modeling. Mechanics and Mechanical Engineering 22, 895–907 (2018).

[6] Nugroho, A. W., and Budiantoro, C.: Improving the tensile properties of 3D printed PLA by optimizing the processing parameter. JEMMME (Journal of Energy, Mechanical, Material, and Manufacturing Engineering) 4(1), 29–36 (2019).

[7] Heidari-Rarani, M., Ezati, N., Sadeghi, P., and Badrossamay, M. R.: Optimization of FDM process parameters for tensile properties of polylactic acid specimens using Taguchi design of experiment method. Journal of Thermoplastic Composite Materials 35(12), 2435–2452 (2020).

[8] Tang, C., Liu, J., Yang, Y., Liu, Y., Jiang, S., and Hao, W.: Effect of process parameters on mechanical properties of 3D printed PLA lattice structures. Composites Part C: Open Access 3, 100076 (2020).

[9] Hsueh, M.-H., Lai, C.-J., Wang, S.-H., Zeng, Y.-S., Hsieh, C.-H., Pan, C.-Y., and Huang, W.-C.: Effect of printing parameters on the thermal and mechanical properties of 3D-printed PLA and PETG, using fused deposition modeling. Polymers 13(11), 1758 (2021).

[10] Harpool, T. D., Alarifi, I. M., Alshammari, B. A., Aabid, A., Baig, M., Malik, R. A., Mohamed Sayed, A., Asmatulu, R., and El-Bagory, T. M.: Evaluation of the infill design on the tensile response of 3D printed polylactic acid polymer. Materials 14(9), 2195 (2021).

[11] Auffray, L., Gouge, P.-A., and Hattali, L.: Design of experiment analysis on tensile properties of PLA samples produced by fused filament fabrication. The International Journal of Advanced Manufacturing Technology 118(11), 4123–4137 (2022).

[12] Le, D., Nguyen, C. H., Pham, T. H. N., Nguyen, V. T., Pham, S. M., Le, M. T., and Nguyen, T. T.: Optimizing 3D printing process parameters for the tensile strength of thermoplastic polyurethane plastic. Journal of Materials Engineering and Performance 32(14), 1–12 (2023).

[13] Huynh, T. T., Nguyen, T. V. T., Nguyen, Q. M., and Nguyen, T. K.: Minimizing warpage for macro-size fused deposition modeling parts. Computers, Materials \& Continua 68(3), 2913–2923 (2021).

[14] Nguyen, B. D., Ton That, N. T., and Nguyen, T. K.: A study on tensile strength of 3D-printed PLA samples by fused deposition modeling. In: Todor, D., Kumar, S., Choi, S. B., Nguyen-Xuan, H., Nguyen, Q. H., Trung Bui, T. (eds) Proceedings of the International Conference on Sustainable Energy Technologies. ICSET 2023. Green Energy and Technology. Springer, Singapore, pp. 349–357 (2024).

[15] Nguyen, T. K., Hwang, C. J., and Lee, B.-K.: Numerical investigation of warpage in insert injection-molded lightweight hybrid products. International Journal of Precision Engineering and Manufacturing 18(2), 187–195 (2017).

[16] Nguyen, T. K., Chau Duc, K., and Pham, A.-D.: Characterization of an FDM-3D printed moldcore in a thermoforming process using taguchi in conjunction with lumped-capacitance method. Arabian Journal for Science and Engineering 48, 11989–12000 (2023).

[17] Nguyen, T. K., Chau, M. Q., Do, T.-C., and Pham, A.-D.: Characterization of geometrical parameters of plastic bottle shredder blade utilizing a two-step optimization method. Archive of Mechanical Engineering 68(3), 253–269 (2021).

[18] Tran, N.-T., and Pham, N. T.-H.: Investigation of the effect of polycarbonate rate on mechanical properties of polybutylene terephthalate/polycarbonate blends. International Journal of Polymer Science 2021, 7635048 (2021).

[19] Pernet, B., Nagel, J. K., and Zhang, H.: Compressive strength assessment of 3D printing infill patterns. Procedia CIRP 105, 682–687 (2022).

[20] Nguyen, T. K., and Lee, B.-K.: Post-processing of FDM parts to improve surface and thermal properties. Rapid Prototyping Journal 24(7), 1091–1100 (2018).

5 Network Optimization Using the Max Product for Multicriteria Decision-Making

Meenakshi Annamalai, O. Mythreyi,
S. Dhanushiya, and Shivangi Mishra J

5.1 INTRODUCTION

Graph theory, as a convenient mathematical tool, has a broad spectrum of uses in computer science, electrical engineering, system analysis, operations research, economics, networking, and transportation. Graphs are visual mathematical representations of practical, real-life problems. A graph is a collection of sets $(\mathbb{V}, \mathbb{E})$ where $\mathbb{V}$ is a non-empty set of vertices connected by $\mathbb{E}$, whose constituents are edges or links. Representing a problem as a graph provides a significant perspective and clarifies the situation.

A network is typically a graph model with a set of nodes connected by edges or links. Networks give us a flexible framework for identifying and observing complex systems. The study of complex social networks is a crucial concept that comprises several disciplines. Complex systems network theory provides techniques for analyzing complex networks that are generally defined by simple graphs that consist of vertices representing the objects under exploration.

Lotfi Zadeh [1] introduced fuzzy set theory in 1965 as a generalization of classical set theory that allows us to represent imprecise and vague phenomena. Fuzzy is an upper version of a crisp set, where every item or element has a varying membership grade [2]. It can illustrate that its elements have distinct membership grades between 1 and 0. Membership degrees are not like probability. Using the fuzzy relation, Kaufmann presented the concept of a fuzzy graph; then, Rosenfeld introduced the concepts of fuzzy paths, fuzzy cycles, fuzzy bridges, fuzzy connectedness, and fuzzy trees to a fuzzy graph and described some of its characteristics. Mathematicians like Rashmanlou and Pal [3, 4] have conducted extensive research on fuzzy graphs and their applications in real-world problems.

Atanassov introduced a new type 1 fuzzy set, the intuitionistic fuzzy set [5]. Type 1 fuzzy sets have only a single membership grade, but the intuitionistic fuzzy set always considers two grades for each element: membership and non-membership.

DOI: 10.1201/9781032635170-5

Shannon and Atanassov [6] described the concept of intuitionistic fuzzy set relationships and graphs for the first time.

Meenakshi et al. [7–11] examined graphs as models to solve problems across various disciplines to optimize network structures. Sheikh Hoseini et al. [12] investigate the idea of the maximum product of graphs in a vague environment, which is a sort of uncertain environment distinguished by imprecise and partial data. Parvathi et al. [13–15] examined various degrees, orders, and sizes of intuitionistic fuzzy graphs that later researchers followed up on [3, 16, 17, 18]. Picture fuzzy graphs may be used to solve problems in fields like decision-making and social networks [19]. Vague and intuitionistic fuzzy graphs represent many real-world problems, but they cannot represent uncertainty due to conflicting or vague real-world information for decision-making [20]. For this reason, experts need new intuitionistic fuzzy graphs that extend above ordinary intuitionistic fuzzy graphs by including a more subtle representation of uncertainty using complex numbers [6, 21]. Aruldoss et al. [22] classified numerous MCDM methodologies, including classic and fuzzy-based approaches. For this chapter, we examined and characterized the maximum products of three intuitionistic fuzzy graphs. Our aim was to find the effective minimal spanning tree weighted network for optimizing a supply chain. In the next section, we define some of the elements of intuitionistic fuzzy graphs.

5.2 PRELIMINARIES

5.2.1 INTUITIONISTIC FUZZY GRAPHS

An intuitionistic fuzzy graph (IFG), $\mathbb{G} = (\mathbb{V}, \mathbb{E})$ where $\mathbb{V} = \{\mathbb{v}_1, \mathbb{v}_2, \dots \mathbb{v}n\}$ is a finite vertex set such that

(i) $\mu_1 : \mathbb{V} \to [0,1]$ and $\gamma_1 : \mathbb{V} \to [0,1]$ denote deterministic and nondeterministic membership, respectively, and $0 \le \mu_1(\mathbb{v}_i) + \gamma_1(\mathbb{v}_i) \le 1$ for every $\mathbb{v}_i \in \mathbb{V}$.

(ii) $\mathbb{E} \subseteq \mathbb{V} \times \mathbb{V}$ where $\mu_2 : \mathbb{V} \times \mathbb{V} \to [0,1]; \gamma_2 : \mathbb{V} \times \mathbb{V} \to [0,1]$ are such that $\mu_2\{(\mathbb{v}_i, \mathbb{v}_j)\} \le \min\{\mu_1(\mathbb{v}_i), \mu_1(\mathbb{v}_j)\}; \gamma_2\{(\mathbb{v}_i, \mathbb{v}_j)\} \ge \max\{\gamma_1(\mathbb{v}_i), \gamma_1(\mathbb{v}_j)\}$ and where $0 \le \mu_2\{(\mathbb{v}_i, \mathbb{v}_j)\}\} + \mu_2\{(\mathbb{v}_i, \mathbb{v}_j)\} \le 1 \forall (\mathbb{v}_i, \mathbb{v}_j) \in \mathbb{E}$.

5.2.2 STRONG ARC

An arc $(\mathcal{U}i, \mathcal{V}j)$ is said to be strong if

$$\mu_2(\mathbb{v}_i, \mathbb{v}_j) = \min\{\mu_1(\mathbb{v}_1), \mu_1(\mathbb{v}_2)\} \text{ and } \gamma_2(\mathbb{v}_i, \mathbb{v}_j) = \max\{\gamma_1(\mathbb{v}_1), \gamma_1(\mathbb{v}_2)\}$$

5.2.3 IFG DEGREES

In an IFG, $\mathbb{G} = (\mathbb{V}, \mathbb{E})$ the degree of a vertex $\mathcal{U}_i$ is defined as the sum of the weight of the strong arcs incident at $\mathcal{U}_i$. The neighbourhood of $\mathcal{U}_i$ is denoted by $N(\mathcal{U}_i) = \{\mathcal{V}j \in \mathbb{V} / (\mathcal{U}i, \mathcal{V}j)$ representing a strong arc. The minimum $\mathbb{G}$ is $\delta(\mathbb{G}) = \min\{d_{\mathbb{G}}(\mathcal{U}i) / \mathcal{U}i \in \mathbb{V}\}$, and the maximum $\mathbb{G}$ is $\Delta(\mathbb{G}) = \max\{d_{\mathbb{G}}(\mathcal{U}i) / \mathcal{U}i \in \mathbb{V}\}$.

5.2.4 Isolated Vertex

A vertex $\mathcal{U}i \in \mathbb{V}$ in an IFG $\mathbb{G} = (\mathbb{V}, \mathbb{E})$ is said to be an isolate vertex if $\mu_2\left(\mathrm{v}_i, \mathrm{v}_j\right) = 0$ and $\gamma_2\left(\mathrm{v}_i, \mathrm{v}_j\right) = 0$.

5.2.5 Cardinality of an IFG

Let $\mathbb{G} = (\mathbb{V}, \mathbb{E})$ be a intuitionistic fuzzy graph. Then the cardinality of $\mathbb{G}$ is defined as

$$|\mathbb{G}| = \left| \sum\nolimits_{\mathrm{V}_i \in \mathbb{V}} \frac{1 + T_{\mathrm{V}_\mathfrak{D}} - F_{\mathrm{V}_{\mathfrak{N}\mathfrak{D}}}}{2} + \sum\nolimits_{\mathrm{V}_i \mathrm{V}_j \in \mathbb{E}} \frac{1 + T_{\mathbb{E}_\mathfrak{D}} - F_{\mathbb{E}_{\mathfrak{N}\mathfrak{D}}}}{2} \right|$$

5.2.6 Dominating Set

Let $\mathbb{G} = (\mathbb{V}, \mathbb{E})$ be an IFG and let $\mathcal{U}i)$ and $\mathcal{V}j \in \mathbb{V}$. Then, say that $\mathcal{U}i$ dominates $\mathcal{V}j$ in $\mathbb{G}$ if there exists a strong arc between them. A subset $\mathbb{V}_d \subseteq \mathbb{V}$ is said to be a dominating set in $\mathbb{G}$ if for every $\mathcal{V}j \in \mathbb{V} - \mathbb{V}_d$, there exists $\mathcal{V}j \in \mathbb{V}_d$ dominates $\mathcal{U}i$.

5.3 MAXIMAL PRODUCT OF THREE INTUITIONISTIC FUZZY GRAPHS

In this section, we discuss the max product of three IFGs:

$$\text{Let } \mathcal{N}_1 = \left(\left(\sigma_{\ell_1}^{\mathcal{N}_1}, \sigma_{\ell_2}^{\mathcal{N}_1}, \sigma_{\ell_3}^{\mathcal{N}_1} \right), \left(\mu_{m_1}^{\mathcal{N}_1}, \mu_{m_2}^{\mathcal{N}_1}, \mu_{m_3}^{\mathcal{N}_1} \right) \right),$$

$$\mathcal{N}_2 = \left(\left(\sigma_{\ell_1}^{\mathcal{N}_2}, \sigma_{\ell_2}^{\mathcal{N}_2}, \sigma_{\ell_3}^{\mathcal{N}_2} \right), \left(\mu_{m_1}^{\mathcal{N}_2}, \mu_{m_2}^{\mathcal{N}_2}, \mu_{m_3}^{\mathcal{N}_2} \right) \right), \text{ and}$$

$$\mathcal{N}_3 = \left(\left(\sigma_{\ell_1}^{\mathcal{N}_3}, \sigma_{\ell_2}^{\mathcal{N}_3}, \sigma_{\ell_3}^{\mathcal{N}_3} \right), \left(\mu_{m_1}^{\mathcal{N}_3}, \mu_{m_2}^{\mathcal{N}_3}, \mu_{m_3}^{\mathcal{N}_3} \right) \right)$$

be three IFGs. The max product of the three graphs is defined as $N_1' = (A_1, B_1)$; $N_2' = (A_2, B_2)$; and $N_3' = (A_3, B_3)$ and denoted by $N_1' \times N_2' \times N_3'$ as follows:

(i) $\forall (l', m', o') \in V_1' \times M\, V_2' \times M\, V_3'$
$(O's_1 * O's_2 * O's_3)((l', m', o')) = \vee\, \{O's_1\, (l'), O's_2\, (m'), Os'_3\, (o')\}$
$(Ts'_1 * Ts'_2 * Ts'_3)((l', m', o')) = \wedge\, \{Ts'_1\, (l'), Ts'_2\, (m'), Ts'_3\, (o')\}$
$(R's_1 * R's_2 * Rs'_3)((l', m', o')) = \wedge\, \{R's_1\, (l'), R's_2\, (m'), Rs'_3\, (o'),$

(ii) $\forall q \in V_2', z \in V_3'$ and $l'm' \in E_1'$
$(O's_1 * O's_2 * O's_3)((l', q, r)(m', q, r)) = \vee\, \{Os'_1\, (l'm'), O's_2\, (q), Os'_3\, (r)\}$
$(Ts'_1 * Ts'_2 * Ts'_3)((l', q, r)(m', q, r)) = \wedge\, \{Ts'_1\, (l'm'), Ts'_2\, (q), Ts'_3\, (r)\}$
$(R's_1 * R's_2 * R's_3)((l', q, r)(m', q, r)) = \wedge\, \{Rs'_1\, (l'm'), R's_2\, (q), Rs'_3\, (r)\},$

(iii) $\forall p \in V_1', q \in V_2'$ and $l'm' \in E_3,$
$(O's_1 * O's_2 * O's_3)((p, q, l')(p, q, m')) = \vee\, \{O's_1\, (p), O's_2\, (q), Os'_3\, (l'm')\}$
$(Ts'_1 * Ts'_2 * Ts'_3)((p, q, l')(p, q, m')) = \wedge\, \{Ts'_1\, (p), Ts'_2\, (q), Ts'_3\, (l'm')\}$
$(R's_1 * R's_2 * R's_3)((p, q, l')(p, q, m')) = \wedge\, \{R's_1\, (p), R's_2\, (q), Rs'_3\, (l'm')\}.$

Let $MP_1 = (\sigma mp_1, \mu mp_1)$ and $MP_2 = (\sigma mp_2, \mu mp_2)$ be two intuitionistic fuzzy networks of the graphs $GMP_1 = (Vmp_1, Emp_1)$ and $GMP_2 = (Vmp_2, Emp_2)$, respectively. Then the maximal product of the graphs MP_1 and MP_2 is denoted by

$$MP_1 * MP_2 = (\sigma mp_1 * \sigma mp_2, \mu mp_1 * \mu mp_2)$$

$$\sigma MP_1 * MP_2 (u_1, v_1) = \sigma MP_1 (u_1) \vee \sigma MP_2 (v_1) \; and$$

$$\forall (u_1, v_1) \in \sigma_{mp} \text{ and } (u_1, v_1)(u_2, v_2) \in \mu_{mp}.$$

$$\mu (MP_1 * MP_2) ((u_1, v_1)(u_2, v_2)) = \sigma MP_1 (u_1, v_1) \vee \sigma MP_2 (u_2, v_2),$$
$$if \; u_1 u_2 \in E_1 \; and \; v_1 = v_2$$

i) $T_{M'}(j_1, j_2) = \max\left(T_{M'_1}(j_1), T_{M'_2}(j_2)\right),$

$\quad F_{M'}(j_1, j_2) = \min\left(F_{M'_1}(j_1), F_{M'_2}(j_2)\right)$

$\quad \forall (j_1, j_2) \in \mathcal{M}'_1 \times \mathcal{M}'_2$

ii) $T_{N'}((j, j_2)(j, k_2)) = \max\left(T_{M'_1}(j), T_{M'_2}(j_2 k_2)\right)$

$\quad F_{N'}((j, j_2)(j, k_2)) = \min\left(F_{M'_1}(j), F_{M'_2}(j_2 k_2)\right)$

We represent $\mathcal{N}_1$, $\mathcal{N}_2$, and $\mathcal{N}_3$ as the three IFGs shown in Figure 5.1, 5.2, and 5.3, respectively, and Figure 5.4 displays the max product of the three graphs.

Efficient time management supports the flow of information from one node to another in the maximized network, and time management of the nodes in collaboration determines the membership or non-membership, as shown in Figure 5.5. With

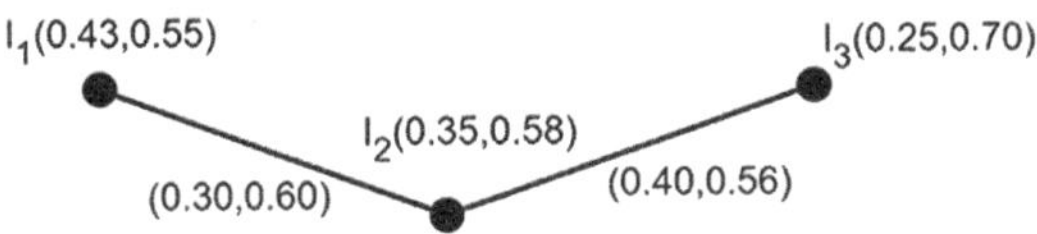

FIGURE 5.1　Intuitionistic graph network $\mathcal{N}_1$.

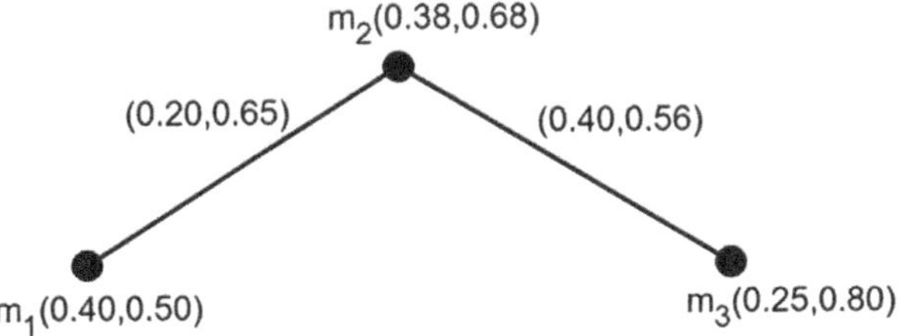

FIGURE 5.2　Intuitionistic graph network $\mathcal{N}_2$.

FIGURE 5.3　Intuitionistic graph network $\mathcal{N}_3$.

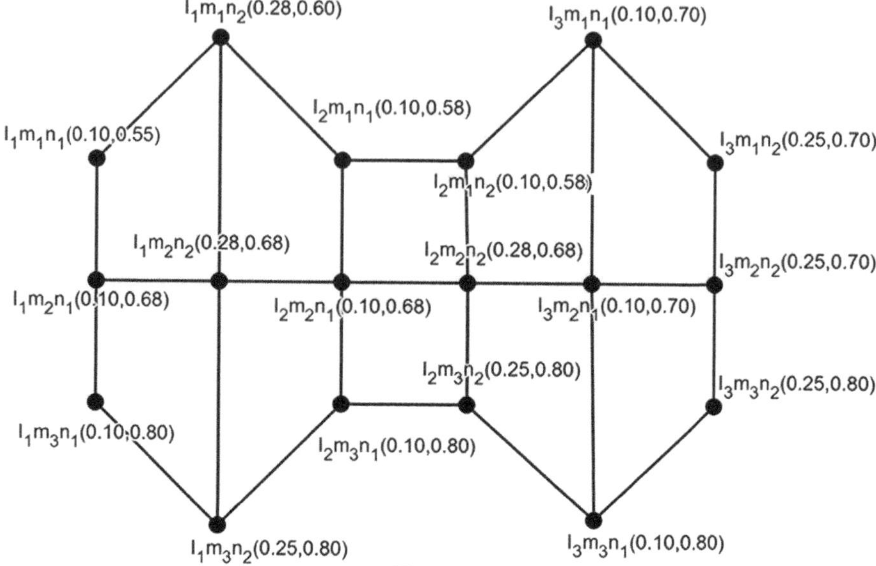

FIGURE 5.4   The maximum product of $\mathcal{N}_1$, $\mathcal{N}_2$, and $\mathcal{N}_3$.

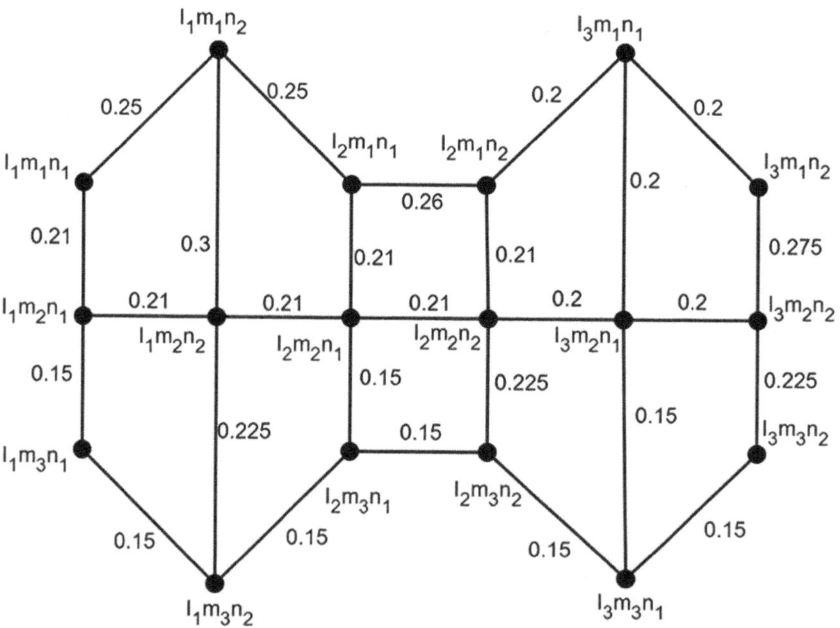

FIGURE 5.5   The maximum products of $\mathcal{N}_1$, $\mathcal{N}_2$, and $\mathcal{N}_3$ with minimum weights.

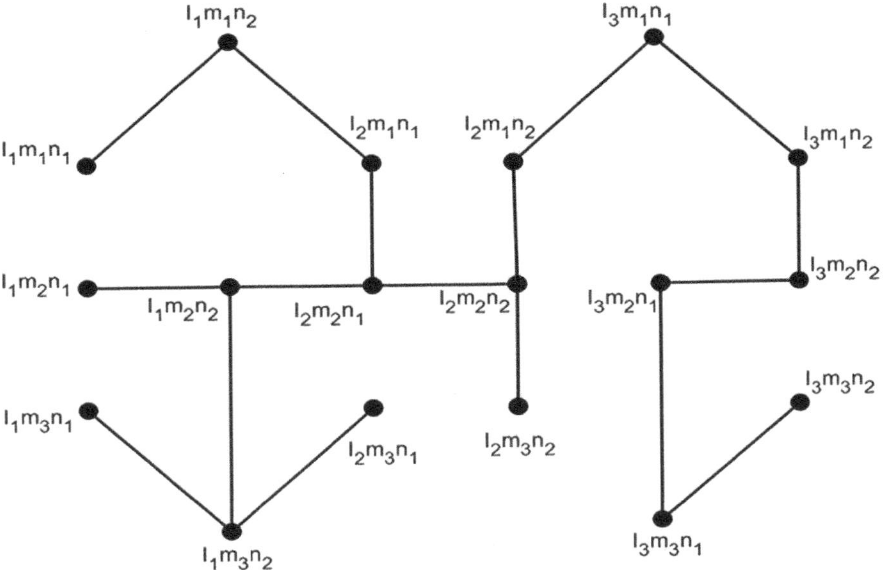

FIGURE 5.6   Minimal spanning tree of $\mathcal{N}_1$, $\mathcal{N}_2$, and $\mathcal{N}_3$.

the IFG model, we can find the minimal spanning tree to make the network more flexible with the minimum possible weights of the edges with score functions found, thereby increasing the minimal cardinality of edges. The minimal spanning tree of the weighted network is shown in Figure 5.6, and its minimum weight of the spanning tree is 3.475.

5.4 APPLICATIONS

Graphs can be incorporated with multicriteria decision-making (MCDM) in a variety of ways to address decision problems involving multiple criteria and alternatives. For instance, in facility location, where the goal is to determine the optimal locations for facilities to serve a set of demand points, MCDM techniques can be applied to evaluate potential locations based on multiple criteria such as transportation costs, customer proximity, and environmental impact.

In network design, operations on graphs with maximization can be used to model various types of networks (e.g., communication, transportation). MCDM helps make decisions about the best network configuration considering cost, reliability, and performance. MCDM can also be applied in project scheduling problems where activities are represented as nodes in a graph and edges represent dependencies; the decision-maker may need to optimize schedules by considering project duration, resource utilization, and cost criteria. Also in networks, MCDM and maximal operation on graphs can model network resource allocation problems and aid in decision-making based on efficiency, fairness, and reliability.

In communication networks or transportation systems, MCDM can be applied to select the optimal route or path; criteria can include factors like latency, bandwidth, and cost. More broadly, network reliability is crucial in many applications, and MCDM can help assess the reliability of different network configurations considering criteria such as fault tolerance, redundancy, and performance. Here, let us assume one such network in which all the nodes play an essential role in completing the tasks in efficient time management. Critical path analysis (CPA) is a project management technique that identifies a project network's critical path. The critical path represents the sequence of tasks that must be completed on time to ensure that the project is completed as planned.

CPA is widely used for scheduling and managing projects, helping project managers identify the most critical tasks that if delayed could extend the overall project duration. Here, we maximized three distinct networks using the max product on graphs where the maximized network played a crucial role in efficient time management. Optimized resource allocation ensures that the most essential tasks receive appropriate attention: By focusing on the critical path, project managers can identify potential delays early, allowing for proactive problem-solving; understanding the critical path helps identify opportunities to compress the schedule by allocating additional resources or adjusting task sequences.

CPA is essential for project managers to effectively plan, schedule, and control projects, especially those with complex task dependencies and time constraints. In these applications, the combination of operations on graphs and MCDM provides a powerful approach to modeling, analyzing, and optimizing complex decision problems where multiple criteria and alternatives need to be considered. The choice of specific MCDM techniques and graph models depends on the nature of the problem and the decision-maker's preferences.

5.5 RESULTS AND DISCUSSION

In this work, we introduced and characterized the max product of three intuitionistic fuzzy graphs, studied its applications, and found the effective minimal spanning tree. Using the maximum product of the three IFGs in a network eases information sharing and expedites understanding data. Maximizing only two networks in the max product of graphs shares information with fewer particular members than using the max product of the three IFGs; three intuitionistic fuzzy networks maximized the information sharing. It is essential to know how to improve the number of edges in networks to increase the flow of information. MCDM provides decision-makers with a powerful approach to modeling, analyzing, and optimizing complex decision problems.

5.6 CONCLUSION

Intuitionistic fuzzy networks enhance the structures of standard fuzzy networks, which helps deal with more ambiguous conditions. For this chapter, we studied the max product of an intuitionistic fuzzy graph structure and discussed the real-world application of the maximized network with a minimum spanning tree algorithm, which we generated to achieve the minimum time to complete tasks in critical project management. In the future, we will extend the study to other operations on graphs.

REFERENCES

[1] Zadeh, L. A., "Fuzzy sets." *Information and Control*, 8(3), 338–353, (1965). https://doi. org/10.1016/S0019-9958(65)90241-X

[2] Akram, M., Waseem, N., & Dudek, W. A., "Certain types of edge m-polar fuzzy graphs." *Iranian Journal of Fuzzy Systems*, (2017). https://doi.org/10.22111/ IJFS.2017.3324

[3] Rashmanlou, H., Samanta, S., Pal, M., & Borzooei, R. A., "Intuitionistic fuzzy graphs with categorical properties." *Fuzzy Information and Engineering*, (2015). https://doi. org/10.1016/j.fiae. 2015.09.005

[4] Mahapatra, T., & Pal, M., "An investigation on m-polar fuzzy threshold graph and its application on resource power controlling system." *Journal of Ambient Intelligence and Humanized Computing*, (2022). https://doi.org/10.1007/s12652-021-02914-6

[5] Koam, A. N., Akram, M., & Liu, P., "Decision-making analysis based on fuzzy graph structures", *Mathematical Problems in Engineering*, (2020). https://doi.org/10.1155/ 2020/6846257

[6] Atanassov, K. T., *On Intuitionistic Fuzzy Sets Theory.* Springer, (2012). https://doi.org/ 10.1007/978-3-7908-1870-3_1

[7] Meenakshi, A., Mythreyi, O., Bramila, M., Kannan, A., & Senbagamalar, J., "Application of neutrosophic optimal network using operations." *Journal of Intelligent & Fuzzy Systems*, (2023). https://doi.org/10.3233/JIFS-223718

[8] Meenakshi, A., & Mythreyi, O., "Applications of neutrosophic social network using max product networks." *Journal of Intelligent & Fuzzy Systems*, (2023). https://doi.org/10. 3233/JIFS-223484

[9] Meenakshi, A., & Mythreyi, O., "Mathematical modeling of social networks using hypergraphs." *First International Conference on Advances in Electrical, Electronics and Computational Intelligence (ICAEECI). IEEE*, (2023). https;//doi.org/10.1109/ICAEECI 58247.2023.10370980

[10] Meenakshi, A., & Mishra, J. S., "The modular product of two cubic fuzzy graph structures." *First International Conference on Advances in Electrical, Electronics and Computational Intelligence (ICAEECI). IEEE,* (2023). https://doi.org/10.1109/ICAEECI58247.2023. 10370796

[11] Meenakshi, A., & Dhanushiya, S., "Fuzzy network using vertex order coloring and efficient domination." *First International Conference on Advances in Electrical, Electronics and Computational Intelligence (ICAEECI). IEEE,* (2023). https://doi.org/10.1109/ ICAEECI58247.2023.10370790

[12] Sheikh Hoseini, B., Akram, M., Sheikh Hosseini, M., Rashmanlou, H., & Borzooei, R. A., "Maximal product of graphs under vague environment." *Mathematical and Computational Applications*, (2020). https://doi.org/10.3390/mca25010010

[13] Parvathi, R., Karunambigai, M. G., & Atanassov, K. T., "Operations on intuitionistic fuzzy graphs." *IEEE International Conference on Fuzzy Systems*, (2009). https://doi.org/10. 1109/FUZZY.2009.5277067

[14] Parvathi, R., & Thamizhendhi, G., "Domination in intuitionistic fuzzy graphs." *Notes on Intuitionistic Fuzzy Sets*, (2010). http://ifigenia.org/wiki/issue:nifs/16/2/39-49

[15] Parvathi, R., & Karunambigai, M. G., "Intuitionistic fuzzy graphs." *Computational Intelligence, Theory and Applications: International Conference 9th Fuzzy Days in Dortmund, Germany*, Springer Berlin Heidelberg, (2006). https://doi.org/10.1007/3-540-34783-6_15

[16] Pasi, G., Yager, R., & Atanassov, K., "Intuitionistic fuzzy graph interpretations of multi-person multi-criteria decision making: Generalized net approach." *2ndInternational IEEE conference on 'Intelligent Systems'*, (2004). https://doi.org/10.1109/ IS.2004.1344787

[17] Shannon, A., & Atanassov, K., "On a generalization of intuitionistic fuzzy graphs." *NIFS*, (2006). http://ifigenia.org/wiki/issue:nifs/12/1/24-29

[18] Shao, Z., Kosari, S., Rashmanlou, H., & Shoaib, M., "New concepts in intuitionistic fuzzy graph with application in water supplier systems." *Mathematics*, (2020). https://doi.org/10.3390/math8081241

[19] Shoaib, M., Mahmood, W., Xin, Q., & Tchier, F., "Certain operations on picture fuzzy graph with application." *Symmetry*, (2021). https://doi.org/10.3390/sym13122400

[20] Islam, S. R., & Pal, M., "Hyper-Wiener index for fuzzy graph and its application in share market." *Journal of Intelligent & Fuzzy Systems*, (2021). https://doi.org/10.3233/JIFS-210736

[21] Yaqoob, N., Gulistan, M., Kadry, S., & Wahab, H. A., "Complex intuitionistic fuzzy graphs with application in cellular network provider companies." *Mathematics*, (2019). https://doi.org/10.3390/math7010035

[22] Aruldoss, M., Lakshmi, T. M., & Venkatesan, V. P., "A survey on multi criteria decision making methods and its applications." *American Journal of Information Systems*, (2013). https://doi.org/10.12691/ajis-1-1-5

6 Optimizing the Surface Roughness of H13 Steel Machined by Wire Electrical Discharge

*Pham Thi Hong Nga, Van Tron Tran, Thanh Tan Nguyen, Xuan Tien Vo, and Van-Thuc Nguyen**

6.1 INTRODUCTION

Wire electrical discharge machining (WEDM) is a method that can machine complex shapes and extremely hard materials that the traditional method cannot process [1–3]. WEDM uses a thin metallic wire made from copper and tungsten as an electrode to discharge the conductive materials. The discharge process erodes the machined part gradually without direct contact with it. The cooling environment removes the thermal energy that appears during the WEDM process; therefore, there is no thermal effect on the heat-treated steel part [4, 5,6].

AISI H13 tool steel is a chromium hot work tool steel that is commonly used in both hot and cold work tooling applications, belonging to group H steel under the AISI classification system. H-13 tool steel has good abrasion resistance at both low and high temperatures, as well as high toughness and ductility [7, 8, 9]. H13 hot work tool steel is used more than any other tool steel in tooling applications due to its exceptional combination of high toughness and fatigue resistance.

However, machining H13 steel is a challenging issue that often requires state-of-the-art tools or methods. For example, Coldwell et al. [10] applied high-speed machining to H13 steel using carbide tooling, and the carbide tool was able to generate 210 holes with 16 mm depth, 30 m/min cutting speed, and a 0.1 mm/rev feed rate. Umbrello et al. [11] presented a model for computer simulation of hard-machining H13 steel. The hardness-based flow stress and fracture models could imitate the effect of hardness on the formation of the chip.

Zhang et al. [12] used physical vapor deposition coating on the cutting tools, reducing the cutting tool temperature. The TiAlN coating produced a higher tool-chip interface temperature than the TiN coating. Do et al. [13] identified the optimal cutting

*Corresponding author: nvthuc@hcmute.edu.vn

DOI: 10.1201/9781032635170-6

97

parameter with the least lubricant for milling H13 steel; they determined that the feed rate has the strongest influence on the surface roughness. Interestingly, Yan et al. [14] investigated the thermo-mechanical model of turning finished H13 hardened steel by examining the machining parameters. The outcomes can be used to improve the design of cutting inserts and optimize process parameters for AISI H13 turning. Zhang et al. [15] studied thermal deformation behavior with thin-walled shapes during WEDM. They machined AISI H13, SKD 11, and Inconel 718 with different water pressures, wire speeds, and pulse-on times, and the thermal deformation during WEDM was strongly impacted by physical properties, yield strength, and thermal conductivity.

Optimizing machining parameters is an effective method of achieving the best results from multiparameter conditions [16, 17, 18]. Sadhana et al. [19] studied the impact of optimizing WEDM parameters on the material removal rate (MRR) and surface roughness (Ra) of AISI H13 steel and found that gap voltage had the strongest impact on MRR and Ra. Chockalingam et al. [20] also discussed the spark gap and dimensional deviation during the WEDM of AISI T-15-HSS steel. They examined the impact of certain machining parameters—reference voltage, pulse-on time (T_{on}), pulse-off time (T_{off}), wire speed, wire tension, and injection pressure mode— and analyzed the results using MATLAB 6.5 gray rational analysis and Lingo 6.0 optimization software. Among these parameters, T_{on}, T_{off}, and wire speed had the greatest impact on MRR, Ra, spark gap, and dimensional deviation.

In particular, T_{off} achieved the highest contribution for all these machining performances. Manjaiah et al. [21] studied the impacts of machining parameters on AISI D2 steel. The authors concentrated on MRR and Ra in the WEDM process, applying Taguchi optimization. Different from Chockalingam et al. [20], Manjaiah et al. found that T_{on} had the most dramatic impact on the MRR and the Ra of the AISI D2 steel.

Meanwhile, Zhang et al. [22] revealed the impact of process parameters on the tungsten tool YG15. They examined the tool's white layer thickness and crack density, and the results indicated that T_{on} and pulse current have a great impact on Ra, white layer thickness, and surface crack density. Rathod et al. [23] also investigated the impact of process parameters on WEDM of H13 steel. They applied the multi-objective JAYA algorithm and the Taguchi method to optimize Ra and MRR, and these authors found that T_{on} and voltage gap had the highest contribution to Ra, while the wire feed was most important for MRR.

Dhobe et al. [24] pinpointed that achieving a good surface finish is a complex issue, especially in WEDM because of the combination of many process parameters and materials. They investigated the WEDM process for cryo-treated AISI D2 tool steel and applied Taguchi Design of Experiments. Dhobe et al. also determined that T_{on} had the most significant impact on Ra, whereas T_{off} had the weakest impact. They provided a regression model for successfully predicting Ra that was consistent with the experimental results.

6.2 PROBLEM STATEMENT

The influence of discharge time on the surface roughness of H13 steel machined by WEDM is rarely discussed and needs further investigation. In this study, we investigated the effects of discharge time on Ra for H13 steel machined by WEDM. We then discuss the microstructure of the machined sample and used the results to optimize the WEDM process for H13 steel.

6.3 MATERIALS AND METHODS

6.3.1 Materials

In this study, we used H13 samples with the dimensions of 26 mm × 45 mm × 12.5 mm, as shown in Figure 6.1. The chemical composition of the H13 steel is presented in Table 6.2. We conducted the WEDM at ACCUTEX DS-430S DM using oil as the dielectric liquid and copper wire with a diameter of 0.2 mm. Following the Taguchi method, we machined nine samples with different T_{on} and T_{off}, as shown in Table 6.3. T_{on} and T_{off} are the on-delay and off-delay times of the discharge process, and we selected them as experimental parameters to follow the possible range of the ACCUTEX DS-430S DM machine and avoid broken wires. We measured

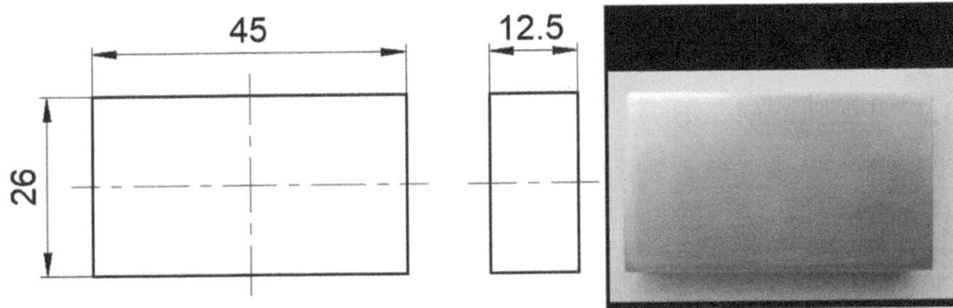

FIGURE 6.1   H13 sample for the WEDM process (in mm).

TABLE 6.1
Properties of the H13 Tool Steel

Properties	Value
Density (kg/m³)	7900
Hardness, Brinell (typical)	290–370
Ultimate Tensile Strength (MPa)	1200–1590
Elastic modulus (GPa)	215
Thermal conductivity (W/m/°K)	28.8
Thermal expansion (°C−1)	3.5×10^{-6} at 20–25 °C

TABLE 6.2
Chemical Composition of the H13 Tool Steel

Elements	C	Mn	Si	Cr	Mo	V	Fe
Percentage (%)	0.32–0.45	0.2–0.6	0.8–1.25	4.75–5.5	1.10–1.75	1.0	Balance

FIGURE 6.2	The Mitutoyo SJ-201 roughness meter to measure surface roughness.

surface roughness using a Mitutoyo SJ-201 roughness meter at the Material Testing Laboratory, HCMC University of Technology and Education, Vietnam, as presented in Figure 6.2. The and examined the microstructure using a JEOL 5410 LV (Japan) scanning electron microscope.

6.3.2 Taguchi Method

The Taguchi technique is a strategy for improving product quality by identifying the elements that influence it and determining the ideal parameters. We executed each parameter three times on the same workpiece with the L9 orthogonal array, for a total of nine workpieces, as shown in Table 6.3. We were looking for the lowest values according to Equation (6.1):

$$\frac{S}{N}(SB) = -10\log_{10}\left(\frac{1}{n}\sum_{u=1}^{n}y_u^2\right) \tag{6.1}$$

where y_u is the result of the uth measurement and n is number of measurements.

TABLE 6.3

Sample and Experimental Parameters Following the Taguchi Design

Sample	T_{on} (μs)	T_{off} (μs)
1	6	10
2	6	14
3	6	18
4	10	10
5	10	14
6	10	18
7	14	10
8	14	14
9	14	18

6.4 RESULTS AND DISCUSSION

6.4.1 MICROSTRUCTURE

The microstructure of H13 steel is presented in Figure 6.3. There are dark carbide particles scattered on the martensite matrix. The carbide phase is micrometer-sized and evenly dispersed throughout the matrix; this small size and well-dispersed structure facilitated the cutting during the WEDM and created better surface roughness. In WEDM, electrical discharges or sparks are the cutting tools that remove the materials; there is no physical contact, and therefore, extremely hard materials can be easily machined by this technique.

The variation in carbide size was approximately 10 µm, which was very modest. The carbide particles in the figure were MoC, VC, and V8C7, which belong to the MC family of carbides, and M23C6, a Cr-rich carbide [25]. These carbide particles have extremely high hardness, helping improve the wear resistance of the steel. Moreover, due to the minor thermal effect of the WEDM process, the microstructure of the H13 steel was better preserved than in the supplied condition. In other words, the microstructure of the H13 steel before and after the WEDM process is similar.

6.4.2 SURFACE ROUGHNESS

Table 6.4 shows the Ra of the H13 steel sample machined by WEDM. For sample 2, T_{on} was 6 µs and T_{off} was 14 µs, achieving the best Ra of 1.91 µm. On the contrary, with sample 8, T_{on} was 14 µs and T_{off} was 14 µs, achieving the highest Ra of 3.35 µm. Rathod et al. [26] also investigated the WEDM of H13 steel and achieved Ra of 2.34–3.22 µm, which was comparable with our results of 1.91–3.35 µm. Giduturi et al. [27] measured Ra of 1.53–4.98 µm with WEDM for H13 steel, which was also

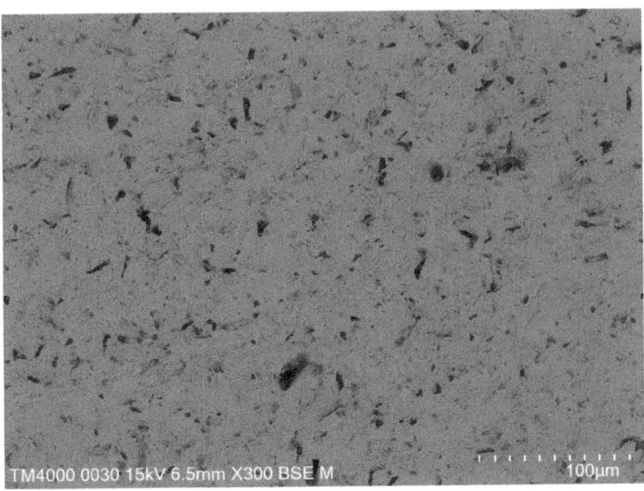

FIGURE 6.3 The microstructure of H13 steel after wire electrical discharge machining.

similar to this study's results, although they found higher rates of deviation than we did. Meanwhile, Rupajati et al. [28] measured Ra ranging from 2.26 to 3.68 μm in WEDM for H13 steel.

To further optimize these results, we applied Taguchi analysis. We calculated the S/N ratio for Ra seeking the smallest ratios; Table 6.5 presents the response table for the S/N ratios. The results show that T_{on} has a stronger effect on Ra than T_{off}, and the findings in Table 6.6 corroborate that effect.

TABLE 6.4
Surface Roughness of the Samples

Sample	T_{on} (μs)	T_{off} (μs)	Surface roughness (Ra, in μm)
1	6	10	2.03
2	6	14	1.91
3	6	18	2.45
4	10	10	2.69
5	10	14	2.54
6	10	18	2.85
7	14	10	3.13
8	14	14	3.35
9	14	18	2.72

TABLE 6.5
Signal-to-Noise Ratios (smaller is better)

Level	T_{on}	T_{off}
1	−6.518	−8.219
2	−8.596	−8.073
3	−9.701	−8.524
Delta	3.183	0.451
Rank	1	2

TABLE 6.6
Means

Level	T_{on}	T_{off}
1	2.130	2.617
2	2.693	2.600
3	3.067	2.673
Delta	0.937	0.073
Rank	1	2

Figure 6.4 presents the influence of EDM parameters on the surface roughness of H13 steel using a main effects plot for means. The optimal values based on the highest point in the graphs are T_{on} = 6 μs, T_{off} = 14 μs, and Ra = 1.91 μm. The regression equation was Equation (6.2):

$$Ra = 1.360 + 0.1171 \times T_{on} + 0.0071 \times T_{off} \tag{6.2}$$

The equation indicates that the ratio of T_{on} is 0.1171, which is considerably higher than the 0.0071 for T_{off}. That is, T_{on} has a stronger influence on surface roughness than T_{off}, which is consistent with the results in Table 6.4. In Table 6.7, the F values corroborate this strong influence of T_{on}.

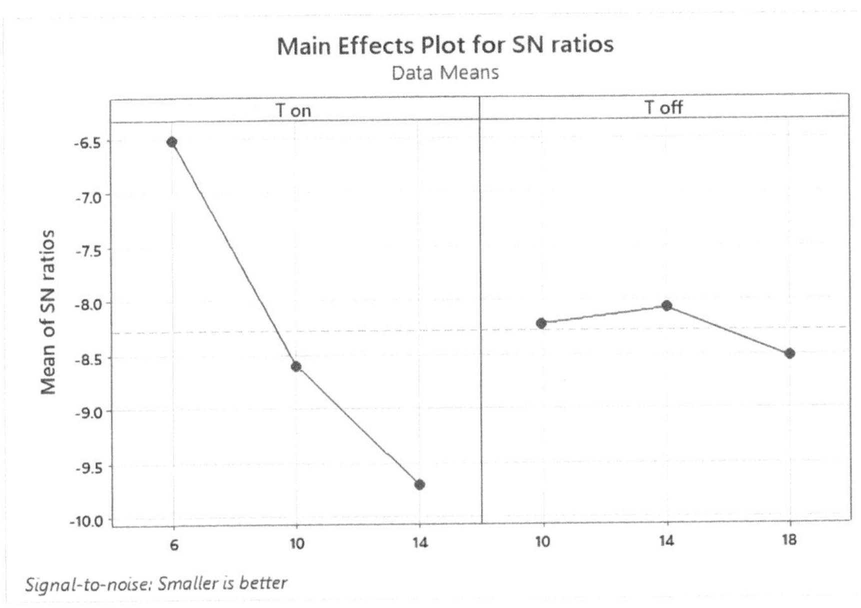

FIGURE 6.4 Influence of EDM parameters on the S/N graph of H13 steel.

TABLE 6.7

ANOVA Results for Surface Roughness

Source	DF	Adj SS	Adj MS	F	P
Regression	2	1.32083	0.66042	9.29	0.015
T on	1	1.31602	1.31602	18.51	0.005
T off	1	0.00482	0.00482	0.07	0.803
Error	6	0.42657	0.07109		
Total	8	1.74740			

6.5　CONCLUSION

For this study, we examined how the discharge duration during wire electrical discharge machining affects the surface roughness of H13 steel and found that the microstructure showed dark carbide particles scattered on the martensite matrix. The best surface roughness, Ra of 1.91 μm, was achieved with Sample 2, where T_{on} was 6 μs and T_{off} was 14 μs. In contrast, for sample 8, Ra = 3.35 μm, with T_{on} and T_{off} of 14 μs as well. According to the Taguchi data, pulse-on time has a greater impact on surface roughness than pulse-off time.

6.6　CONFLICTS OF INTEREST

The authors state that they have no known competing financial interests or personal relationships that could have appeared to influence the work reported in this article.

One of the authors (Van-Thuc Nguyen) is a member of this book's editorial board but was not involved in the editorial review or the decision to publish this article. The authors state that the work has no potential conflict of interest. All authors have read and agreed to the published version of the manuscript.

6.7　QUESTIONS

1. Between T_{on} and T_{off}, which has a stronger impact on the surface roughness of the H13 steel after the WEDM process?
2. Please describe the microstructure of H13 steel.
3. Please explain the meaning of "Signal-to-Noise Ratios (smaller is better)" in the title of Table 6.5.

6.8　ACKNOWLEDGEMENT

We would like to express our deep gratitude to Ho Chi Minh City University of Technology and Education for sponsoring the machines and equipment for the experiments.

REFERENCES

[1] Ho, K. H., Newman, S. T., Rahimifard, S., & Allen, R. D. (2004). State of the art in Wire Electrical Discharge Machining (WEDM). *International Journal of Machine Tools and Manufacture, 44*(12–13), 1247–1259. https://doi.org/10.1016/j.ijmachtools.2004.04.017

[2] Abbas, N. M., Solomon, D. G., & Bahari, M. F. (2007). A review on current research trends in Electrical Discharge Machining (EDM). *International Journal of Machine Tools and Manufacture, 47*(7–8), 1214–1228. https://doi.org/10.1016/j.ijmachtools.2006.08.026

[3] Mahapatra, S. S., & Patnaik, A. (2006). Parametric optimization of Wire Electrical Discharge Machining (WEDM) process using Taguchi method. *Journal of the Brazilian Society of Mechanical Sciences and Engineering, 28*, 422–429. https://doi.org/10.1590/S1678-58782006000400006

[4] Dhobe, M. M., Chopde, I. K., & Gogte, C. L. (2013). Investigations on surface characteristics of heat treated tool steel after wire electro-discharge machining. *Materials and Manufacturing Processes, 28*(10), 1143–1146. https://doi.org/10.1080/10426914.2013.822976

[5] Choi, K. K., Nam, W. J., & Lee, Y. S. (2008). Effects of heat treatment on the surface of a die steel STD11 machined by W-EDM. *Journal of Materials Processing Technology, 201*(1–3), 580–584. https://doi.org/10.1016/j.jmatprotec.2007.11.156

[6] Gowthaman, P. S., Gowthaman, J., & Nagasundaram, N. J. M. T. P. (2020). A study of machining characteristics of AISI 4340 alloy steel by wire electrical discharge machining process. *Materials Today: Proceedings, 27,* 565–570. https://doi.org/10.1016/j.matpr.2019.12.037

[7] Wang, J., Liu, S., Fang, Y., & He, Z. (2020). A short review on selective laser melting of H13 steel. *The International Journal of Advanced Manufacturing Technology, 108,* 2453–2466. https://doi.org/10.1007/s00170-020-05584-4

[8] Cormier, D., Harrysson, O., & West, H. (2004). Characterization of H13 steel produced via electron beam melting. *Rapid Prototyping Journal, 10*(1), 35–41. https://doi.org/10.1108/13552540410512516

[9] Guanghua, Y., Xinmin, H., Yanqing, W., Xingguo, Q., Ming, Y., Zuoming, C., & Kang, J. (2010). Effects of heat treatment on mechanical properties of H13 steel. *Metal Science and Heat Treatment, 52,* 393–395. https://doi.org/10.1007/s11041-010-9288-4

[10] Coldwell, H., Woods, R., Paul, M., Koshy, P., Dewes, R., & Aspinwall, D. (2003). Rapid machining of hardened AISI H13 and D2 moulds, dies and press tools. *Journal of Materials Processing Technology, 135*(2–3), 301–311. https://doi.org/10.1016/S0924-0136(02)00861-0

[11] Umbrello, D., Rizzuti, S., Outeiro, J. C., Shivpuri, R., & M'Saoubi, R. (2008). Hardness-based flow stress for numerical simulation of hard machining AISI H13 tool steel. *Journal of Materials Processing Technology, 199*(1–3), 64–73. https://doi.org/10.1016/j.jmatprotec.2007.08.018

[12] Zhang, J., Liu, Z., & Du, J. (2015). Modelling and prediction of tool-chip interface temperature in hard machining of H13 steel with PVD coated tools. *International Journal of Machining and Machinability of Materials, 17*(5), 381–396. https://doi.org/10.1504/IJMMM.2015.073148

[13] Do, T. V., & Hsu, Q. C. (2016). Optimization of minimum quantity lubricant conditions and cutting parameters in hard milling of AISI H13 steel. https://doi.org/10.3390/app6030083

[14] Yan, H., Hua, J., & Shivpuri, R. (2005). Numerical simulation of finish hard turning for AISI H13 die steel. *Science and Technology of Advanced Materials, 6*(5), 540. https://doi.org/10.1016/j.stam.2005.04.002 tempering. https://doi.org/10.3390/met7030070

[15] Zhang, Y., Zhang, Z., Huang, H., Huang, Y., Zhang, G., Li, W., & Liu, C. (2018). Study on thermal deformation behavior and microstructural characteristics of wire electrical discharge machining thin-walled components. *Journal of Manufacturing Processes, 31,* 9–19. https://doi.org/10.1016/j.jmapro.2017.10.024

[16] Gang, W. A. N. G., Li, W. L., Fan, R. A. O., He, Z. R., & Yin, Z. P. (2019). Multi-parameter optimization of machining impeller surface based on the on-machine measuring technique. *Chinese Journal of Aeronautics, 32*(8), 2000–2008. https://doi.org/10.1016/j.cja.2018.09.005

[17] Kumar, R., Katyal, P., Kumar, K., & Singh, V. (2022). Multiresponse optimization of end milling process parameters on ZE41A Mg alloy using Taguchi and TOPSIS approach. *Materials Today: Proceedings, 56,* 2497–2504. https://doi.org/10.1016/j.matpr.2021.08.271

[18] Li, Z. J., Hong, M. S., Su, H., & Wei, Y. L. (2003). Machining accuracy analysis for step multi-element varying-parameter vibration drilling of laminated composite materials. *The International Journal of Advanced Manufacturing Technology, 21,* 760–768. https://doi.org/10.1007/s00170-002-1391-2

[19] Sadhana, A. D., Prakash, J. U., Ananth, S., Juliyana, S. J., & Rubi, C. S. (2022). Effect of process parameters for wire EDM of AISI H13 tool steel. *Materials Today: Proceedings, 52,* 1870–1874.

[20] Chockalingam, K., Jawahar, N., Muralidharan, N., & Jeyaraj, K. L. (2019). Material subtraction study of AISI T-15-HSS by wire cut Electrical Discharge Machining (CNC-wire cut EDM) based on Taguchi grey relational analysis. *International Journal of Machining and Machinability of Materials*, *21*(3), 139–168. https://doi.org/10.1504/IJMMM.2019.099481

[21] Manjaiah, M., Laubscher, R. F., Kumar, A., & Basavarajappa, S. (2016). Parametric optimization of MRR and surface roughness in Wire Electro Discharge Machining (WEDM) of D2 steel using Taguchi-based utility approach. *International Journal of Mechanical and Materials Engineering*, *11*(1), 1–9. https://doi.org/10.1186/s40712-016-0060-4

[22] Zhang, Z., Ming, W., Huang, H., Chen, Z., Xu, Z., Huang, Y., & Zhang, G. (2015). Optimization of process parameters on surface integrity in wire electrical discharge machining of tungsten tool YG15. *The International Journal of Advanced Manufacturing Technology*, *81*, 1303–1317. https://doi.org/10.1007/s00170-015-7266-0

[23] Rathod, L., Poonawala, N. S., & Rudrapati, R. (2020, March). Multi response optimization in WEDM of H13 steel using hybrid optimization approach. In *IOP Conference Series: Materials Science and Engineering*, *814*(1), 012015). IOP Publishing. https://doi.org/10.1088/1757-899X/814/1/012015

[24] Dhobe, M. M., Chopde, I. K., & Gogte, C. L. (2014). Optimization of wire electro discharge machining parameters for improving surface finish of cryo-treated tool steel using DOE. *Materials and Manufacturing Processes*, *29*(11–12), 1381–1386. https://doi.org/10.1080/10426914.2014.930890

[25] Ning, A., Mao, W., Chen, X., Guo, H., & Guo, J. (2017). Precipitation behavior of carbides in H13 hot work die steel and its strengthening during tempering. *Metals*, *7*(3), 70. https://doi.org/10.3390/met7030070

[26] Rathod, L., Poonawala, N. S., & Rudrapati, R. (2020, March). Multi response optimization in WEDM of H13 steel using hybrid optimization approach. In *IOP Conference Series: Materials Science and Engineering*, *814*(1), 012015. IOP Publishing. https://doi.org/10.1088/1757-899X/814/1/012015

[27] Giduturi, S., & Kuma, A. (2018). Parameter optimization of wire EDM for H-13 tool steel. *International Journal of Current Engineering and Technology*, *8*, 120–127. https://doi.org/10.14741/ijcet/v.8.1.24

[28] Rupajati, P., Soepangkat, B. O. P., Pramujati, B., & Agustin, H. C. K. (2014). Optimization of recast layer thickness and surface roughness in the wire EDM process of AISI H13 tool steel using Taguchi and fuzzy logic. *Applied Mechanics and Materials*, *493*, 529–534. https://doi.org/10.4028/www.scientific.net/AMM.493.529

7 The Impact Toughness of PBT/PA6 Composite Reinforced with Glass Fibers

*Nguyen Tran Dan Truong, Nguyen Cong Dat,
Nguyen Quang Tuan, Pham Thi Hong Nga*,
Van-Thuc Nguyen, Nguyen Vinh Tien,
Pham Quan Anh, Nguyen Thanh Tan,
and Ho Thi My Nu*

7.1 INTRODUCTION

According to World Bank statistics, over two billion tons of solid waste each year are generated worldwide, with many components such as plastic, fabric, and rubber; plastic waste accounts for up to 10% of the total waste and is extremely difficult to decompose or treat. Therefore, reusing or recycling plastic materials is an issue of great interest to researchers (Wang et al., 2017). While popular plastics such as polypropylene and polyvinyl chloride receive significant attention, there are still plastics that receive little attention, such as polybutylene terephthalate (PBT), while the recycling source is vast, which are toothbrush bristles (Nga et al., 2021). The reason is that PBT has limitations in mechanical properties.

PBT is a semi-crystalline engineering plastic material belonging to the polyester polymer family. Based on the advantages of mechanical properties such as durability, high hardness, good electrical insulation, and water resistance (Xiao et al., 2006), PBT has excellent potential when applied in industries such as automobiles, electricity, and electronics (Wei et al., 2011). However, disadvantages remain, such as high shrinkage, easy to uneven shrinkage deformation, poor notched impact toughness, and low deformation temperature.

PBT plastic is rugged in impact or high-temperature environments (An et al., 1996), but some researchers have improved on its mechanical properties. Researchers studied the efficiency of recycling PBT/PC/ABS composite with or without glass fiber (GF) combination and found that the flexural strength and impact toughness of the recycled PBT/PC/ABS-GF composite were greater than those of the original PBT/PC/ABS-GF

*Corresponding author: hongnga@hcmute.edu.vn

DOI: 10.1201/9781032635170-7

composite (Kuram et al., 2014). Rawat et al. (2022) added thermoplastic polyurethane to improve the impact toughness of PBT and found that the toughness had increased nearly three times. We conducted this study to attempt to improve PBT's flexural and impact toughness by mixing it with polyamide 6 (PA6) and GF. From there, PBT can be applied more widely in automobile manufacturing and household electrical appliances.

Mixing is the most widely used method to overcome the disadvantages of plastics. This method helps improve the limitations and retains the existing advantages (Bruce Brown, 2023). Dekkers et al. (1991) researched improving the impact toughness of PBT, polycarbonate (PC), and polyphenylene ether (PPE) mixtures by adding styrene-ethylene/butadiene-styrene (SEBS); under the same conditions of 25 °C, the PBT/PC/PPE/SEBS mixture (46/10/30/14) had an impact toughness of 600 J/m, 12 times higher than the PBT/PPE/SEBS blend (46/30/14) only reached 50 J/m. Maiti et al. (2007) studied the effects of acrylonitrile-butyl acrylate-styrene on the mechanical properties of PBT and found that the PBT's flexural strength and impact toughness had increased significantly.

PA6 is a polymer line used mainly in technical applications with high physical and mechanical property requirements, such as automobile manufacturing, electrical and electronic industries, precision machinery, and medical equipment. PA6 has outstanding advantages in durability and impact toughness, extremely high corrosion resistance, and high melting temperature. However, its limitations are its resistance to acid and inability to be used in many acidic environments and its poor dimensional deformation (Chih-Rong et al., 1996).

Not only do polyamide and polyester have excellent combination capabilities that many researchers are interested in, but they also expect this to be a promising option for creating high-performance polymer materials in the future. In particular, PBT is considered the optimal solution for mixing with PA6 from the contrasting mechanical properties. Researchers tested the compatibility of PBT and PA6 by adding E-44 epoxy resin to the mixture and found excellent compatibility (An et al., 1996), and flexural strength also improved when a small amount of E-44 epoxy was added (Hongrui et al., 2017). The compatibility of PBT/PA6 was also studied by adding poly (n-butyl acrylate)/poly (methyl methacrylate-co-methacrylic acid). The results show that the impact toughness of the new mix was 25.66 kJ/m^2, four times higher than the original mix. At the same time, the unique blend absorbed only 1.3% water content, down 53.6% from PA6 and 26.6% from the original blend (Hongrui et al., 2017). The impact toughness of the PBT/PA6 mixture also increases by adding ethylene vinyl acetate-g-maleic anhydride when made by injection molding (Kim et al., 2003).

Injection molding is a popular method when manufacturing PBT products reinforced with GFs because the resulting products have stable and highly accurate dimensions, short molding cycles, and low cost (Kelly et al., 1965). Reinforced PBT is widely used in electrical engineering, electronics, automobile component manufacturing, etc., so it is necessary to research more possibilities for strengthening PBT to meet additional needs (Zhang et al., 2014). Researchers have investigated the burning ability of PBT reinforced with GFs because of its high applicability, increasing the ability to create fire-resistant plastic. Ramani et al. (2014) studied GF-reinforced fire-resistant PBT and found that the GF did not change the degradability of PBT. Other researchers confirmed that GF-reinforced PBT had a 2.5 times lower mass than the original PBT (Kempel et al., 2012) and improved the fire resistance of

GF-reinforced PBT by adding 2,4,6-tris(2',4',6'-tribromophenoxy)-1,3,5-triazine (Chen et al., 2006). In addition, polyethylene terephthalate's impact toughness increases with the addition of GF and other additives (Monti et al., 2021). The results showed that both notched and unnotched impact toughness increased, up to 52% higher than the original. Based on these results, we determined that we could use GF to improve impact toughness in this research.

Currently, reinforcing compounds have attracted much attention in many fields, such as automobiles and electronics, thanks to the significantly superior mechanical properties of the material after reinforcement (Thattaiparthasarathy et al., 2008). Common materials used for reinforcement are GF, carbon fiber, and thermoplastics. Synthetic mixes of PA6 and GFs are widely used in research and industrial fields (Zuo et al., 2013). Because PA6 has good mechanical properties, low cost, and high-temperature resistance, a GF coating applied to produce PA6-based composite panels achieves high performance (Kiss et al., 2021). SEM results show that PA6 adheres tightly to GFs. In addition, the plastic strength of synthetic panels is up to 500 MPa, much higher than that of industrial containers.

From these studies on PBT, PA6, and GF mixtures, especially the findings from Hamlaoui et al. (2022), we examined PBT's mechanical properties using injection molding after GF reinforcement and found that after being reinforced with GF, PBT's mechanical properties increased significantly. For example, Young's modulus increased to 77.33%, the tensile strength increased to 63.28%, and the yield strength increased to 65.12% when the ratio of GF was 30%. Based on these results, GF significantly improves the mechanical properties of a PBT/PA6 mixture, especially unnotched Izod impact toughness.

7.2 MATERIALS AND METHODS

7.2.1 MATERIALS

The PBT was supplied by Toan Dai Hung made with Lanxess TC BN60 (Germany); the PBT-30GF B3235 was also supplied by Lanxess. The PA6 was supplied by MoSuCo originating in Hong Kong from LiBoLon (Germany). Table 7.1 shows the

TABLE 7.1
The PBT/PA6/GF Mixing Ratios

		Percentage (wt.%)		
Sample	Code	PBT	PA6	GF
PBT/PA6/0%GF	0GF	50	50	0
PBT/PA6/3%GF	3GF	48.5	48.5	3
PBT/PA6/6%GF	6GF	47	47	6
PBT/PA6/9%GF	9GF	45.5	45.5	9
PBT/PA6/12%GF	12GF	44	44	12
PBT/PA6/15%GF	15GF	42.5	42.5	15
PBT/PA6/18%GF	18GF	41	41	18

percentages of GF in the PBT/PA6 mixtures at multiple ratios, with our aim of determining the most appropriate GF ratio for optimizing the PBT/PA6 mixture.

7.2.2 MEASUREMENT METHODS

First, we injection molded and produced samples. Next, we selected five random samples to measure the unnotched Izod impact toughness according to the ASTM D256 standard. The sample in Figure 7.1 is $63.5 \pm 2 \times 12.7 \pm 0.2$ mm. To measure impact toughness, we used the Tinius Olsen IT504 impact toughness tester shown in Figure 7.2, including digital controller 104; this controller allows quick and accurate calculation and meets most experimental needs. The primary impact energy of the machine is 2.82 J with a standard height of 610 mm; the corresponding impact speed is 3.46 m/s. However, the aerodynamic design allows lowering the height of the hammer, meeting the need for impact testing smaller than 2.82 J, which is 2.75 J to 2 J, with speeds of 3.14 m/s and 2.91 m/s, respectively. When mounted on the hammer, the impact energy can easily vary up to 25 J using different load sets. The machine has a solid structure with an aerodynamic design that ensures impact accuracy and energy loss caused by wind. The samples were broken after the impact toughness test, as shown in Figure 7.3. The breaking positions are not homogeneous because the of the samples' different GF mixing ratios.

7.3　RESULTS AND DISCUSSION

7.3.1 THE IMPACT TOUGHNESS OF PBT/PA6/GF BLENDS

Table 7.2 shows the data of each ratio after five measurements. Figure 7.4 and the table show that the impact toughness differed little for the different ratios, with the PBT/PA6/0%GF sample showing the lowest toughness at 22.23 kJ/m^2 and sample PBT/PA6/18%GF showing the highest at 26.45 kJ/m^2.

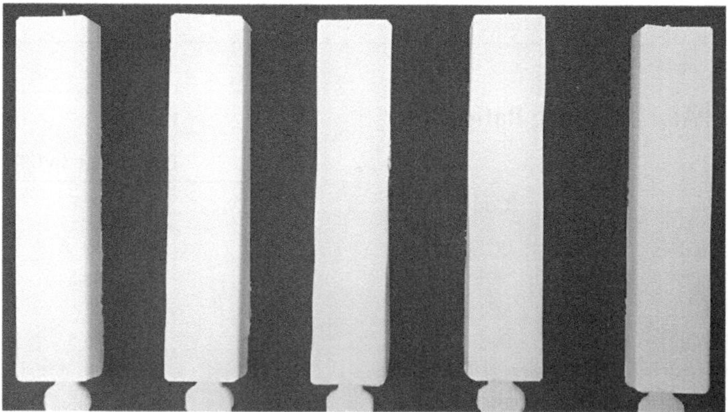

FIGURE 7.1　Sample after injection molding.

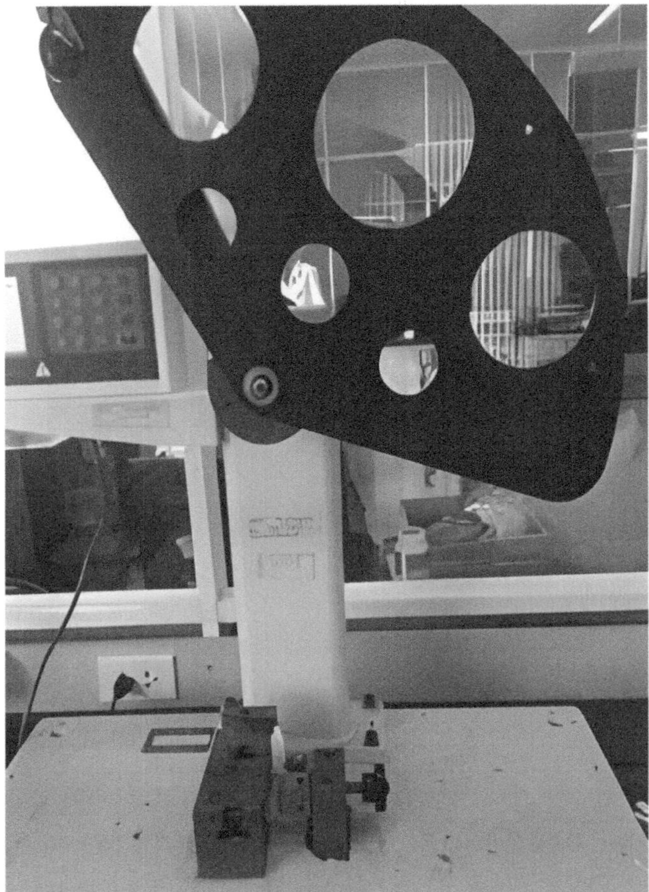

FIGURE 7.2   Tinius Olsen IT504 impact toughness tester.

Table 7.2 and Figure 7.4 show that reinforcing a PBT/PA6 composite with GF increased the unnotched impact toughness, confirming Cheng et al.'s (2020) finding that the higher the GF content in the mixture, the better its mechanical properties.

7.3.2　The Sample Microstructures

After measurement, we took samples to capture the microscopic structures at the fracture surfaces for observation (Figure 7.5).

Figure 7.6 shows the microstructures of the PBT/PA6/GF samples with each GF ratio: 0, 3, 6, 9, 12, 15, and 18% GF. Observing sample PBT/PA6/0%GF, we see small round dots that are PA6 plastic components evenly dispersed on the surface of the PBT substrate. Next, the PBT/PA6/9%GF sample shows cylindrical fibers that are the GF component; at this time, the small round dots gradually disappear, and the PBT/PA6 mixture becomes smoother. We finally observe that the samples with the higher proportions of GF show increasingly even GF distribution on the PBT/PA6 substrate.

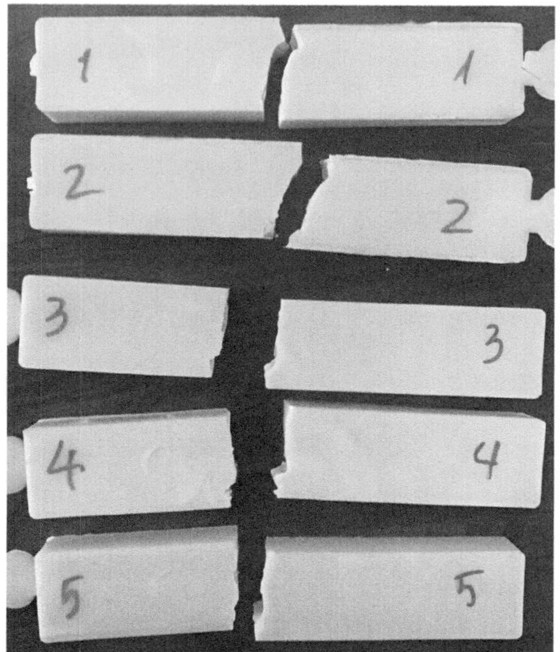

FIGURE 7.3   Sample after measurement.

TABLE 7.2

The Unnotched Izod Impact Toughness of PBT/PA6/GF Samples

	Unnotched Impact Toughness (kJ/m²)						
Sample	0GF	3GF	6GF	9GF	12GF	15GF	18GF
1	22.30	22.64	21.80	25.59	26.76	29.54	25.89
2	22.09	26.21	26.19	22.87	24.66	24.19	25.92
3	22.89	21.16	22.58	24.40	23.44	24.40	27.40
4	25.59	24.40	25.59	26.53	27.91	25.89	28.91
5	18.26	22.89	22.64	23.84	24.19	26.08	24.16
Average	22.23	23.46	23.76	24.85	25.39	26.02	26.45
Standard Deviation	2.62	1.92	1.98	1.44	1.87	2.14	1.79

7.4 POLYNOMIAL REGRESSION ANALYSIS

If x is the GF content, and y is the impact toughness, the quadratic regression equation has the form:

$$y = a_{11}x^2 + a_1 x + a_0 \tag{7.1}$$

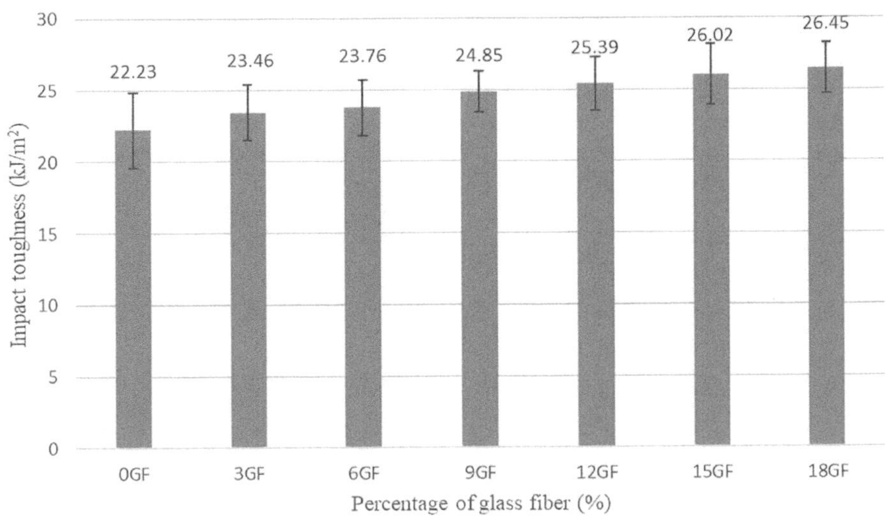

FIGURE 7.4   Average unnotched Izod impact toughness of the PBT/PA6/GF blends.

FIGURE 7.5   Fracture surfaces of the samples.

Finding the regression equation required the specific data from Table 7.3.

Applying the parabolic curve quadratic regression method, we get the following system of equations:

$$
\begin{cases}
a_0 \cdot n + a_1 \cdot \sum_{k=1}^{n} x_k + a_{11} \cdot \sum_{k=1}^{n} x_k^{\,2} = \sum_{k=1}^{n} y_k \\[2mm]
a_0 \cdot \sum_{k=1}^{n} x_k + a_1 \cdot \sum_{k=1}^{n} x_k^{\,2} + a_{11} \cdot \sum_{k=1}^{n} x_k^{\,3} = \sum_{k=1}^{n} x_k y_k \\[2mm]
a_0 \cdot \sum_{k=1}^{n} x_k^{\,2} + a_1 \cdot \sum_{k=1}^{n} x_k^{\,3} + a_{11} \cdot \sum_{k=1}^{n} x_k^{\,4} = \sum_{k=1}^{n} x_k^{\,2} y_k
\end{cases}
$$

$$
\begin{cases}
7a_0 + 63a_1 + 819a_{11} = 172.16 \\
63a_0 + 819a_1 + 11907a_{11} = 1607.67 \\
819a_0 + 11907a_1 + 184275a_{11} = 21159.81
\end{cases}
\rightarrow
\begin{cases}
a_0 = 22.309285714 \\
a_1 = 0.313214286 \\
a_{11} = -0.004563492
\end{cases}
\tag{7.2}
$$

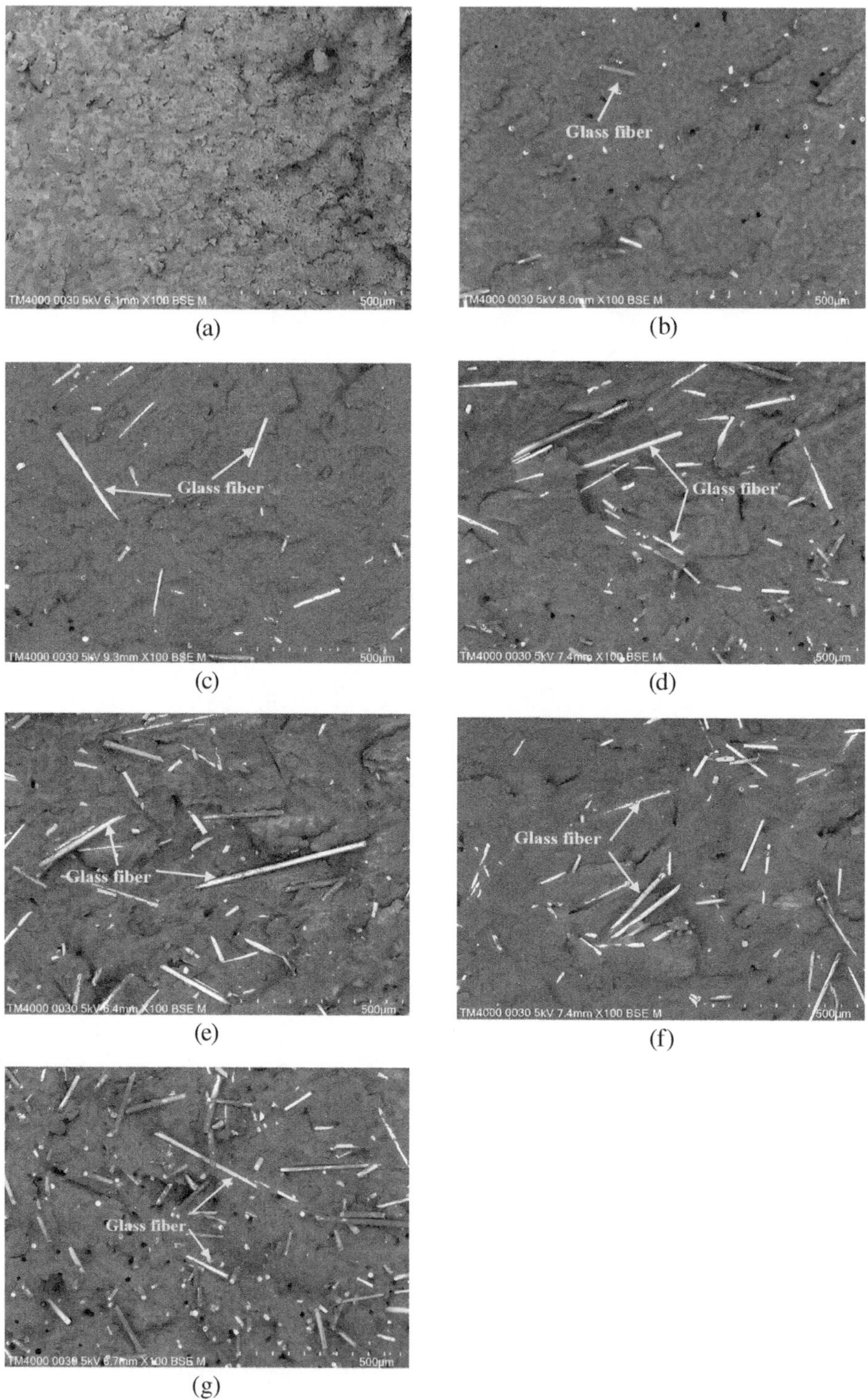

FIGURE 7.6 The microstructures of the PBT/PA6/GF samples: a, 0% GF; b, 3% GF; c, 6% GF; d, 9% GF; e, 12% GF; f, 15% GF; g, 18% GF.

TABLE 7.3

Experimental Unnotched Izod Impact Toughnesses of the PBT/PA6/GF Blends

No	x	y	xy	$x^2 y$	x^2	x^3	x^4
1	0	22.23	0	0	0	0	0
2	3	23.46	70.38	211.14	9	27	81
3	6	23.76	142.56	855.36	36	216	1296
4	9	24.85	223.65	2012.85	81	729	6561
5	12	25.39	304.86	3656.16	144	1728	20736
6	15	26.02	390.30	5854.50	225	3375	50625
7	18	26.45	476.10	8569.80	324	5832	104976
Σ	63	172.16	1607.67	21159.81	819	11907	184275

Substituting Equation (7.1) into Equation (7.2) gives the second-order regression Equation (7.3):

$$y = -0.004563492x^2 + 0.313214286x + 22.309285714 \tag{7.3}$$

We checked the regression equation results by drawing graphs in Excel and found nearly identical expressions from the two methods, with differences only due to rounding; that is, the formula accurately expresses the influence of GF on the impact toughness of a PBT/PA6 mixture. Figure 7.7 shows the regression model of the correlations between the average unnotched impact toughness of the sample and the GF ratio. The graph shows that toughness gradually increases as the GF content in the mixture increases.

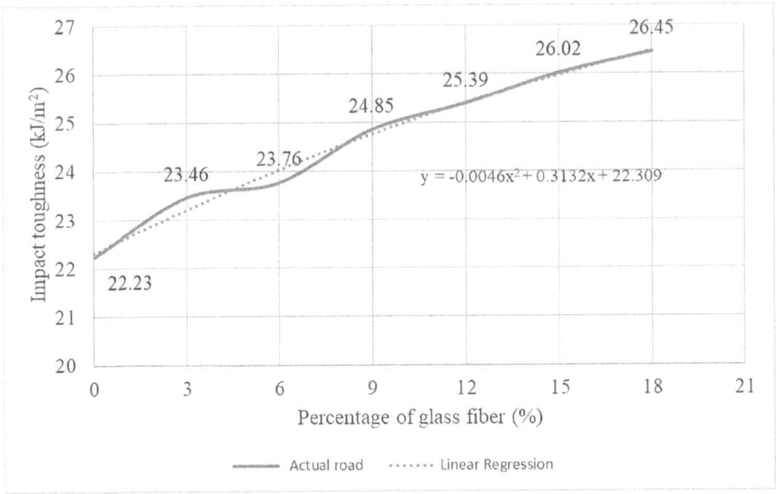

FIGURE 7.7 Graphic display of the quadratic regression equation for the unnotched Izod impact toughness of the PBT/PA6/GF mixtures prepared in Excel.

The figure shows that impact toughness increased from 22.23 kJ/m^2 at 0% GF to 26.45 kJ/m^2 at 18% GF. This could have been because of the overlapping bonds of the GFs as the GF content increased; we can closely observe in Figure 7.6 g that the GFs overlap each other. According to Fung et al. (2006), who studied the influence of GF on the mechanical properties of polymer composites, GF in higher quantities acts as a reinforcing agent for composite materials.

7.5   CONCLUSION

We studied the unnotched impact toughness of GF-reinforced PBT/PA6 composites with seven GF ratios: 0%, 3%, 6%, 9%, 12%, 15%, and 18%. GF demonstrably increased the toughness in all cases, although the difference between the highest toughness for sample PBT/PA6/18%GF and the lowest, for PBT/PA6/0%GF, was only 15.95%. This result opens a new research direction for improving the impact toughness and other mechanical properties of PBT/PA6 composite by reinforcing with 18% GF. Factories and plastic manufacturers can increase the impact toughness of recycled PBT/PA6 blends by reinforcing them with GF, which can help with addressing environmental pollution.

7.6   QUESTIONS

1. Why choose GF to increase the mechanical properties of the PBT/PA6 mixture and not other polymer components?
2. Why do we choose the regression equation as a quadratic equation of a parabolic curve and not a first-order equation?
3. Through structural observation, why do the PA6 dots gradually disappear and become smoother as the mixture's GF percentage increases?

7.7   ACKNOWLEDGMENT

This work belongs to the project in 2024 funded by Ho Chi Minh City University of Technology and Education grant number SV2024-225.

7.8   CONFLICTS OF INTEREST

The authors state that they have no known competing financial interests or personal relationships that could have appeared to influence the work reported in this article.

One of the authors (Van-Thuc Nguyen) is a member of this book's editorial board but was not involved in the editorial review or the decision to publish this article. All authors have read and agreed to the published version of the manuscript.

REFERENCES

Bruce Brown, S. "Reactive compatibilization of polymer blends", *Polymer Blends Handbook*, 2023, https://doi.org/10.1007/0-306-48244-4_5

Balaji Thattaiparthasarathy, K., Pillay, S., Ning, H., Vaidya, U.K. "Process simulation, design and manufacturing of a long fiber thermoplastic composite for mass transit application",

Composites Part A: Applied Science and Manufacturing, 2008, https://doi.org/10.1016/j.compositesa.2008.05.017

Chih-Rong Chiang, F. C. Chang, "Polymer blends of polyamide-6 (PA6) and Polyphenylene Oxide (PPO) compatibilized by Styrene-Maleic Anhydride (SMA) copolymer", *Polymer*, 1997, https://doi.org/10.1016/S0032-3861(96)00015-8

Chin-Ping Fung, Jiun-Ren Hwang, Chun-Chia Hsu, "The effect of injection molding pro-cess parameters on the tensile properties of short glass fiber-reinforced PBT", *Polymer Plastics Technology and Engineering*, 2006, https://doi.org/10.1081/PPT-120016335

Dekkers, M. E. J., Hobbs, S. Y., Watkins, V.H. "The morphology and deformation behavior of toughened blends of polybutylene terephthalate/polycarbonate and polyphenylene ether", *Polymer*, 1991, https://doi.org/10.1016/0032-3861(91)90039-L

Emel Kuram, Babur Ozcelik, Faruk Yilmaz, Gokhan Timur, Zeynep Munteha Sahin, "The effect of recycling number on the mechanical, chemical, thermal, and rheological properties of PBT/PC/ABS ternary blends: With and without glass-fiber", *Polymer Composites*, 2014, https://doi.org/10.1002/pc.22869

Florian Kempel, Bernhard Schartel, Gregory T. Linteris, Stanislav I. Stoliarov, Richard E. Lyon, Richard N. Walters, Anja Hofmann, B. Schartel, Stanislav I. Stoliarov, Richar N. Walters, "Prediction of the mass loss rate of polymer materials: Impact of residue formation", *Combustion and Flame*, 2012, https://doi.org/10.1016/j.combustflame.2012.03.012

Hamlaoui, O., Klinkova, O., Tawfiq, I., Elleuch. R. "Effect of the glass fiber content of a poly-butylene terephthalate reinforced composite structure on physical and mechanical char-acteristics", *Polymers (Basel)*, 2022, https://doi.org/10.3390/polym14010017

Hongrui Li, Guohua Li, Yanqiu Lu, Jinwei Wang, Nongyue Wang, Qingxin Zhang, Xiongwei Qu, "Preparation of core-shell structured particle and its application in toughening PA6/PBT blends", *Polymers Advanced Technologies*, 2017, https://doi.org/10.1002/pat.3969

Jun An, Jingying Ge, Youxi Liu, "Special effect of epoxy resin E-44 on compatibility and mec-hanical properties of polybutylene terephthalate/polyamide-6 blends", *Journal of Applied Polymer Science*, 1996, https://doi.org/10.1002/(sici)1097-4628(19960613)60:11<1803::aid-app3>3.0.co;2-p

Jun Chen, Dongyu Cai, Weixi Jin, Fei Wu, Xianyi Chen, "Effect of a type of trizaine on the properties and morphologies of polybutylene terepthalate composites", 2006, https://doi.org/10.1002/app.23404

Jungfeng Xiao, Yuan Hu, Ling Yang, et al. "Fire retardant synergism between melamine and triphenyl phosphate in polybutylene terephthalate", *Polym. Degrad. Stabil.*, 2006, https://doi.org/10.1016/j.polymdegradstab.2006.01.018

Ke-qing Han, Zheng-jun Liu, Mu-huo Yu, "Preparation and mechanical properties of long glass fiber reinforced PA6 composites prepared by a novel process", Materials and Engineering, 2005, DOI:10.1002/mame.200500051

Kelly, A. Tyson, W.R. "Tensile properties of fibre-reinforced metals: copper/tungsten and cop-per/ molybdenum", *Journal of the Mechanics and Physics of Solids*, 1965, https://doi.org/10.1016/0022-5096(65)90035-9

Maiti, S. N., Neetu Tomar, "Mechanical properties of PBT/ABAS blends", *Applied Polymer Science*, 2007, https://doi.org/10.1002/app.25831

Marco Monti, Maria Teresa Scrivani, Irene Kociolek, Age G. Larsen, Kjell Olafsen, Vito Lambertini, "Enhanced impact strength of recycled PET/Glass Fiber composites", *Polymers*, 2021, https://doi.org/10.3390/polym13091471

Nga Pham Thi Hong, "Characterization of virgin polybutylene terephthalate and recycled Polybutylene terephthalate from toothbrush bristles," *IOSR Journal of Mechanical and Civil Engineering, 2021, (IOSRJMCE)*, https://doi.org/10.9790/1684-1802025056

Peter Kiss, Joachim Schoefer, Wolfgang Stadlbauer, Christoph Burgstaller, Wasiliki-Maria Archodoulaki, "An experimental study of glass fibre roving sizings and yarn finishes in high-performance GF-PA6 and GF-PPS composite laminates", *Composites Part B: Engineering*, 2021, https://doi.org/10.1016/j.compositesb.2020.108487

Qian Cheng, Han Jiang, Yonghua Li, "Effect of fiber content and orientation on the scratch behavior of short glass fiber reinforced PBT composites", Tribology International, 2020, https://doi.org/10.1016/j.triboint.2020.106221

Ramani, A., Dahoe, A. E., "On the performance and mechanism of brominated and halogen free flame retardants in formulations of glass fibre reinforced polybutylene terephthalate", *Polymer Degradation and Stability*, 2014, https://doi.org/10.1016/j.polymdegradstab. 2014.03.021

Rawat, R. S., Tyagi, M., Talwar, M., Tyagi, V. K., Das, M., Kumar, M., Roy, P. "Morphological, mechanical and thermal properties of polybutylene tetrephthlate-polyester based thermoplastic polyurethane blends", 2022, https://doi.org/10.1002/slct.202202947

Seon-Joon Kim, Dong-Keun Kim, Won-Jei Cho, Chang-Sik Ha, "Morphology and properties of PBT/nylon 6/EVA-g-MAH ternary blends prepared by reactive extrusion", *Polymer Engineering and Science (PES)*, 2003, https://doi.org/10.1002/pen.10110

Wang Baolong, Wu Di, Zhu Lien, Jin Zheng, Zhao Kai, "High-density polyethylene-based ternary blends toughened by PA6/PBT core—Shell particles", *Polymer-Plastics Technology and Engineering*, 2017, https://doi.org/10.1080/03602559.2017.1295314

Wen-bing Liu, Wen-faa Kuo, Chiu-jung Chiang, Feng-chiu Chang, "Eur. In situ compatibilization of PBT/PPO blends", *Polym. J.*, 1996, https://doi.org/10.1016/0014-3057(95)00115-8

Wei Yang, Hongdian Lu, Qilong Tai, Zhihua Qiao, "Flame retardancy mechanisms of poly (1,4-butylene terephthalate) containing microencapsulated ammonium polyphosphate and melamine cyanurate", *Polym. Adv. Technol.*, 2011, https://doi.org/10.1002/pat.1735

Xiaoling Zuo, Kaizhou Zhang, Yang Lei, Shuhao Qin, Zhi Hao, Jiangbing Guo, "Influence of thermooxidative aging on the static and dynamic mechanical properties of long-glass fiber-reinforced polyamide 6 composites", *Journal of Applied Polymer Science*, 2013, https://doi.org/10.1002/app.39594

Zhang Shuidong, Tan Lingcao, Liang Jizhao, Huang Hanxiong, Jiang Guo, "Relationship between structure and properties of reprocessed glass fiber reinforced flame retardant poly(butylene terephthalate)", Polymer Degradation and Stability, 2014, https://doi.org/10.1016/j.polymdegradstab.2014.04.009

8 The Effect of Chamber Temperature on the Flexural Strength of Thermoplastic Polyurethane Plastic via FDM Technology

Truong Hoang Phuc, Huynh Nguyen Vinh Phuc, Le Nguyen Trung Nam, Pham Thi Hong Nga, Nguyen Thanh Tan, Nguyen Vinh Tien, Le Duong, Van-Thuc Nguyen, and Ho Thi My Nu*

8.1 INTRODUCTION

Additive manufacturing (AM), also known as 3D printing, has become the most widely used technology to rapidly create parts with complex geometries, which are difficult to produce using traditional methods (Rasiya et al., 2020). Over the years, AM technology has seen many advancements based on developments in material science and manufacturing technologies (Swift et al., 2013), in particular fused deposition modeling (FDM) and fused filament fabrication, the most widespread 3D printing technologies.

FDM was developed in 1989 by Scott Crump, cofounder of Stratasys Inc., and Stratasys first brought it to the market in 1991 (Gurr et al., 2012) and has since been used extensively in various industries such as automotive, health care, education, and aerospace. FDM is a cost-effective, flexible, and user-friendly technology that has revolutionized the manufacturing industry. With the current FDM technology, printed parts can be quickly produced and precisely sized with complex geometries, smooth surface roughness, and good mechanical properties. However, the mechanical characteristics are affected by printing process parameters such as printing speed, infill partner, layer height, infill density, nozzle temperature, and nozzle diameter. Depending on each specific application, it is essential to adjust the process parameters accordingly (Azar Equbal et al., 2017; Mohamed et al., 2015; Hasdiansah et al., 2023; Dragos et al., 2023; Sahu et al., 2023).

*Corresponding author: hongnga@hcmute.edu.vn

DOI: 10.1201/9781032635170-8

"

Many researchers have investigated the effects of various process parameters on the mechanical properties of each 3D-printed material using the FDM method. Adan Rasheed investigated the influence of infill density, number of layers, printing speed, and bed temperature on the mechanical properties to determine the optimal parameters for better tensile strength for a bilayered polylactic acid (PLA) and acrylonitrile butadiene styrene composite (Rasheed et al., 2023). The authors used Taguchi's method for the parameter optimization and used ANOVA to examine the tensile strength test data; the result of the experiment showed that infill density was the most influential parameter, and an infill density of 75%, 30 layers per part, a printing speed of 20 mm/s, and a bed temperature of 100 °C gave the best tensile strength.

Rahman et al. (2023) conducted a study on the effects of layer height, nozzle temperature, printing speed, and bed temperature on a cubic lattice structure printed from PLA filament; they determined that the best process parameters were layer height of 0.1 mm, nozzle temperature of 210 °C, printing speed of 30 mm/s, and bed temperature of 60 °C for the highest compressive strength. In addition, higher nozzle temperature enhanced the tensile strength of PLA-printed parts, with the highest values at 220 °C (Karim et al., 2022). Gohar and colleagues showed the effects of raster angle on the mechanical properties of thermoplastic polyurethane (TPU) honeycomb sandwich structures; they confirmed that 0/90° gave better mechanical performance than −45/45° (Gohar et al., 2023).

Venkat et al. (2023) highlighted the effect of infill patterns in determining the mechanical characteristics of 3D-printed parts; specifically, the zig-zag infill consistently exhibited better tensile and flexural strength than the gyroid infill. Other researchers determined that at raster angle 0° and nozzle temperature 200 °C, fused filament fabrication achieved high strength for TPU filament (Arifvianto et al., 2021). In another study on achieving the lowest surface roughness, authors recommended an infill of 20%, layer height of 0.12 mm, nozzle temperatures of 210 °C for Rz and 220 °C for Ra, and printing speed of 90 mm/s, and layer height had the most significant impact on the surface roughness of printed parts (Tuncel et al., 2023). Other researchers have calculated the parameter values to produce maximum tensile strength for TPU to be 210 °C, 45° infill angle, 100% infill density, 0.18 mm layer thickness, and an infill angle of 45° has a better stacking structure than an infill angle of 90° (Duong Le et al., 2023).

Although many researchers have investigated the optimal parameters for FDM printing, few have studied the effect of chamber temperature. FDM 3D printing often takes place in ambient room temperature conditions, and plastic's chemical and mechanical properties are usually strongly influenced by temperature, particularly high temperatures. The temperature inside the printing chamber induces molecular diffusion and entanglement in the linking of layers, improving the interlayer bonding and the mechanical properties of FDM-printed parts. Increasing chamber temperature over 40 °C improved the interlayer bonding in PLA-printed products, which was informed by the increased fracture toughness (Supaphorn Thumsorn et al., 2023).

Abraheem Hadeeyah showed that printing with the closed and opened printer at the same ambient temperature has equal dimensional accuracy. However, the closed printer provided better surface roughness, while the open printer produced higher tensile strength (Hadeeyah et al., 2023). Therefore, we chose an initial printing chamber temperature of 30 °C (corresponding to room temperature) and gradually increased the temperature to test the flexural strength of the printed samples by using FDM technology with TPU filament. Our aim was to determine the optimal chamber temperature for achieving the best flexural strength.

8.2 THE EXPERIMENT

TPU has many properties, including elasticity, transparency, and resistance to oil, grease, and abrasion. It can be formulated as different materials, from soft and flexible elastomers to rigid plastics with high modulus. TPU's flexibility and durability make it suitable for various applications such as medical devices, sports equipment, robotics, and functional prototypes. For this study, we used TPU filament provided by Kingroon Tech Co., Ltd. (Shenzhen, China) with the properties shown in Table 8.1.

The printed sample for the tensile strength test was designed using 3D design software with dimensions shown in Figure 8.1.

We produced the printed samples shown in Figure 8.2 with an FDM X500 printer and chose the printing parameters based on previous studies with three different chamber temperatures, as shown in Table 8.2. Figure 8.2 shows the samples after the FDM printing.

TABLE 8.1

Original Properties of TPU Filament

Performance	Parameters	Unit
Filament diameter	1.75	mm
Nozzle temperature	210–230	°C
Heat bed temperature	70–80	°C
Print speed	25–30	mm/s
Density	1.25 ± 0.05	g/cm³
Tensile strength	39	MPa
Hardness	95	Shore A
Elongation at break	580	%

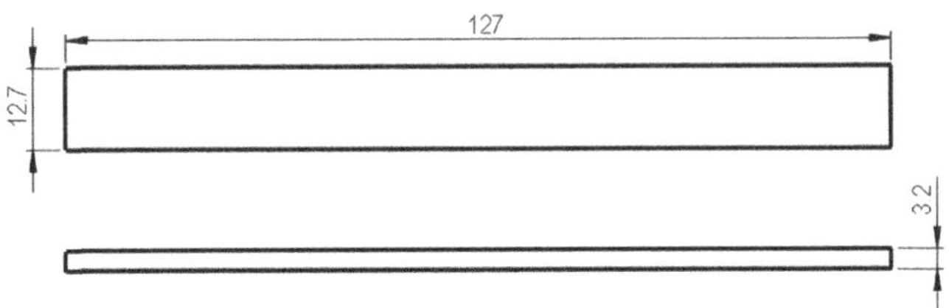

FIGURE 8.1 Design of the 3D-printed sample.

(a)

(b)

FIGURE 8.2 3D printing and testing machine: (a) FDM X500 printing machine and (b) Testing machine.

TABLE 8.2
3D Printing Parameters

Layer Height (mm)	Infill Percentage (%)	Nozzle Temperature (°C)	Heat Bed (°C)	Printing Speed (mm/s)	Outline Shells	Chamber Temperature (°C)
0.25	100	230	60	40	3	30
0.25	100	230	60	40	3	45
0.25	100	230	60	40	3	60

8.3 RESULTS AND DISCUSSION

The test result obtained from the stress–strain curve of each sample is shown in Figure 8.3, and the average flexural stress values are shown in Table 8.3. From there, we can compare and determine each printed sample's flexural strength at different chamber temperatures. The flexural stress or strength of a material is the maximum stress a material can withstand during flexural testing. The highest point of the stress curve in the stress–strain curve is the flexural strength achieved by the 3D sample printed at different chamber temperatures. Figure 8.4 graphically presents the average flexural stresses we calculated for the three different printing chamber temperatures.

Figure 8.4 shows that at the chamber temperature of 30 °C, the flexural stress was highest at 1.80 MPa, and it was lowest, 1.21 MPa, at 45 °C; at 60 °C, it increased to 1.37 MPa. The values in Table 8.3 show that the flexural stress we measured in the samples printed at 30 °C were stable and had the least deviation, and the figure shows that flexural strength correlates with chamber temperature. At high

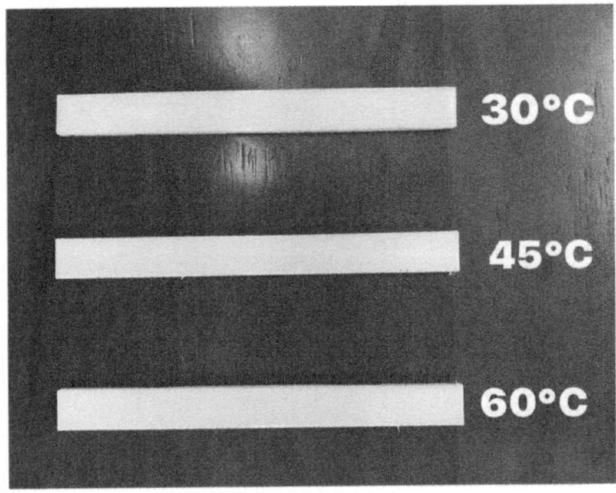

FIGURE 8.3 The stress–strain curve of each sample printed at different chamber temperatures.

TABLE 8.3
Average Flexural Stresses of the Printed Samples

Flexural Stress (MPa)

No.	Chamber Temperature		
	30°C	45°C	60°C
Sample 1	1.61	—	1.19
Sample 2	1.80	1.10	—
Sample 3	1.91	1.20	1.50
Sample 4	1.78	1.28	1.78
Sample 5	1.89	1.27	1.02
Average Flexural Stress (MPa)	1.80	1.21	1.37
Standard Deviation (MPa)	0.12	0.08	0.34

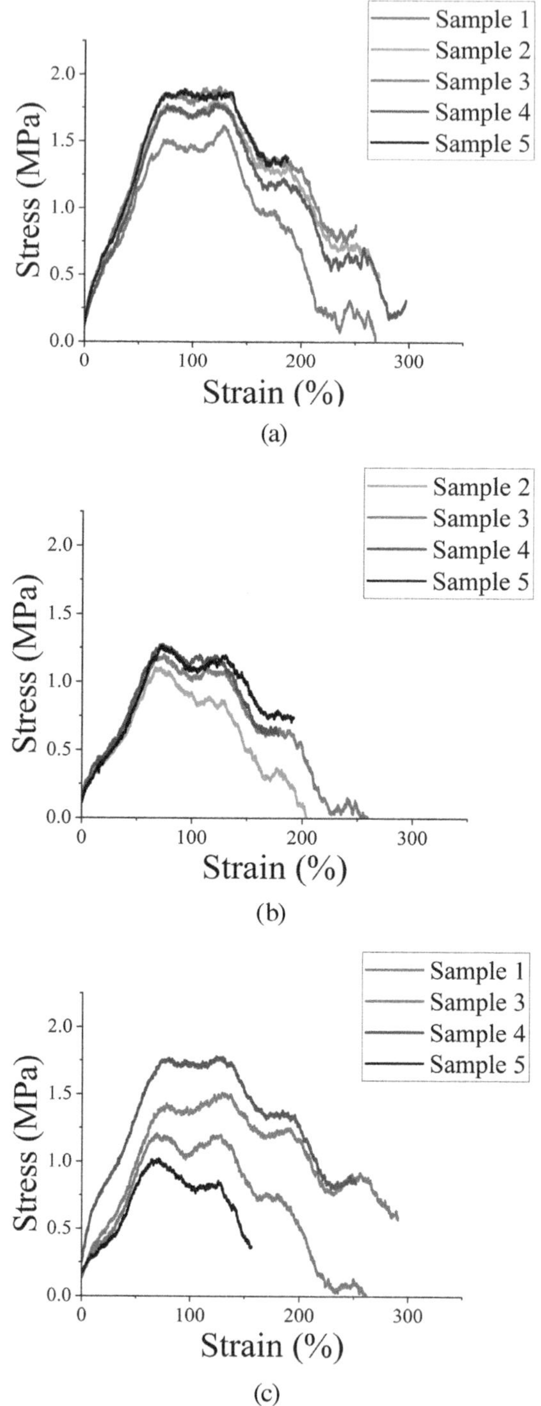

FIGURE 8.4   The influence of different chamber temperatures on flexural stress.

temperatures, the fracture strength of the TPU sample increased, indicating that high temperature is conducive to layer adhesion, which improved the mechanical properties of the printed samples (Lu et al., 2018).

However, flexural stress decreased as chamber temperature increased. Still, the tensile stress strongly decreases at 45 °C, which means that strong interlayer adhesion when printing at high temperatures does not increase flexural stress. In contrast, it reduces the flexural strength. Abraheem Hadeeyah also showed that tensile strength decreases the higher the temperature (Hadeeyah et al., 2023). In short, the chamber temperature affects the flexural strength of 3D-printed samples.

8.4 POLYNOMIAL REGRESSION ANALYSIS

We used polynomial regression (Equation (8.1)) to determine the effects of chamber temperature method; Table 8.4 shows our regression results based on the flexural stress values shown in Table 8.3 at confidence interval (CI) = 95%.

The regression equation is:

$$\text{Flexural Stress} = 5.205 - 0.1633 \text{ Chamber Temperature} + 0.001657 \text{ Chamber Temperature}^2 \tag{8.1}$$

Analyzing the variance and regression between the obtained quantities gives the following results:

$$\text{Correlation of determination: } R^2 = 0.6675, P = 0.004 < 0.05$$

$$\text{Flexural strength equation: } y = 5.205 - 0.1633x + 0.001657x^2 \tag{8.2}$$

The Student's t regression coefficient for the correlation between temperature and durability was negative at $r = -0.60$: As chamber temperature increased, flexural stress decreased. Figure 8.5 shows that the optimal chamber temperature is 30 °C with a maximum flexural strength of ymax = 1.91 based on Table 8.3.

8.5 CONCLUSION

The study's results demonstrate a significant effect of chamber temperature on the mechanical properties of TPU using FDM technology: Flexural strength decreases as chamber temperature increases from 30 to 60 °C. The results show that to obtain the best flexural strength (1.80 MPa) of TPU material, the chamber temperature should be at 30 °C.

8.6 CONFLICTS OF INTEREST

The authors state that they have no known competing financial interests or personal relationships that could have appeared to influence the work reported in this article. One of the authors (Van-Thuc Nguyen) is a member of this book's editorial board but was not involved in the editorial review or the decision to publish this article. The

TABLE 8.4

Polynomial Regression Findings for Flexural Stress versus Chamber Temperature

Model Summary

S	R²	R²(adj)
0.204262	66.75%	60.10%

Analysis of Variance

Source	DF	SS	MS	F	P
Regression	2	0.83745	0.418723	10.04	0.004
Error	12	0.41723	0.041723		
Total	14	1.25468			

Sequential Analysis of Variance

Source	DF	SS	MS	F	P
Linear	1	0.454157		6.24	0.030
Quadratic	1	0.383290		9.19	0.013

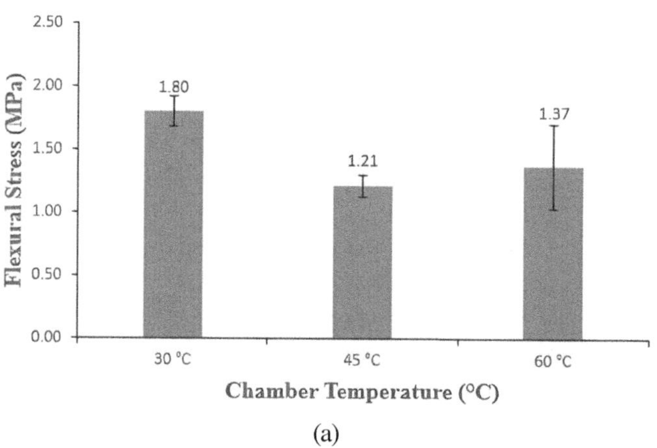

FIGURE 8.5   The relationship between flexural stress and chamber temperature: a) bar graph; b) line graph.

authors state that the work has no potential conflict of interest. All authors have read and agreed to the published version of the manuscript.

8.7 QUESTIONS

1. Which chamber temperature gives the maximum flexural strength?
2. What printer did the authors of this study use?
3. How many chamber temperature levels were tested in this study?

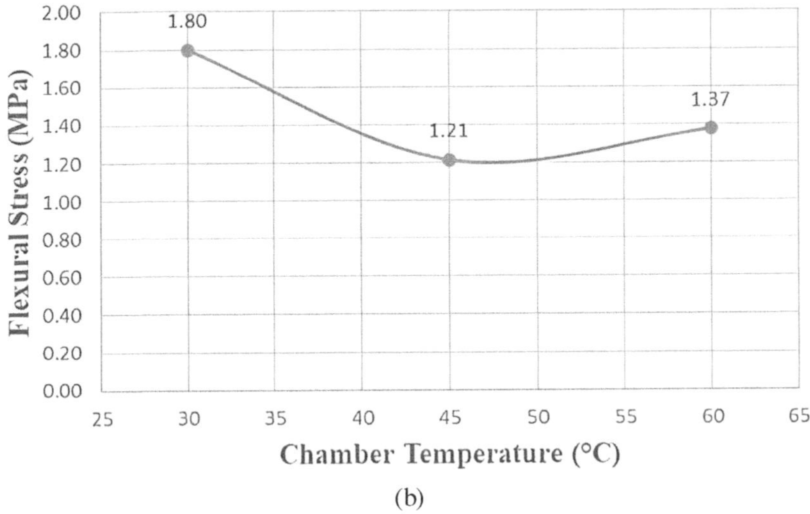

FIGURE 8.5   (Continued)

8.8   ACKNOWLEDGMENT

This work belongs to the project in 2024 funded by Ho Chi Minh City University of Technology and Education grant number SV2024-227.

REFERENCES

Abraheem Hadeeyah, Amir Kessentini, Ibrahim Ali Farj Emhemed, Fouzi Alhadar, Mondher Wali, Neila Khabou Masmoudi (2023). The effect of ambient temperature on the quality of three-dimensional printer products in FDM technology for ABS material. https://www.researchgate.net/publication/370770766

Adnan Rasheed, Muhammad Hussain, Shafi Ullah, Zeeshan Ahmad, Hasnain Kakakhail, Asim Ahmad Riaz, Imran Khan, Sajjad Ahmad, Waseem Akram, Sayed M Eldin, Ilyas Khan (2023). Experimental Investigation and Taguchi optimization of FDM process parameters for the enhancement of tensile properties of Bi-layered printed PLA—ABS. http://doi.org/10.1088/2053-1591/acf1e7

Azar Equbal, Anoop Kumar Sood, Abdul Razzaq Ansari, Md. Asif Equbal (2017). Optimization of process parameters of FDM part for minimizing its dimensional inaccuracy. https://www.researchgate.net/publication/316077743_Optimization_of_process_parameters_of_FDM_part_for_minimiizing_its_dimensional_inaccuracy

Budi Arifvianto, Benidiktus Tulung Prayoga, Rini Dharmastiti, Teguh Nur Iman (2021). Tensile properties of the FFF-processed Thermoplastic Polyurethane (TPU) elastomer. http://doi.org/10.21203/rs.3.rs-299979/v1

Dragos Gabriel Zisopol, Mihail Minescu, Dragos Valentin Iacob (2023). A theoretical-experimental study on the influence of FDM parameters on the dimensions of cylindrical spur gears made of PLA. https://www.etasr.com/index.php/ETASR/article/view/5733/3059

Duong Le, Canh Ha Nguyen, Minh Tai Le, Thi Hong Nga Pham, Thanh Tan Nguyen, Son Minh Pham, Van Thuc Nguyen (2023). Optimization 3D printing process parameters for the tensile strength of thermoplastic polyurethane plastic. https://doi.org/10.1007/s11665-023-07892-8

Elhattab Karim, Sarit B Bhaduri, Prabaha Sikder (2022). Influence of fused deposition modeling nozzle temperature on the rheology and mechanical properties of 3D β-Tricalcium Phosphate (TCP)/Polylactic Acid (PLA) composite. https://doi.org/10.3390/polym14061222

Gulnaaz Rasiya, Abhinav Shukla, Karan Saran (2021). Additive manufacturing-a review. https:// doi.org/10.1016/j.matpr.2021.05.181

Gohar, S., Hussain, G., Ali, A., Ahmad H. (2023). Mechanical performance of honeycomb sandwich structures built by FDM printing technique. http://doi.org/10.1177/0892705721997892

Gurr, M., Mülhaupt R. (2012). "Rapid prototyping," *Reference Module in Material Science and Materials Engineering*, vol. 8, pp. 77–79. http://doi.org/10.1016/B978-0-444-53349- 4.00202-8

Hasdiansah Hasdiansah, Yaqin Rizqi Ilmal, Pristiansyah Pristiansyah, Umar Mega Lazuardi, Priyambodo Bambang Hari (2023). FDM-3D printing parameter optimization using Taguchi approach on surface roughness of thermoplastic polyurethane parts. http://doi.org/10.1007/s12008-023-01304-w

Jitendra Kumar Sahu, Ranjeet Kumar Sahu, Jitendra Kumar Katiyar, P Sai Kiran (2023). Optimization of process parameters for dimensional stability in FDM. https://doi.org/10.1177/09544089231206800

Lu, C.-H., Yin, J., Du, M., Zhu Y.-B. (2018). Influence of 3D printing parameters on the interlayer bonding strength for TPU soft materials. http://doi.org/10.11777/j.issn1000-3304.2017.17146

Omar Ahmed Mohamed, Syed Hasan Masood, Jahar Lal Bhowmik (2016). Optimization of fused deposition modeling process parameters for dimensional accuracy using I-optimality criterion. https://doi.org/10.1016/j.measurement.2015.12.011

Oguz Tuncel, Mumin Tutar (2023). Optimization of FDM-3D printing parameters on surface roughness of TPU parts using taguchi technique. https://www.researchgate.net/publication/374948267

Rahman Md Mazedur, Sultana Jakiya, Rayhan Saiaf Bin, Ahmed Ammar (2023). Optimization of FDM manufacturing parameters for the compressive behavior of cubic lattice cores: An experimental approach by Taguchi method. http://doi.org/10.1007/s00170-023-12342-9

Swift, K.G., Booker J.D. (2013). Manufacturing process selection handbook. https://www.sci-encedirect. com/book/9780080993607/manufacturing-process-selection-handbook

Supaphorn Thumsorn, Wattanachai Prasong, Akira Ishigami, Takashi Kurose, Yutaka Kobayashi, Hiroshi Ito (2023). Influence of ambient temperature and crystalline structure on frac-ture toughness and production of thermoplastic by enclosure FDM 3D printer. https:// doi.org/10.3390/jmmp7010044

Venkat Reddy Yadavalli, Amar Kumar Myadam, Sriranga Babu Telu (2023). FDM 3D-print on Thermoplastic Polyurethane (TPU) with different process parameters using gyroid and zigzag infill patterns. https://www.scirp.org/journal/paperinformation?paperid=131360

9 Enhancing Underwater Imagery Using Multicriteria Decision-Making with Machine Learning Techniques

Ganesh Khekare, Shivani Kerai,
Anil V Turukmane, Urvashi Khekare,
Rahul Sharma, and Rahul Agrawal

9.1 INTRODUCTION

The realm of underwater image analysis is fraught with inherent challenges, stemming from the intricate interplay of environmental factors such as water turbidity, variable lighting conditions, and the attenuation of colors in different spectral ranges. These challenges impede the effectiveness of traditional computer vision and image analysis techniques, necessitating innovative strategies to extract meaningful information from underwater scenes. In this research chapter, we delve into the transformative impact of two pivotal image enhancement methodologies—histogram equalization and manual white balancing—on the intricacies of underwater imagery.

Histogram equalization, a widely employed technique in image processing, seeks to overcome the limitations of low contrast in underwater scenes. By redistributing pixel intensities to achieve a more uniform distribution, this method enhances the overall contrast of the image. In underwater environments, where visual details are often obscured by water properties, histogram equalization proves instrumental in revealing finer features and nuances that are crucial for subsequent analytical stages.

Complementing this, manual white balancing rectifies the color distortions inherent in underwater imaging. The aquatic milieu introduces shifts in color tones due to the absorption and scattering of light at different depths, and manual white balancing addresses this challenge by compensating for these variations, resulting in a more accurate and visually faithful representation of the underwater scene. This nuanced adjustment not only aids in discerning objects of interest but also significantly contributes to the reliability of subsequent computer vision tasks. Amalgamating histogram equalization with manual white balancing comprehensively lays a robust foundation for advanced image analysis models; with

DOI: 10.1201/9781032635170-9

"

this research, we endeavor to showcase the symbiotic relationship between these techniques, illustrating how their judicious integration enhances the interpretability of underwater scenes.

9.2　RATIONALE FOR THE STUDY

Khekare et al. (2023a) tackled the problems with absorption and scattering effects in underwater photographs, which impede the identification of marine life and the detection of objects; the authors combined a dehazing algorithm with contrast enhancement based on least information loss and histogram distribution prior (Yun et al., 2020). They optimized one version for display with genuine color and natural appearance, and the other, with heightened contrast and brightness, was conducive to detailed information extraction, and their method surpassed several state-of-the-art techniques in visual quality (Yeh et al., 2022), information value, and color restoration (Khekare et al., 2022). The dual output approach and superior performance demonstrated in various underwater scenarios contribute significantly to advancing the field of underwater imaging (Xiutiao et al., 2022), providing a valuable foundation for underwater object detection and image captioning.

Deluxni et al. (2023) tackled the underwater photography difficulties of color distortion, blurriness, and low contrast brought on by light absorption and scattering, and Khekare et al. (2023b) systematically investigated current underwater picture enhancement and restoration techniques, specifically spatial-domain (Xiang et al., 2023) and transform-domain enhancement (Wu et al., 2023) and color constancy, retinex, and restoration. The authors highlighted the technical challenges posed by wavelength-dependent absorption and scattering (Wentian et al., 2021), as well as varying illumination in underwater environments (Liu, C. et al., 2022); they also examined data sets used for testing algorithms and discussed quality evaluation methods, both objective and subjective (Khekare et al., 2023b). They concluded by acknowledging the complexity of underwater image processing and advocating for further research in this field (Liu, J. et al., 2022), positioning the review as a valuable resource for understanding the current state (Majeed et al., 2020), challenges, and advancements in underwater image enhancement and restoration techniques (Khekare et al., 2023c).

Li et al. (2019) addressed the crucial need for enhancing low-light underwater images in the context of deep-sea exploration. The paper emphasizes the challenges posed by low light, color distortions, and noise in capturing high-quality underwater images (Arisa et al., 2023). The authors first reviewed dark channel prior and histogram equalization, underscoring the limitations of generic approaches for underwater imaging and highlighting the necessity for specialized techniques. Li et al. (2023) introduced an innovative bright channel model (Panetta et al., 2022), transmission map refinement using guided filtering, deep convolutional neural networks for denoising, and a spectral characteristics-based approach for color correction (Bai et al., 2020). The experimental results demonstrated the effectiveness of the method on real-world deep-sea images, showcasing improvements in noise reduction, exposure enhancement, and color correction over existing methods. The paper's significance for underwater object detection and image captioning lies in its focus on addressing the challenges specific to low-light underwater environments, providing a valuable resource for research aiming to improve algorithmic performance in these scenarios (Li et al., 2019).

Wang G. et al. (2023) outlined a novel approach, optimal contrast and attenuation difference (OCAD), for enhancing underwater images plagued by challenges such as color casts and blurring caused by light scattering and absorption. OCAD is a two-step process involving transmission map estimation and refinement and including dark channel prior (Saleem et al., 2023) and guided filter methods (Song et al., 2021). Additionally, the method addresses color correction (Tan et al., 2021) and deblurring by estimating veiling light based on the differential attenuation of red, green, and blue (RGB) light underwater.

Wang T. et al. (2023) demonstrated that using OCAD to enhance established underwater image data sets was superior to state-of-the-art and classical enhancement techniques based on improved objective metrics such as peak signal-to-noise ratio (Ueda et al., 2023), structural similarity index, and underwater color image quality evaluation. The authors established that OCAD effectively tackled color distortion and blurring issues, offering significant advantages over existing methods (Vinyals et al., 2015). In their study, OCAD enhanced image quality and holds substantial promise for improved accuracy in object detection and more precise identification and description of underwater scenes for image captioning applications.

Feng et al. (2021) proposed correcting the red channel in underwater images to enhance underwater object detection, addressing the issue of red channel attenuation, which is a characteristic challenge in underwater imaging (Tsung et al., 2014). The method involved weighting and correcting the red channel, constructing a quaternion with RGB components, and employing the phase spectrum of quaternion Fourier transform (PQFT) for salient object detection (Haque et al., 2021). Experimental results showed that correcting the red channel outperformed the traditional spectral residual and PQFT algorithms in accuracy and error rate for detecting visually significant underwater objects (Im et al., 2023).

Voronin et al. (2018) introduced a novel algorithm that combined local and global image processing in the frequency domain, using logarithmic transform histogram matching with spatial equalization (Jiang et al., 2021) to improve underwater image quality. Their approach (Kandala et al., 2022) entailed processing different image blocks and integrating them through weighted means optimized by the measure of enhancement (Kastner et al., 2021). The authors demonstrated the effectiveness of this method through experiments on real underwater images and showed improvements over classical contrast enhancement and adaptive histogram equalization.

Zhang et al. (2023) presented an innovative approach for detecting marine organisms using image processing and machine learning; specifically, they combined underwater image enhancement and domain generalization with an improved YOLOv7 architecture (Galassi et al., 2021). The authors proposed a double-domain data augmentation strategy that integrated self-attention operations and convolutional computation. The improved YOLOv7 network incorporated the SIoU loss function for better convergence and regression accuracy, and the authors tested the method on multiple specific data sets. The method gave more precise detection than any existing methods.

9.3 MATERIALS AND METHODS

For the study for this chapter, the image enhancement was a crucial process, seamlessly integrated as a preprocessing step before we input the images into the feature extraction model. This strategic placement was designed to optimize the condition of the images, laying the groundwork for more effective object detection and subsequent caption generation. By enhancing the images before they the feature extraction, we aimed to ensure that the model would receive input in an optimal state, primed to extract relevant information. The anticipated outcome was improved object detection accuracy in the clearer, enhanced underwater images that captured richer images with intricate details. This image enhancement was necessary to enhance the overall performance of the integrated approach to underwater object detection and image captioning for advanced marine exploration.

9.3.1 IMAGE READING

The first step in the process was reading the input image using the OpenCV library with the 'cv2.imread' function. OpenCV is a widely used computer vision library that provides various tools and functions for image processing tasks. In cv2.imread, we loaded the images into the computational environment, making it accessible for subsequent analysis and manipulation.

9.3.2 COLOR SPACE CONVERSION

Following the image reading, we adapted each image to a more suitable color representation. The default color space in OpenCV is BGR (blue, green, red). However, it is frequently preferable to convert an image to the RGB color space for a variety of reasons, such as natural color representation and interoperability with other tools and libraries. This conversion involves rearranging the color channels from BGR to RGB, ensuring that the color information is correctly interpreted and displayed. RGB is a widely used color model in computer vision and image processing, making it a standard choice for various applications.

9.3.3 SPLITTING RGB IMAGES

Next, we split the R, G, and B channels in the RGB images. This division was essential because it permitted each channel to be processed independently, allowing for focused modifications to be made to color components.

9.3.4 HISTOGRAM EQUALIZATION

The next step in the process was applying histogram equalization to each of the individual RGB channels independently. The goal of the contrast enhancement machine learning technique known as histogram equalization is to modify the distribution of pixel intensities within an image.

An image's pixel intensity distribution is represented graphically by a histogram that displays the frequency of each intensity level. A grayscale image has two axes: the y-axis, which shows the frequency of each intensity, and the x-axis, which shows the intensity, which ranges from 0 to 255. The cumulative distribution function is calculated from the histogram, and it ranges from 0 to the total number of pixels in the image and represents the cumulative sum of the histogram values.

The next step involved creating a transformation function based on the cumulative distribution function. This function maps the original intensity values to new values; the goal is to spread out the intensity values such that the resulting histogram is as uniform as possible. Then we mapped each pixel in an image to its new intensity using the transformation function. This process effectively redistributes the pixel intensities, enhancing the overall contrast of the image with more evenly distributed pixel intensities.

Histogram equalization is particularly useful for images with uneven lighting or a limited range of pixel intensities. It is commonly applied to grayscale images but can be extended to color images by applying the technique independently to each color channel in RGB images as we just described. In this case, we used OpenCV cv2.equalizeHist to perform the equalization on each channel separately. The process begins with computing the histogram, a distribution of pixel intensities, for the given image. The histogram P(i,j) is determined by counting the frequency of each intensity level in the image:

$$P(i, j) = \text{Frequency of intensity level } (i, j)$$

Subsequently, we calculated the cumulative distribution function, denoted as $C(i,j)$, representing the cumulative probability of occurrence for each intensity level as shown in Equation (9.1):

$$C(i, j) = \sum_{k=0}^{i} \sum_{l=0}^{j} P(k, l) \tag{9.1}$$

Then we computed the enhanced pixel values (I_{eq}) using Equation (9.2):

$$I_{eq}(x, y) = \frac{L-1}{W \times H} \sum_{i=0}^{x} \sum_{j=0}^{y} P(i, j) \tag{9.2}$$

where W and H are, respectively, the width and height of the image; L is the number of intensity levels; and P(i, j) is the histogram of the image. This formula achieved the highest accuracy with valid data sets.

9.3.5 MANUAL WHITE BALANCING

Manual white balancing is a crucial step in refining the color representation of underwater images, mitigating distortions induced by complex lighting conditions inherent in aquatic environments. This technique is used to adjust the colors in an image to appear more natural and visually appealing. The equalized RGB image must first be converted into the LAB color space using OpenCV's cv2.COLOR_RGB2LAB conversion function. LAB separates the image into three channels: L (luminance), A (green-red), and B (blue-yellow). This separation is advantageous for color correction tasks as it allows for independent adjustment of luminance and color information.

Once in the LAB color space, L is equalized separately. This involves enhancing and balancing the contrast and brightness of the luminance component of the image. After the luminance equalization, the LAB channels are merged back together to reconstruct the color image. Using the cv2.COLOR_LAB2RGB conversion function, the processed LAB image is finally returned to the regular RGB format, suitable for additional examination or display.

We performed this adjustment by scaling each pixel value in a color channel (c) based on the ratio of the average intensity of the current image ($Avg_{reference}$) to the average intensity of the current image (Avg_{image}). The formulas for computing the white-balanced pixel value at position (x, y, c) is given as shown in Equation (9.3):

$$I_{wb}(x, y, c) = I(x, y, c) \times \frac{Avg_{reference}}{Avg_{image}} \tag{9.3}$$

where I(x, y, c) is the pixel value at position (x, y) in channel c and $Avg_{reference}$ and Avg_{image} are the average intensities of corresponding channels in a reference white-balanced image and the current image, respectively. These meticulous adjustments effectively correct color distortions, leading to a more accurate and visually pleasing representation of the underwater scene. Figure 9.1 shows enhanced images using histogram equalization and manual white balancing. The enhanced image clearly provides more clarity, allowing for easier captioning.

9.4 RESULT AND DISCUSSION

We applied the proposed image enhancement methodologies, histogram equalization, and manual white balancing to a diverse set of underwater images and measured the quantitative results using established metrics and multicriteria decision-making into consideration.

FIGURE 9.1 Image enhancement using histogram equalization and manual white balancing.

9.4.1 CONTRAST IMPROVEMENT

To quantify the effectiveness of histogram equalization, we calculated the enhancement in contrast (Ceq) using Equation (9.4):

$$C_{eq} = \frac{Mean\,intensity\,of\,enhanced\,image}{Mean\,intensity\,of\,original\,image} \tag{9.4}$$

9.4.2 COLOR ACCURACY

To quantify the improvement in color accuracy achieved by the manual white balancing, we measured the color accuracy using Equation (9.5):

$$CA = 1 - \frac{Mean\,squared\,difference\,between\,white\,\_\,balanced\,image\,and\,reference\,image}{Maximum\,possible\,mean\,squared\,difference} \tag{9.5}$$

9.4.3 OVERALL IMAGE QUALITY EVALUATION

We evaluated overall image quality using the peak signal-to-noise ratio (PSNR) and the structural similarity index (SSI). SSI is computed using a combination of luminance, contrast, and structure measures, providing a comprehensive evaluation of image quality, and PSNR is computed as Equation (9.6):

$$PSNR = 10 \cdot log_{10}\left(\frac{Max\,intensity^2}{Mean\,squared\,difference\,between\,original\,and\,enhanced\,images}\right) \tag{9.6}$$

Our PSNR analysis of enhanced images is shown in Figure 9.2.

9.5 CONCLUSION

Our introduction in this chapter of combining novel image enhancement methodologies with machine learning techniques, namely histogram equalization and manual white balancing, represents a substantial advancement in underwater imaging. The discernible enhancements with multicriteria decision-making in contrast, color accuracy, and overall image quality address the intrinsic challenges of underwater scenes by increasing the accuracy of object detection and the precision of image captioning in the realm of marine exploration. By systematically tackling the nuances of underwater imagery, we have not only improved the interpretability of detected objects but also enriched the descriptive capabilities of image captioning models. The findings of this study are poised to propel further innovations in underwater imaging techniques, offering valuable insights that have practical implications for both researchers and practitioners, ultimately contributing to the ongoing evolution

Original Image	Grayscale Original Image	Enhanced Image	Grayscale Enhanced Image	PSNR (dB)	Grayscale PSNR (dB)
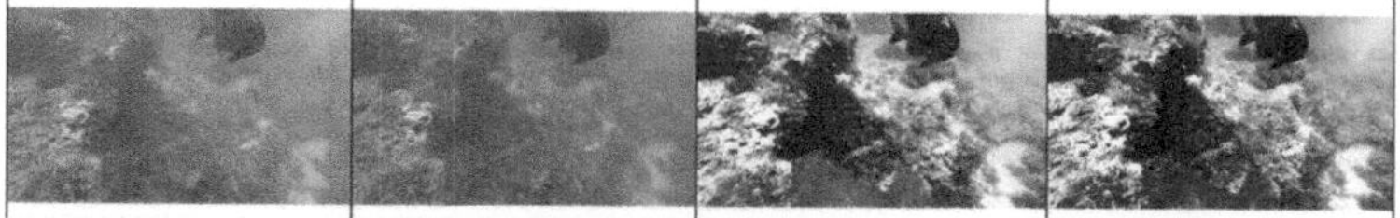				12.26	14.14
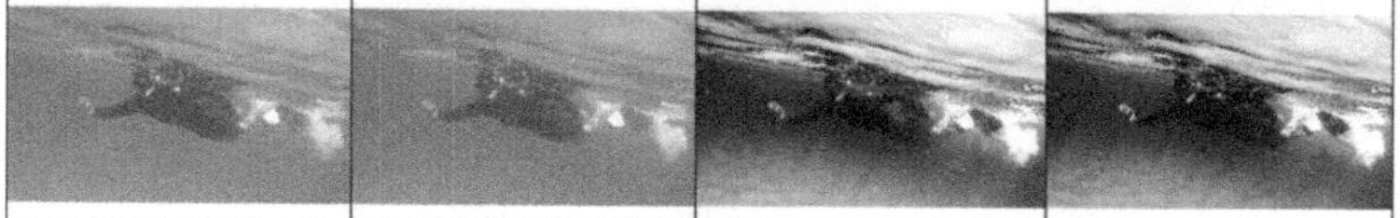				12.29	27.64
				13.31	27.39
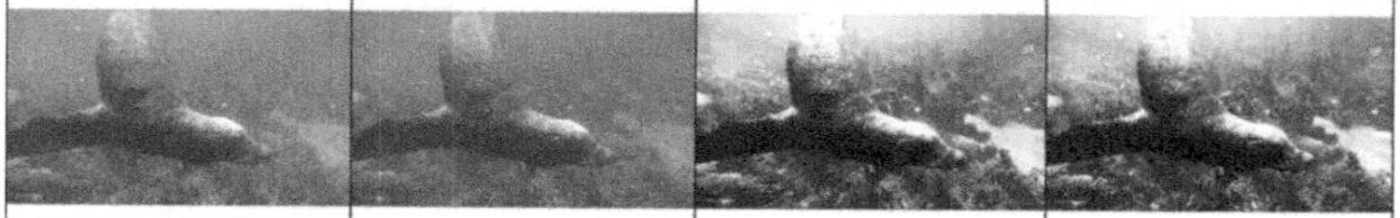				14.43	16.05
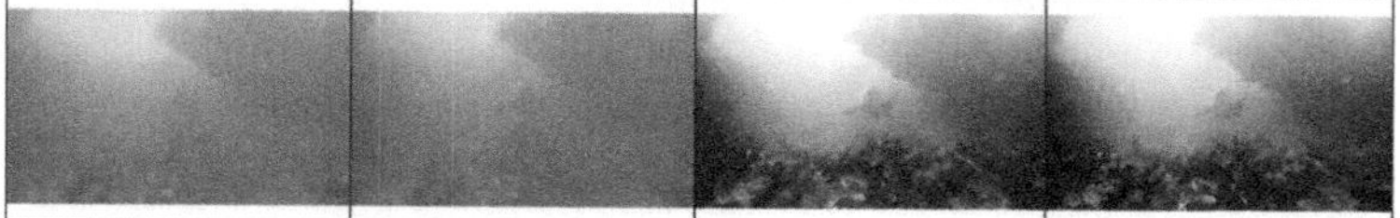				13.32	14.30
				13.15	14.23
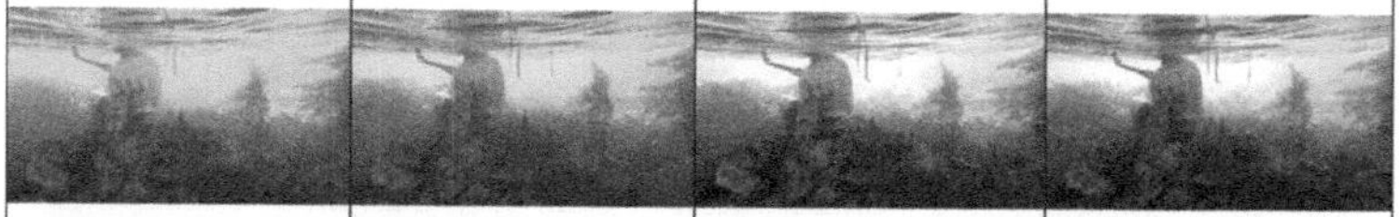				18.48	25.53
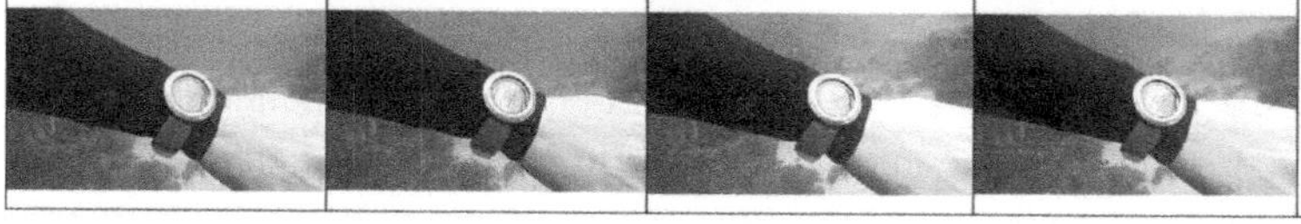				19.30	24.93

FIGURE 9.2 PSNR analysis of enhanced images.

of imaging technologies in the marine exploration domain. Captioning these images will remain as a future scope.

REFERENCES

Arisa, U., Wei, Y. & Komei, S., 2023. Switching Text-Based Image Encoders for Captioning Images With Text. IEEE Access, pp. 55706–55715.

Bai, L., Zhang, W., Pan, X. & Zhao, C., 2020. Underwater Image Enhancement Based on Global and Local Equalization of Histogram and Dual-Image Multi-Scale Fusion. IEEE Access, Volume 8, pp. 128973–128990.

Deluxni, N., Sudhakaran, P., Kitmo & Ndiaye, M. F., 2023. A Review on Image Enhancement and Restoration Techniques for Underwater Optical Imaging Applications. IEEE Access, Volume 11, pp. 111715–111737.

Feng, H., Xu, L., hong Yin, X. & hui Chen, Z., 2021. Underwater Salient Object Detection based on Red Channel Correction. 2021 IEEE 2nd International Conference on Big Data, Artificial Intelligence and Internet of Things Engineering (ICBAIE). s.l.:s.n., pp. 446–449.

Galassi, A., Lippi, M. & Torroni, P., 2021. Attention in Natural Language Processing. IEEE Transactions on Neural Networks and Learning Systems, 32(10), pp. 4291–4308.

Haque, A. U., Ghani, S. & Saeed, M., 2021. Image Captioning With Positional and Geometrical Semantics. IEEE Access, pp. 160917–160925.

Im, S.-K. & Chan, K.-H., 2023. Context-Adaptive-Based Image Captioning by Bi-CARU. IEEE Access, pp. 84934–84943.

Jiang, J. et al., 2021. Enhancements of Attention-Based Bidirectional LSTM for Hybrid Automatic Text Summarization. IEEE Access, Volume 9, pp. 123660–123671.

Kandala, H., Saha, S., Banerjee, B. & Zhu, X. X., 2022. Exploring Transformer and Multilabel Classification for Remote Sensing Image Captioning. IEEE Geoscience and Remote Sensing Letters, pp. 1–5.

Kastner, M. A. et al., 2021. Imageability- and Length-Controllable Image Captioning. IEEE Access, pp. 162951–162961.

Khekare, G., Gambhir, S., Abdulrahman, I. S., Kumar, C. M. S. & Tripathi, V., 2023a. D2D Network: Implementation of Blockchain Based Equitable Cognitive Resource Sharing System. 2023 3rd International Conference on Advance Computing and Innovative Technologies in Engineering (ICACITE), Greater Noida, India, pp. 908–912. https://doi.org/10.1109/ICACITE57410.2023.10182834.

Khekare, G., Kumar, K. P., Prasanthi, K. N., Godla, S. R., Rachapudi, V., Al Ansari, M. S. & El-Ebiary, Y. A. B., 2023b. Optimizing Network Security and Performance Through the Integration of Hybrid GAN-RNN Models in SDN-based Access Control and Traffic Engineering. International Journal of Advanced Computer Science and Applications (IJACSA), 14(12). http://dx.doi.org/10.14569/IJACSA.2023.0141262.

Khekare G. & Midhunchakkravarthy, 2023c. Smart Image Recognition System for The Visually Impaired People. 2023 International Conference on Energy, Materials and Communication Engineering (ICEMCE), Madurai, India, pp. 1–6. https://doi.org/10.1109/ICEMCE57940.2023.10434130.

Khekare, G., Verma, P. & Raut, S., 2022. The Smart Accident Predictor System using Internet of Things. Cloud IoT. Chapman and Hall/CRC, pp. 163–175.

Liu, C. et al., 2022. Focus Your Attention: A Focal Attention for Multimodal Learning. IEEE Transactions on Multimedia, Volume 24, pp. 103–115.

Liu, J., Liu, S., Xu, S. & Zhou, C., 2022. Two-Stage Underwater Object Detection Network Using Swin Transformer. IEEE Access, Volume 10, pp. 117235–117247.

Li, Y. et al., 2019. Low-Light Underwater Image Enhancement for Deep-Sea Tripod. IEEE Access, Volume 7, pp. 44080–44086.

Li, Y. et al., 2023. Vision-Based Target Detection and Positioning Approach for Underwater Robots. IEEE Photonics Journal, 15(1), pp. 1–12.

Majeed, S. H. & Isa, N. A. M., 2020. Iterated Adaptive Entropy-Clip Limit Histogram Equalization for Poor Contrast Images. IEEE Access, Volume 8, pp. 144218–144245.

Panetta, K., Kezebou, L., Oludare, V. & Agaian, S., 2022. Comprehensive Underwater Object Tracking Benchmark Dataset and Underwater Image Enhancement With GAN. IEEE Journal of Oceanic Engineering, 47(1), pp. 59–75.

Saleem, A. et al., 2023. A Non-Reference Evaluation of Underwater Image Enhancement Methods Using a New Underwater Image Dataset. IEEE Access, Volume 11, pp. 10412–10428.

Song, Y., He, B. & Liu, P., 2021. Real-Time Object Detection for AUVs Using Self-Cascaded Convolutional Neural Networks. IEEE Journal of Oceanic Engineering, 46(1), pp. 56–67.

Tan, H. et al., 2021. KT-GAN: Knowledge-Transfer Generative Adversarial Network for Text-to-Image Synthesis. IEEE Transactions on Image Processing, Volume 30, pp. 1275–1290.

Tsung-Yi, L. et al., 2014. Microsoft Coco: Common Objects in Context. Zurich, Switzerland, Springer.

Ueda, A., Yang, W. & Sugiura, K., 2023. Switching Text-Based Image Encoders for Captioning Images With Text. IEEE Access, 11(10.1109/ACCESS.2023.3282444), pp. 55706–55715.

Vinyals, O., Toshev, A., Bengio, S. & Erhan, D., 2015. Show and Tell: {A} Neural Image Caption Generator. CoRR, pp. 3156–3164.

Voronin, V. et al., 2018. Underwater Image Enhancement Algorithm Based on Logarithmic Transform Histogram Matching With Spatial Equalization. 2018 14th IEEE International Conference on Signal Processing (ICSP). s.l.:s.n., pp. 434–438.

Wang, G. et al., 2023. Polarization-Enhanced Underwater Detection Method for Multiple Material Targets Based on Deep-Learning. IEEE Photonics Journal, 15(6), pp. 1–6.

Wang, T. et al., 2023. Underwater Image Enhancement Based on Optimal Contrast and Attenuation Difference. IEEE Access, Volume 11, pp. 68538–68549.

Wang, Y. et al., 2019. An Experimental-Based Review of Image Enhancement and Image Restoration Methods for Underwater Imaging. IEEE Access, Volume 7, pp. 140233–140251.

Wentian, Z., Xinxiao, W. & Jiebo, L., 2021. Cross-Domain Image Captioning via Cross-Modal Retrieval and Model Adaptation. IEEE Transactions on Image Processing, pp. 1180–1192.

Wu, F. et al., 2023. Fish Target Detection in Underwater Blurred Scenes Based on Improved YOLOv5. IEEE Access, Volume 11, pp. 122911–122925.

Xiang, D., Wang, H., He, D. & Zhai, C., 2023. Research on Histogram Equalization Algorithm Based on Optimized Adaptive Quadruple Segmentation and Cropping of Underwater Image (AQSCHE). IEEE Access, Volume 11, pp. 69356–69365.

Xiutiao, Y. et al., 2022. A Joint-Training Two-Stage Method For Remote Sensing Image Captioning. IEEE Transactions on Geoscience and Remote Sensing, pp. 1–16.

Yeh, C.-H.et al., 2022. Lightweight Deep Neural Network for Joint Learning of Underwater Object Detection and Color Conversion. IEEE Transactions on Neural Networks and Learning Systems, Volume 33, pp. 6129–6143.

Yun, J., Xu, Z. & Gao, G., 2020. Context-Driven Image Caption With Global Semantic Relations of the Named Entities. IEEE Access, pp. 143584–143594.

Zhang, J. et al., 2023. Marine Organism Detection Based on Double Domains Augmentation and an Improved YOLOv7. IEEE Access, Volume 11, pp. 68836–68852.

10 Selecting Optimal Electric Vehicle Charging Station Sites Based on Analytic Hierarchy and VIKOR

Mousumi Karmakar and Subhajit Bhattacharyya

10.1 INTRODUCTION

Economic stability across the globe has improved steadily because of the rapidly growing demand for industrialization in our cities, but as a result, we are depleting many of our natural resources and causing severe environmental degradation (Bilgen, 2014) and (Zhao et al., 2014). However, if we stop industrialization to save a green environment, then we delay or lose our economic improvement. Therefore, it is necessary to develop ways to balance economic advancement by reducing fossil fuel consumption.

Although as of 2023, India's Climate Change Performance Index (CCPI) had improved to eighth place worldwide (Puneet et al., 2022), India's world carbon footprint rank still ranks third after only China and the United States. CCPI is calculated based on four criteria, greenhouse gas emissions, renewable energy, energy use, and environmental policy. In India, nearly 13% of CO_2 emissions are caused by the transportation sector (Yingwen et al., 2023), but the aggressive adoption of EVs can drastically reduce the emission of harmful greenhouse gases, although ongoing research is needed to boost EV battery life and range (Xiuhong et al., 2023). The Indian government is promoting EVs by offering subsidies to customers. With support from the government and attention from the public, we can make a sustainable, carbon-emission-free transportation for our future.

Building on the rapidly growing demand for and public acceptance of EVs, the optimal selection of EV charging station (EVCS) sites plays a significant role in green transportation systems. A sufficient number of EVCSs can eliminate range anxiety among EV users, and the public will be attracted to EVs if the charging stations are economical, well-organized, and easy to access. The charging station site determines its success rate; therefore, the method of selecting the sites needs to be well organized for choosing optimal locations.

In this paper, to choose an optimal site for EVCS, we considered five conflicting criteria: population, air quality, total construction cost, the total number of EVs served per day, and distance of the site from the main road. We propose a hybrid multicriteria decision-making (MCDM) model that combined the analytical hierarchy

DOI: 10.1201/9781032635170-10

process (AHP) for computing the relative weights of the different criteria followed by VIKOR MCDM ranking to rank the site alternatives. Finally, we selected the optimal EVCS site based on the VIKOR rank.

The rest of the paper is organized as follows. First, we present a literature review on EVCS site selection. Next, we explain the basic concepts of traditional MCDM methods followed by discussing the basic theories of AHP and VIKOR. In the third section, we discuss and apply the working AHP-VIKOR hybrid model, and in the fourth section, we give the results. Lastly, we conclude the proposed research work.

10.1.1 Literature Review

Selecting optimal sites for EVCSs is a real-life multicriteria decision-making challenge because the relevant criteria conflict with each other (Sen et al., 2015), yet doing so is critical for the evolution of the green industry. Authors have conducted a number of investigations on EVCS sites, for instance Hookah et al.'s (2015) study on EVCS planning and management. Zhang (2018) proposed an EV charging station planning model using a distribution network with the aim of minimizing the charging cost while maximizing customer service capacity and satisfaction. Erbaş et al. (2018) used GIS-based fuzzy AHP multicriteria decision analysis and TOPSIS (technique for order preference by similarity to ideal solution) to select optimal EVCS sites in Istanbul, Turkey. Aman et al. (2021) also integrated fuzzy AHP with TOPSIS to improve commercially feasible EV charging stations based on considering different environmental, societal, economic, location, and other factors.

Abapour et al. (2015) maximized the benefits of a distribution system manager to determine the optimal size and location for a plug-in hybrid electric vehicle charging station; their technique not only reduced active power losses and CO_2 emission but also improved supply energy in full load condition. Hosseini et al. (2019) proposed a Bayesian network (BN) model for optimal site selection considering various quantitative and qualitative factors. Based on sensitivity analysis and propagation analysis, they validated the proposed BN model for selecting the optimal EVCS site among all alternatives.

Yao et al. (2014) proposed a multi-objective planning model for integrated power distribution as well as for EV charging systems; they also faced multiple conflicting objectives: reducing the final annual investment cost, minimizing energy consumption, and maximizing the annual traffic flow attained. Yao et al. computed multiple multi-objective evolutionary algorithms to validate the proposed model's feasibility and effectiveness. Based on electro-mobility and energy demand, Baouche et al. (2014) proposed an optimized model for properly allocating EVCS sites in the city of Lyon, France. The proposed model not only satisfied all the mobility energy demands but also minimized trip energy consumption and total location expenditure. To get the vehicle consumption data, Baouche et al. tested the model on real trips and conducted data sensitivity analysis to validate the model.

Wu et al. (2016) developed a PROMETHEE (preference ranking organization method for enrichment evaluations)-based cloud model for choosing optimal sites for EVCSs in Beijing. The proposed cloud-based model integrated the uncertainty of information criteria and their mutual influence. In this model, PROMETHEE enhanced the evaluation and ranking of all alternatives, and Wu et al. performed a sensitivity analysis to select the final reasonable charging site.

Guo et al. (2020) used VIKOR to evaluate the performance of renewable energy power plants based on several criteria such as energy-saving, environmental impact, and economic feasibility. Tzeng et al. (2005) compared the performance of VIKOR with AHP, TOPSIS, and simple additive weighting and used a case study to show that VIKOR was the most effective for solving complex decision-making problems. In another study, Kaya et al. (2011) proposed a fuzzy version of VIKOR that could handle uncertain and imprecise information in solving complex decision-making problems. Wang et al. (2020) suggested an improved version of VIKOR that could handle interval data and hesitant fuzzy information in decision-making. The authors used a case study to indicate that the modified VIKOR method can effectively rank alternatives based on multiple criteria.

Researchers have applied VIKOR in other fields as well, for instance in finance to calculate the performance of mutual funds (Zhang et al., 2013) and identify optimal stock portfolios (Gupta et al., 2018), and in the field of supply chain management, AHP has been used to integrate and evaluate the model objectives in group decision-making processes for the development of supply network systems (Tavana et al., 2012). In engineering, VIKOR has been adopted to resolve supplier selection (Chen et al., 2011), project management (Gao et al., 2019), and product design (Chen et al., 2014), and in environmental management, VIKOR has been used to choose a suitable location for a waste disposal site (Duzakin et al., 2011) and to evaluate the sustainability of construction projects (You et al., 2015).

Overall, VIKOR has been widely used and shown promising results in solving MCDM problems in various fields. However, it is a significant condition that VIKOR requires carefully weighting the criteria and selecting appropriate compromise solutions in any real-life applications. For the study for this chapter, we propose a hybrid AHP-VIKOR model for selecting an optimal EVCS site based on population, air quality, cost, number of EVs served, and distance to the main city. The proposed model determined the criteria weights and selected the best alternative for EVCS.

10.2　TRADITIONAL METHODS OF MCDM

Traditional MCDM techniques have long been a cornerstone in decision analysis and have played a vital role in helping individuals and organizations make complex decisions. These methods encompass a range of well-established approaches, such as AHP, analytic network process (ANP), TOPSIS, and VIKOR. These techniques offer a structured framework for evaluating and prioritizing alternatives when faced with multiple conflicting criteria. They enable decision-makers to systematically assess the trade-offs between different options, taking into account various qualitative and quantitative factors. While these traditional MCDM methods have been widely used and trusted for many years, they are now being complemented and, in some cases, supplanted by more advanced computational and data-driven approaches as technology and data availability continue to evolve. Nonetheless, traditional MCDM techniques remain a valuable resource for decision-makers seeking to make informed and transparent choices in a wide range of domains.

10.2.1   AHP

Thomas Saaty (1980) created the decision-making framework known as the analytic hierarchy process in the 1970s; it is a methodical strategy for handling difficult choices including numerous criteria and options. With AHP, a choice is divided into a hierarchy of factors (criteria) (Vaidya et al., 2006; Cabala, 2010) and subfactors before each level is given a weight based on its relative importance. This process is usually done through pairwise comparisons, where each level is compared against the other levels in terms of their importance. The beauty of AHP is there is a provision for consistency checking to minimize inconsistencies in beliefs. Once the weights have been determined, the decision maker can evaluate each alternative against each criterion and compute a total score for every option. Next, the alternative with the best score is picked as the preferred choice. AHP is broadly used in different fields, including business, engineering, and government. It is particularly useful for decision-making scenarios that involve trade-offs between multiple conflicting criteria.

Some mathematical steps are required for AHP final weight calculations. The steps are as follows:

I. First an (m X m) pairwise comparison matrix is formed.
    Let B be the comparison matrix of size (m X m) and is represented by Equation (10.1):

$$B = \begin{bmatrix} b11 & b12 & \cdots & b1m \\ & \vdots & \ddots & \vdots \\ bm1 & bm2 & \cdots & bmm \end{bmatrix} \tag{10.1}$$

Any random matrix item may be described by Bij, where i,j =1,2, ...,m.
Bij = 1 when i = j and
Bij = 1/Bji when i ≠ j.
M = Number of comparison factors/attributes.

The diagonal elements of matrix B always equal 1 because a criterion is always equally important to itself. When the ith factor has a y-fold greater importance than the jth factor, then Bij = y and Bji = 1/y, usually calculated on a 1–9 scale (Table 10.1), where 1 means that the two sub-problems are equally important and 9 means that one is more important than the other.

II. In the second step, the normalized weight matrix is formed. To calculate the normalized weight matrix, every element of the jth column is multiplied by the inverse of the grand total of all elements of the jth column. After normalization, the sum of every column element must be one. Normalize pairwise comparison matrix is calculated as Equation (10.2):

$$Bw = \begin{bmatrix} \dfrac{b11}{\sum bi1} & \cdots & \dfrac{b1m}{\sum bim} \\ \vdots & \ddots & \vdots \\ \dfrac{bm}{\sum bi1} & \cdots & \dfrac{bmm}{\sum bim} \end{bmatrix} \tag{10.2}$$

TABLE 10.1

The Scale for the Pairwise Comparison of Criteria Weights

Intensity of Importance	Definition	Explanation
1	Equally significant	The property is equally dependent on two factors.
3	Medium significant of thing to another	Perceptions as well as expertise only slightly prefer one alternative to the other.
5	Substantial or vital significance	Perceptions as well as expertise substantially prefer one alternative to the other.
7	Truly significant	An element is given strong preference, and its dominance is shown in action.
9	Very High significant	One of the strongest possible orders of confirmations that prefer one element to the other.
2,4,6,8	Values in the middle of two consecutive opinions.	It is necessary to reconcile two opinions.
Reciprocals	If activity i is assigned with one nonzero number from 1 to 9 when differentiated from activity j, then j has the reciprocal value when differentiated from i.	
Rational	Ratios appearing from forcing stability of awareness	

Source: Saaty (1980).

III. In the third step, the eigenvector of matrix B is calculated; the eigenvector is the average of each row element of matrix Bw and is represented by Di. Obviously, Di is a column vector of size m × 1; it is calculated as Equation (10.3):

$$
Di = \begin{bmatrix} D1 \\ D2 \\ . \\ . \\ Dm \end{bmatrix} = \begin{bmatrix} \dfrac{\frac{b11}{\sum bi1}}{m} + \cdots + \dfrac{\frac{b1m}{\sum bim}}{m} \\ \vdots \quad \ddots \quad \vdots \\ \dfrac{\frac{bm1}{\sum bi1}}{m} \quad \cdots \quad \dfrac{\frac{bmm}{\sum bim}}{m} \end{bmatrix} \tag{10.3}
$$

IV. In the fourth step, the eigenvalue of original matrix B is calculated, which is represented by λmax. To calculate λmax requires another column vector which we get by multiplying matrix B by D (Equations (10.4) and (10.5)):

$$
B \times D = \begin{bmatrix} b11 & b12 & \cdots & b1m \\ \vdots & & \ddots & \vdots \\ bm1 & bm2 & \cdots & bmm \end{bmatrix} \times \begin{bmatrix} D1 \\ D2 \\ . \\ . \\ Dm \end{bmatrix} = \begin{bmatrix} y1 \\ y2 \\ . \\ . \\ ym \end{bmatrix} \tag{10.4}
$$

$$\lambda \max = \sum_{i=1}^{m} \frac{yi}{Ci} \tag{10.5}$$

V. In the fifth step, we calculate the consistency index (CI) and consistency ratio (CR); CR is calculated by dividing CI by the random consistency index (RI), a number based on the size of the matrix (Equation (10.6)) (Zou et al., 2008).

$$CI = \frac{\lambda \max - m}{m - 1} \text{ and } CR = \frac{CI}{RI} \tag{10.6}$$

RI is different for different m, and m can range between 1 and 10; Table 10.2 displays the corresponding RIs. If the CR is greater than 0.1, the matrices need to be revised (Cay et al., 2013; Chakraborty et al., 2006).

10.2.2 The VIKOR Method

A literature review on VIKOR reveals that the method has gained significant popularity among researchers due to its ability to handle decision-making problems with multiple criteria. Researchers have applied VIKOR across a broad range of disciplines, including engineering, management, finance, environmental studies, and health care.

One of the key advantages of VIKOR is its ability to balance multiple criteria, allowing decision-makers to select the most effective alternative that meets the requirements of all stakeholders. The method ranks choices according to their final performance and allows for sensitivity analysis to be carried out to evaluate the robustness of the decision. However, researchers have also identified some limitations of the method; for example, the method supposes that all the criteria are equally important and the criteria's weights are predefined, but this may not always reflect the actual preferences of decision-makers, leading to biased results. Additionally, the approach disregards risk and uncertainty in decision-making. Despite these limitations, VIKOR remains a popular MCDM method because of its clarity, simplicity in adoption, and capacity to handle a mass number of criteria. Researchers continue to explore new applications of the method and develop extensions to overcome its limitations.

The steps to calculate the final rank using the VIKOR technique are as follows:

I. Construct a decision matrix: The decision matrix contains the performance ratings of each choice (alternative) in relation to each factor (criterion). The matrix should have m alternatives and n criteria (see Equation (10.7)).

TABLE 10.2

Random Indices

No. of Features (m)	1	2	3	4	5	6	7	8	9	10
RI	0.00	0.00	0.58	0.90	1.12	1.24	1.32	1.41	1.45	1.49

$$C_1 \quad C_2 \quad .. \quad C_n$$

$$X = \begin{bmatrix} A_1 \\ A_2 \\ .. \\ A_m \end{bmatrix} \begin{bmatrix} I_{11} & I_{12} & .. & I_{1m} \\ I_{21} & I_{22} & .. & I_{2n} \\ .. & .. & .. & .. \\ I_{m1} & I_{m2} & .. & I_{mn} \end{bmatrix} \tag{10.7}$$

II. Calculate the weight of each criterion using a weighting method such as AHP (Singh et al., 2022).

III. Determine the ideal and undesirable values for each criterion based on the available data. The ideal value is the most favorable value for that criterion, while the undesired value is the most unfavorable value.

For example, let's say we have a decision-making problem that involves three criteria: cost, quality, and delivery time. We have four alternatives to choose from, and the data for each criterion are given in Table 10.3.

To determine the best and worst values for every criterion, we look at the data for each criterion and identify the highest and lowest values. The best value is characterized by g_j^+ and the worst value is characterized by g_j^- for each criterion.

- For cost (C1), the best value is 90 (A3) and the worst value is 120 (A2).
- For quality (C2), the best value is 90 (A2) and the worst value is 70 (A3).
- For delivery time (C3), the best value is 3 (A4) and the worst value is 6 (A3).

Once we have identified the ideal and undesirable value for each criterion, we can move on to calculating S, R, and Q for each.

IV. Calculate the maximum gap of improvement for both favorable and detrimental criteria: In VIKOR, each criterion can be either beneficial or nonbeneficial depending on whether higher or lower values of the criterion are preferred.

Beneficial criteria are those for which higher values are desirable, such as fuel efficiency, maximum speed, and passenger capacity for motor vehicles. The formula for a maximum gap of improvement for beneficial criteria is Equation (10.8)

$$B_{ij} = w_j \times \frac{g_{j^+} - g_{ij}}{g_{j^+} - g_j} \tag{10.8}$$

TABLE 10.3

Criteria for All Alternatives

Alternative	Cost (C1) (in USD)	Quality (C2) (in %)	Delivery Time (C3) (in Hours)
A1	100	80	5
A2	120	90	4
A3	90	70	6
A4	110	85	3

Nonbeneficial criteria are those that are desirable in lower values, for example cost, weight, and emissions for cars. The formula for maximum gap of improvement for nonbeneficial criteria is Equation (10.9):

$$\text{Bij} = \text{wj} \times \frac{g_{ij} - g_j-}{g_j - g_j-} \tag{10.9}$$

Bij = normalized value of ith alternative for j^{th} criteria.

VIKOR allows decision-makers to consider both favorable and detrimental criteria in evaluating alternatives and considers both the positive and negative aspects of each criterion. By considering both types of criteria, supporting more informed decisions.

V.   Next is calculating utility (Si; Equation (10.10)) and regret (Ri; Equation (10.11)) for every choice:

$$si = \sum\nolimits_{j=1}^{n} S_{ij} \tag{10.10}$$

$$Ri = \max(S_{ij}) \text{ for } j = 1, 2, \ldots n \tag{10.11}$$

VI.   The VIKOR index (Qi) is then calculated for each choice using Equation (10.12):

$$Qi = p \times \frac{S_i - S_{i\#}}{S_{i*} - S_{i\#}} + (1-p) \times \frac{R_i - R_{i\#}}{R_{i*} - R} \tag{10.12}$$

where S_{i*} is the highest S_i and $S_{i\#}$ is the lowest, R_{i*} is the highest R_i and $R_{i\#}$ is the lowest, and p is a weighting factor that reflects the decision-maker's preference for Q or R. A value of p = 0.5 gives equal weight to both values.

VII.   Rank the alternatives according to Si, Ri, and Qi (Yildirim et al., 2022), with the alternative having the highest value being ranked first. VIKOR also requires considering two other conditions ranking decisions: acceptable advantage and acceptable stability.

VIKOR acceptable advantage refers to the difference in utility between the two best-ranked alternatives, for instance Q (Rank1) and Q (Rank2). If the difference between Q (Rank1) and Q (Rank2) is greater than or equal to 1/(n−1), where n is the number of alternatives, then the acceptable advantage condition is satisfied. If the difference between Q (Rank1) and Q (Rank2) is less than 1/(n−1), then the first and second alternatives may not be sufficiently different in terms of their utility. In this case, a compromise solution can be proposed such as selecting the alternatives ranked first and second if only condition 2 is unsatisfied. The acceptable advantage condition ensures that the top-ranked alternatives are significantly better than the others, and it prevents the ranking from being based on small differences that are not meaningful.

Acceptable stability ensures that the chosen alternative is stable and consistent with the decision-maker's preferences even when different criteria and weights are used; it requires that the alternative ranked first by utility Q also be the best-ranked alternative according to at least one of the other decision criteria used in the analysis, such as cost or risk. This ensures that the top-ranked alternative is not only good in terms of utility but also satisfies the other criteria that are important for the decision-maker.

If the first-ranked alternative does not meet the acceptable stability condition, a set of compromise solutions may be proposed, such as selecting the alternatives that rank first and second based on Q, or selecting the alternatives that are ranked among the top k alternatives based on Q if only the permissible advantage condition is unsatisfied. Overall, the acceptable stability condition ensures that the selected alternative is not only the best in terms of utility but also meets the other criteria that are important for the decision-maker, thereby providing a more robust and stable decision-making process.

10.3 THE PROPOSED HYBRID MODEL

Here, we describe our analytic hierarchy model for selecting EVCS sites. Figure 10.1 depicts that there are three levels in the hierarchy we proposed: level I, the goal of the decision, to determine the optimal site for an EVCS among a selection of four; level II, consider the main attributes that impact the choice of the optimal site: population, air quality, cost, distance to the main city, and number of EVs served per day; and

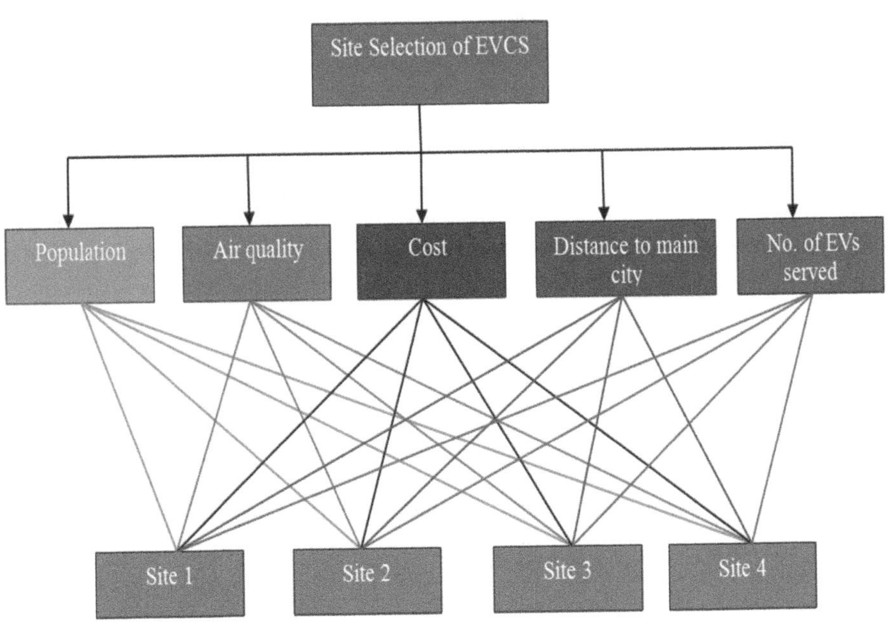

FIGURE 10.1 The hierarchy model for selecting an optimal EVCS site.

level III, consider the four alternative solutions that potentially met the objective of the task. The proposed hybrid model is an efficacious MCDM technique that is a combination of AHP and VIKOR-MCDM techniques.

Every MCDM method can solve problems (Roy, 2018), but AHP increases decision-making accuracy very well from the beginning with the help of a comparison matrix. AHP models evaluate the weights of each criterion, and VIKOR is a highly useful ranking method for selecting optimal alternatives. Combining the two was highly effective for selecting the optimal site for an electric vehicle charging station. The proposed AHP-VIKOR model entailed the following steps:

Step 1: Recognize the criteria for the pairwise comparison matrix.
Step 2: Formulate the weight matrix using AHP.
Step 3: Normalize the weight matrix and test consistency
Step 4: Use VIKOR to compute the ranks of all alternatives.

10.4   RESULTS AND DISCUSSION

For this research work, we were evaluating four locations in India, Kolkata, Chinsurah, Durgapur, and Bankura, to choose the optimal site for an electric vehicle charging station based on five conflicting parameters: population, air quality, total construction cost, the total number of EVs served per day, and distance of the site from the main road are identified to choose an optimal site for EVCS.

We applied AHP to create the pairwise criteria matrix in Table 10.4 and compute criteria weights before calculating the maximum eigenvalue (λm) of the matrix. Table 10.5 shows that distance to the main city had the highest weight in the decision (0.43) and that air quality had the lowest (0.5). Next, we calculated CI, RI, and CR following Step V for AHP (Section 10.2.1); we know that an acceptable CR is less than 0.1. Figure 10.2 plots the different attribute values for each criterion and each of the four locations.

Finally, we applied VIKOR to choose the optimal EVCS site following the steps in Section 10.2.2 by calculating the rank evaluation matrix in Table 10.6. We considered all the attributes for each alternative, and Table 10.6 shows that VIKOR ranked Kolkata the highest followed by Durgapur, Chinsurah, and Bankura, in order. Therefore, we concluded that Kolkata was the optimal location for the EVCS site.

TABLE 10.4

Pairwise Comparisons of the Criteria

	Population	Air Quality	Cost	Distance to the Main City	No. of EVs Served
Population	1	3	0.5	0.25	0.33
Air Quality	0.33	1	0.25	0.14	0.2
Cost	2	4	1	0.33	0.5
Distance to the Main City	4	7	3	1	2
No. of EVs Served/Day	3	5	2	0.5	1

TABLE 10.5

The Weights of Individual Criteria and Consistency Testing

	Criteria Weights	Consistency Test
Population	0.10	Consistency index = $(\lambda m - n)/(n - 1)$ = 0.019
Air Quality	0.05	Random index = 1.12
Cost	0.16	Consistency ratio = CI/RI = 0.017
Distance to the Main City	0.43	
No. of EV Served/Day	0.26	

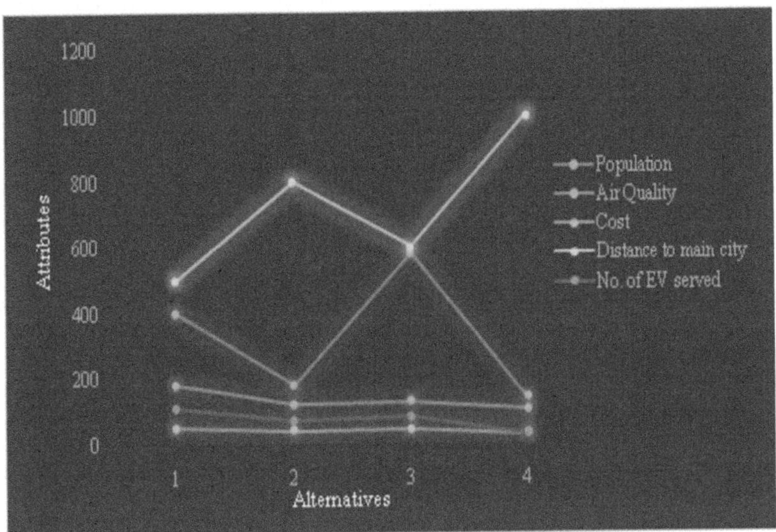

FIGURE 10.2   The different attributes for all four alternatives: A1, Kolkata; A2, Chinsurah; A3, Durgapur; and A4, Bankura.

Notes: population: thousands; air quality (PM2.5): $\mu g/m^2$; cost: lakhs; distance to main city: meters.

TABLE 10.6   THE RANKS OF THE FOUR ALTERNATIVES

Attribute/ Alternative	Population	Air Quality	Cost	No. of EVServed/Day	Distance to Main Road	Si	Ri	Rank
Kolkata	0.0409	0.05	0.16	0	0	0.2509	0.16	1
Chinsurah	0.0909	0.010256	0.08	0.215	0.156	0.5522	0.215	3
Duragpur	0.0000	0.016026	0.12	0.16125	0.052	0.3493	0.161	2
Bankura	0.1000	0	0	0.43	0.26	0.7900	0.43	4
				Min	Si*,Ri*	0.2509	0.16	
				Max	Si#,Ri#	0.79	0.43	

10.5 CONCLUSION

For this research work, we proposed a hybrid MCDM model for selecting an optimal EVCS site from four cities in India that combined AHP and VIKOR determined based on multiple conflicting criteria that are crucial in choosing charging sites. The conventional AHP calculates the weights of each attribute, and VIKOR finds the net rank of all alternatives. In more detail, AHP coordinates customer views and employs a pyramid shape for dissolving complicated challenges into easier forms, and VIKOR helps to resolve complex decision-making problems that have contradictory criteria and choices by ranking alternatives; customers can arrive at prompt resolution with minimum attempts. The new well-structured proposed hybrid model was not only effective for choosing an optimal location site for an electric vehicle charging station but can in fact be used to solve complex real-life challenges. The proposed model does not include societal risk criteria; in the future, we may extend the model by including these criteria to compare different MCDM techniques.

REFERENCES

Aman Dhingra, Avnish Jareda, Himanshu Choudhary, Dr. Saurabh Agrawal, "Selection of optimal electric vehicle charging station location using AHP-fuzzy TOPSIS approach", ICIDSSD, New Delhi, India, 2021. http://dx.doi.org/10.4108/eai.27-2-2020.2303237

Abapour Saeed, Abapour Mehdi, Khalkhali Kazem, Moghaddas Tafreshi, Seyed Masoud, "Application of data envelopment analysis theorem in plug-in hybrid electric vehicle charging station planning", IET Generation, Transmission & Distribution, 9(7), 666–676, 2015. https://doi.org/10.1049/iet-gtd.2014.0554

Baouche Fouad, Billot Romain, Trigui Rochdi, El Faouzi, Nour-Eddin, "Efficient allocation of electric vehicles charging stations: Optimization model and application to a dense urban network", IEEE Intelligent Transportation Systems Magazine, 6(3), 33–43, 2014. http://dx.doi.org/10.1109/MITS.2014.2324023

Bilgen, S. "Structure and environmental impact of global energy consumption", Renewable and Sustainable Energy Reviews, 38, 890–902, 2014. https://doi.org/10.1016/j.rser.2014.07.004

Bartosz Paradowski, Zdzislaw Szyjewski, "Comparative analysis of muti-criteria methods in supplier selection problem", Procedia of Computer Science, 207, 4593–4602, 2022. https://doi.org/10.1016/j.procs.2022.09.523

Cabala Pawel, "Using the analytic hierarchy process in evaluating decision alternatives", Operations Research and Decision, 20(1), 5–23, RePEc:wut:journl:v:1:y:2010:p:1–23, 2010.

Cay Tayfun, Mevlut Uyan, "Evaluation of reallocation criteria in land consolidation studies using the Analytic Hierarchy Process (AHP)", Land Use Policy, 541–548, 2013. http://dx.doi.org/10.1016/j.landusepol.2012.04.023

Chakraborty, S. Banik, D. "Design of a material handling equipment selection model using analytic hierarchy process", International Journal of Advanced Manufacturing Technology, 28, 1237–1245, 2006. https://doi.org/10.1007/s00170-004-2467-y

Chen, C. T., Lin, C. T., Huang, S. F. "A fuzzy approach for supplier evaluation and selection in supply chain management", International Journal of Production Economics, 128(1), 289–298, 2011. https://doi.org/10.1016/j.ijpe.2005.03.009

Chen, Y. J., Li, X., Li, H. "Multicriteria decision making for product design under uncertainty", Journal of Mechanical Design, 136(6), 061004, 2014.

Duzakin, E. R., Aras, H., Tavli, B. "Multi-criteria hazardous waste landfill site selection using geographic information systems: A case study in Istanbul, Turkey", Waste Management & Research, 29(8), 867–878, 2011.

Erbaş Mehmet, Kabak Mehmet, Özceylan Eren, Çetinkaya Cihan, "Optimal Siting of Electric Vehicle Charging Stations: A GIS-Based Fuzzy Multi-Criteria Decision Analysis", Energy, 2018. https://doi.org/10.1016/j.energy.2018.08.140

Hookah Khooi, Rashad Yazdanifard, "How to plan and strategically manage an Electric Vehicle (EV) charging station," Global Journal of Management and Business Research, 15(4). 1–5, 2015.

Hosseini Seyedmohsen, M. D. Sarder, "Development of a Bayesian network model for optimal site selection of electric vehicle charging station", International Journal of Electrical Power & Energy Systems, 105, 110–122, 2019. https://doi.org/10.1016/j.ijepes.2018.08.011

Gao Zhicheng, Liang, Robert Y., Xuan Tiemin, "VIKOR method for ranking concrete bridge repair projects with target-based criteria", Results in Engineering, 3, 100018, 2019. https://doi.org/10.1016/j.rineng.2019.100018

Guo Zhong Zheng, Xiao Wang, "The comprehensive evaluation of renewable energy system schemes in tourist resorts based on VIKOR method", Energy, 193, 116676, ISSN 0360-5442, 2020.

Gupta, P., Yadav. S. P., "Selection of optimal portfolio using VIKOR and TOPSIS methods", International Journal of Emerging Markets, 13(1), 142–157, 2018.

Gwo-Hshiung Tzeng, Cheng-Wei Lin, Serafim Opricovic, "Multi-criteria analysis of alternative fuel buses for public transportation", Energy Policy, 33(11), 1373–1383, 2005. https:// doi.org/10.1016/j.enpol.2003.12.014

Kaya, I., Kahraman, C. "A fuzzy VIKOR approach for supplier selection: A case study in the cable industry", Expert Systems with Applications, 38(5), 5967–5975, 2011. https://doi.org/10.1088/1757-899X/394/4/042126

Mingsheng Zhang, "Location planning of electric vehicle charging station", IOP Conference Series: Materials Science and Engineering, 394(4). 042126, 2018.

Puneet Kamboj, Ankur Malyan, Harsimran Kaur, Himani Jain, Vaibhav Chaturvedi, "India transport energy outlook", New Delhi: Council on Energy, Environment and Water, 2022.

Roy, S. "Comparative study of some MCDM techniques for ECommerce applications", International Journal of Advanced Research in Computer Science", 9(1), 440–443, 2018. http://dx.doi.org/10.26483/ijarcs.v9i1.5361

Saaty, T. L. "The Analytic Hierarchy Process", McGraw-Hill International, New York, NY, USA, 1980.

Singh, S., Kumar, R., Kumar, R., Chohan, J. S., Ranjan, N., Kumar, R. "Aluminum metal composites primed by fused deposition modeling-assisted investment casting: Hardness, surface, wear, and dimensional properties", Proc. Inst. Mech. Eng. Part L J. Mat. Des. Appl., 236, 674–691, 2022. https://doi.org/10.1177/14644207211054143

Sen Guo, Huiru Zhao, "Optimal site selection of electric vehicle charging station by using fuzzy TOPSIS based on sustainability perspective", Applied Energy, 158, 390–402, 2015. https://doi.org/10.1016/j.apenergy.2015.08.082

Tavana, M., Fazlollahtabar, H., Hajmohammadi, H. "Supplier selection and order allocation with process performance index in supply chain management", International Journal of Information and Decision Sciences, 4(4), 329, 2012. https://doi.org/10.1504/ijids.2012.050379

Vaidya, O. S., Kumar, S. "Analytic hierarchy process: An overview of applications", European Journal of Operational Research, 169(1), 1–29, 2006. https://doi.org/10.1016/j.ejor.2004.04.028

Wang, Y., Zhang, H., Zhang, X., Cao, Y. "A new multi-criteria decision-making method for interval data and hesitant fuzzy information", Soft Computing, 24(8), 5695–5711, 2020.

Wu Yunna, Yang Meng, Zhang Haobo, Chen Kaifeng, Wang Yang, "Optimal site selection of electric vehicle charging stations based on a cloud model and the PROMETHEE method", Energies, 9(3), 157, 2016. https://doi.org/10.3390/en9030157

Xiuhong He, Yingying Hu, "Optimal mileage of electric vehicles considering range anxiety and charging times", World Electric Vehicle Journal, 14(1), 21, 2023. https://doi.org/10.3390/wevj14010021

You Peng-Sheng, Hsieh Yi-Chih, "A hybrid heuristic approach to the problem of the location of vehicle charging stations", Computers & Industrial Engineering, 70, 195–204, 2014. https://doi.org/10.1016/j.cie.2014.02.001

Yao Weifeng, Zhao Junhua, Wen Fushuan, Dong Zhaoyang, Xue Yusheng, Xu Yan, Meng Ke, "A multi-objective collaborative planning strategy for integrated power distribution and electric vehicle charging systems", IEEE Transactions on Power Systems, 29(4), 1811–1821, 2014. https://doi.org/10.1109/TPWRS.2013.2296615

Yingwen Wu, Fu Gu, Yangjian Ji, Shaochao Ma, Jianfeng Guo, "Electric vehicle adoption and local PM2.5 reduction: Evidence from China", Journal of Cleaner Production, 396, 136508, 2023. https://doi.org/10.1016/j.jclepro.2023.136508

You Xiao-Yue, You Jian-Xin, Liu Hu-Chen, Zhen Lu, "Group multi-criteria supplier selection using an extended VIKOR method with interval 2-tuple linguistic information", Expert Systems with Applications, 42(4), 1906–1916, 2015. https://doi.org/10.1016/j.eswa.2014.10.004

Yildirim, B., Kuzu Yildirim, S. "Evaluating the satisfaction level of citizens in municipality services by using picture fuzzy VIKOR method: 2014–2019 period analysis", Decision Making: Applications in Management and Engineering, 5, 50–66, 2022. https://doi.org/10.31181/dmame181221001y

Zhao Hui-ru, Guo Sen, Fu Li-wen, "Review on the costs and benefits of renewable energy power subsidy in China", Renewable and Sustainable Energy Reviews, 37, 538–549, 2014. https://doi.org/10.1016/j.rser.2014.05.061

Zhang Nian, Wei Guiwu, "Extension of VIKOR method for decision-making problem based on hesitant fuzzy set", Applied Mathematical Modelling", 37(7), 4938–4947, 2013. https://doi.org/10.1016/j.apm.2012.10.002

Zou X, Li D. A., "Multidisciplinary GIS-based approach for the potential evaluation of land consolidation projects: a model and its application", In Proceedings of the 7th WSEAS International Conference on Applied Computer and Applied Computational Science. China; p. 551–6, 2008.

11 Optimum Topological Indices for Intuitionistic Fuzzy Graphs

J. Senbagamalar and V. Gomathi

11.1 INTRODUCTION

A fuzzy set gives the degree of membership of an object in a given set (Zadeh, 1965). An intuitionistic fuzzy graph (IFG) contains two-valued logic (two dimensions) where the values lie between [0, 1]. The IFG edge values, the minimum and maximum, are graphed based on the vertex. IFGs represent only membership and non-membership value. In 1974, basic theoretic concepts and fuzzy graph operations were viewed by Rosenfield. K. Y. Atanassov introduced intuitionistic fuzzy sets in 1986 (Atanassov, 1986) and interval-valued intuitionistic fuzzy sets in 1999 (Atanassov, 1999). Fuzzy, intuitionistic fuzzy, and neutrosophic graphs are one-dimensional (only membership), two-dimensional (membership and non-membership), and three-dimensional (membership, non-membership and indeterminacy), respectively. Battacharya introduced the center and eccentricity of fuzzy graph and also given comparison study of crisp set and fuzzy set (Bhattacharya, 1987). IFGs extend the fuzzy graphs, and neutrosophic graphs extend IFGs (Szmidt and Kacprzyk, 2000). The intuitionistic fuzzy bridge, cut vertices, cycle and trees in intuitionistic fuzzy graph were introduced by Akram (Akram and Alshehehri, 2014). In 2012, graph operations on strong intuitionistic fuzzy graph were discussed (Akram and Davvaz, 2012) and studied the connectivity concepts in intuitionistic fuzzy graph (Alzoubi and Alnaser, 2021). A real-life application was related with human trafficking and illegal immigration using Wiener fuzzy graph (Binu et al., 2019). In 2016, Akram introduced an idea about operations on intuitionistic fuzzy graph structures (Akram Muhammad and Akmal, 2018). For this chapter, we characterize several topological indices for IFGs: first Zagreb, second Zagreb, forgotten, Randić, and harmonic (Javeria Dinar, 2023) and discuss algorithms and applications.

11.2 PRELIMINARIES

11.2.1 INTUITIONISTIC FUZZY GRAPHS

IFGs are extensions of classical fuzzy graphs that incorporate intuitionistic fuzzy sets. Fuzzy graph theory deals with graphs whose edges and vertices are labeled with degrees of membership in the interval [0,1] (Rosenfeld, A., 1975). Intuitionistic fuzzy sets extend the idea by including a third parameter called the non-membership degree, representing the degree to which does not to the set. An intuitionistic fuzzy graph uses both membership and non-membership degrees to model the uncertainty associated with the presence or absence of edges and vertices. This allows for a more flexible representation of uncertainty in graph-based systems. Nagoorgani (2016) characterized double domination on intuitionistic fuzzy graphs.

11.2.1.1   Definition

A fuzzy graph G = (S, α, μ) is a triple consisting of a non-empty set S together with a pair of functions α: S → [0, 1] and $\mu\colon T \to$ [0, 1] such that

$$\mu(xy) \le \alpha(x) \wedge \alpha(y) \ \forall \ xy \in T.$$

11.2.1.2   Definition

Let $G = (S, T)$ be an IFG if

   (1) $S = \{v_1, v_2, v_3, \ldots, v_n\}$ such that $\alpha_1\colon S \to [0, 1]$, $\mu_1\colon S \to [0, 1]$ denote the degree of membership and non-membership of the element $v_i \in S$, respectively, and $0 \le \alpha_1(v_i) + \mu_1(v_i) \le 1$ for every $v_i \in S$.
   (2) $T \subseteq S \times S$ where $\alpha_2 : S \times S \to [0,1]$ and $\mu_2 : S \times S \to [0,1]$ are such that

$\alpha_2(v_i, v_j) \le \min\left[\alpha_1(v_i), \alpha_1(v_j)\right], \mu_2(v_i, v_j) \le \max\left[\mu_1(v_i), \mu_1(v_j)\right]$ denotes the degree of membership and non-membership of the edge $(v_i,\ v_j) \in T$ respectively, where $0 \le \alpha_2(v_i, v_j) + \mu_2(v_i, v_j) \le 1$ for every $(v_i,\ v_j) \in T$.

11.2.1.3   Definition

$G = (S, T)$ is said to be a strong IFG if

$$\mu_T(xy) = \min\left(\mu_S(x), \mu_S(y)\right) \text{ and}$$

$$\sigma_T(xy) = \max\left(\sigma_S(x), \sigma_S(y)\right) \ \forall \ xy \in T.$$

11.2.1.4   Definition

$G = (S, T)$ is said to be complete and is denoted by $C_m(G)$ if $\mu_T(xy) = \min\left(\mu_S(x), \mu_S(y)\right)$ and

$$\sigma_T(xy) = \min\left(\sigma_S(x), \sigma_S(y)\right) \ \forall \ xy \in T.$$

11.2.1.5   Definition

The first Zagreb index in fuzzy graphs was introduced by Kalaithan. S et al. (2019):
$$M(G) = \sum_{i=1}^{n} \sigma(u_i) d_{u_i}^2, \ \forall u_i \in S.$$

11.2.1.6   Definition

The second Zagreb index in fuzzy graphs was also introduced by Kalaithan, S et al. (2019): $M^*(G) = \dfrac{1}{2} \displaystyle\sum_{ij \in T(G)} \sigma(u_i) d_{u_i} \sigma(v_i) d_{v_j}, \ \forall \ i \ne j.$

11.2.1.7   Definition

The forgotten index in fuzzy graphs is $F(G) = \sum_{i=1}^{n} \sigma(u_i) d_{u_i}^3$, $\forall u_i \in S$.

11.2.1.8   Definition

The harmonic index in fuzzy graphs was introduced by Kalaithan et al. as well:

$$H(G) = \frac{1}{2} \sum_{ij \in T(G)} \frac{1}{\sigma(u_i)d(u_i) + \sigma(v_j)d(v_j)}, \ \forall \ i \neq j.$$

11.2.1.9   Definition

Kalaithan et al. also introduced the Randić index for fuzzy graphs:

$$R(G) = \frac{1}{2} \left[\frac{1}{\sqrt{\sum_{i=1}^{n} \sigma(u_i)d(u_i) \cdot \sigma(v_j)d(v_j)}} \right], \ \forall \ i \neq j.$$

11.3   INTUITIONISTIC TOPOLOGICAL INDICES

11.3.1   First Zagreb Index

Let $G = (S, T)$ be the IFG shown in Figure 11.1, where S is a vertex set and T is the edge set on the graph. The first Zagreb index is denoted by $M(G)$ and defined as

$$M(G) = \sum_{i=1}^{n} [T_S(u_i), F_S(u_i)] d_{u_i}^2$$

$G = (S, T)$ with the vertex set S = $\{u_1, u_2, u_3, u_4\}$ such that $(T_S, F_S)(u_1) = (0.3, 0.5)$, $(T_S, F_S)(u_2) = (0.2, 0.4)$, $(T_S, F_S)(u_3) = (0.1, 0.3)$, and $(T_S, F_S)(u_4) = (0.4, 0.6)$. The edge set contains $(T_T, F_T)(u_1, u_2) = (0.2, 0.5)$, $(T_T, F_T)(u_2, u_3) = (0.1, 0.4)$, $(T_T, F_T)(u_1, u_4) = (0.3, 0.6)$, and $(T_T, F_T)(u_3, u_4) = (0.1, 0.6)$. This gives the following:

FIGURE 11.1   (a) A fuzzy graph; (b) the complement fuzzy graph G.

$$d(u_1) = (0.2 + 0.3, 0.5 + 0.6) = (0.5, 1.1)$$

$$d(u_2) = (0.2 + 0.1, 0.5 + 0.4) = (0.3, 0.9)$$

$$d(u_3) = (0.1 + 0.1, 0.6 + 0.4) = (0.2, 1)$$

$$d(u_4) = (0.3 + 0.1, 0.6 + 0.6) = (0.4, 1.2)$$

$$d^2(u_1) = (0.04 + 0.09, 0.25 + 0.36) = (0.13, 0.61)$$

$$d^2(u_2) = (0.04 + 0.01, 0.25 + 0.16) = (0.05, 0.41)$$

$$d^2(u_3) = (0.01 + 0.01, 0.36 + 0.16) = (0.02, 0.52)$$

$$d^2(u_4) = (0.09 + 0.01, 0.36 + 0.36) = (0.1, 0.72)$$

$$d^3(u_1) = (0.008 + 0.027, 0.125 + 0216) = (0.035, 0.341)$$

$$d^3(u_2) = (0.008 + 0.001, 0.125 + 0.064) = (0.009, 0.189)$$

$$d^3(u_3) = (0.001 + 0.001, 0.216 + 0.064) = (0.002, 0.28)$$

$$d^3(u_4) = (0.027 + 0.001, 0.216 + 0.216) = (0.028, 0.432)$$

$$M_1(G) = (0.3, 0.5)(0.13, 0.61) + (0.2, 0.4)(0.05, 0.41) + (0.1, 0.3)(0.02, 0.52)$$
$$+ (0.4, 0.6)(0.1, 0.72)$$

$$= (0.039 + 0.305) + (0.01 + 0.164) + (0.002 + 0.156) + (0.04 + 0.432)$$

$$= 0.344 + 0.174 + 0.158 + 0.0472$$

$$M_1(G) = 0.7232$$

11.3.2 Theorem

Let G be the regular IFG. Then we have

$$M(G) = c^2 \sum_{i=1}^{n} [T_S(u_i) + F_S(u_i)], \ \forall u_i \in S \text{ where } \sum_{v \neq u} T_S(v,u) = c, \sum_{v \neq u} F_S(v,u) = c$$

Proof: Given the degree of the definition of each points, $d(v) = \left[d_T(v), d_F(v) \right]$

$$d(v) = \left[\sum_{\substack{v \in V \\ v \neq u}} T_S(v,u), \sum_{\substack{v \in V \\ v \neq u}} F_S(v,u) \right].$$

On the other hand, for regular intuitionistic fuzzy graph, we know that $\sum_{v \neq u} T_S(v,u) = c, \sum_{v \neq u} F_S(v,u) = c$

$$d(v) = \left[d_T(v), d_F(v) \right] = (\text{c, c})$$

Now by applying the formula in the first Zagreb index, we will get the results.

11.3.3 SECOND ZAGREB INDEX

The second Zagreb index is denoted by $M^*(G)$ and defined as

$$M^*(G) = \frac{1}{2}\sum_{i=1}^{n}\left[\left(T_S(u_i), F_S(u_i)\right)d(v_i)\right]\left[\left(T_S(u_i), F_S(u_i)\right)d(v_j)\right] \forall\ i \neq j.$$

If G is the same IFG in Figure 11.1 and Section 11.3.1, we have

$$M^*(G) = \frac{1}{2}\begin{bmatrix}(0.3,0.5)(0.5,1.1)\times(0.2,0.4)(0.3,0.9)\\ +(0.3,0.5)(0.5,1.1)\times(0.4,0.6)(0.4,1.2)\\ +(0.1,0.3)(0.2,1)\times(0.2,0.4)(0.3,0.9)\\ +(0.1,0.3)(0.2,1)\times(0.4,0.6)(0.4,1.2)\end{bmatrix}$$

$$M^*(G) = \frac{1}{2}\begin{bmatrix}(0.15+0.55)\times(0.06+0.36)\\ +(0.15+0.55)\times(0.16+0.72)\\ +(0.02+0.3)\times(0.06+0.36)\\ +(0.02+0.3)\times(0.16+0.72)\end{bmatrix}$$

$$M^*(G) = \frac{1}{2}\begin{bmatrix}(0.7\times0.42)+(0.7\times0.88)+(0.32\times0.42)\\ +(0.32\times0.88)\end{bmatrix}$$

$$M^*(G) = \frac{1}{2}\left[0.294+0.616+0.1344+0.2816\right]$$

$$M^*(G) = \frac{1}{2}\left[1.326\right]$$

$$M^*(G) = 0.663$$

11.3.4 FORGOTTEN INDEX

The forgotten index is denoted by $F(G)$ and defined as $F(G) = \sum_{i=1}^{n}\left[T_S(u_i), F_S(u_i)\right]d_{u_i}^3,$ $\forall\ i \neq j.$

If G is the IFG in Figure 11.1 and Section 11.3.1, we have

$$F(G) = (0.3,\ 0.5)(0.035,\ 0.341) + (0.2,\ 0.4)(0.009,\ 0.189) + (0.1,\ 0.3)(0.002,\ 0.28)$$
$$+ (0.4,\ 0.6)(0.028,\ 0.432)$$
$$= (0.0105 + 0.1705) + (0.0018 + 0.0756) + (0.0002 + 0.084) + (0.0112$$
$$+ 0.2592)$$
$$= 0.181 + 0.0774 + 0.0842 + 0.2704$$
$$= 0.613.$$

11.3.5 HARMONIC INDEX

The harmonic index is denoted by $M^*(G)$ and defined as

$$H(G) = \frac{1}{2}\left[\sum_{ij \in T(G)}\left[\frac{1}{\left(T_S(u_i), F_S(u_i)\right)d(u_i) + \left(T_S(u_i), F_S(u_i)\right)d(v_i)}\right]\right] \quad \forall \; i \neq j$$

If G is the IFG in Figure 11.1 and Section 11.3.1, we have

$$H(G) = \frac{1}{2}\left[\begin{array}{c}\dfrac{1}{(0.3,0.5)(0.5,1.1) + (0.2,0.4)(0.3,0.9)} \\[2mm] + \dfrac{1}{(0.3,0.5)(0.5,1.1) + (0.4,0.6)(0.4,1.2)} \\[2mm] + \dfrac{1}{(0.1,0.3)(0.2,1) + (0.2,0.4)(0.3,0.9)} \\[2mm] + \dfrac{1}{(0.1,0.3)(0.2,1) + (0.4,0.6)(0.4,1.2)}\end{array}\right]$$

$$= \frac{1}{2}\left[\begin{array}{cc}\dfrac{1}{(0.15+0.55)+(0.06+0.36)} + \dfrac{1}{(0.15+0.55)+(0.16+0.72)} \\[2mm] + \dfrac{1}{(0.02+0.3)+(0.06+0.36)} + \dfrac{1}{(0.02+0.3)+(0.16+0.72)}\end{array}\right]$$

$$= \frac{1}{2}\left[\frac{1}{0.7+0.42} + \frac{1}{0.7+0.88} + \frac{1}{0.32+0.42} + \frac{1}{0.32+0.88}\right]$$

$$= \frac{1}{2}\left[\frac{1}{1.12} + \frac{1}{1.58} + \frac{1}{0.74} + \frac{1}{1.2}\right]$$

$$= \frac{1}{2}[3.7105]$$

$$= 1.8553$$

11.3.6 Randić Index

The Randić index is denoted by $R(G)$ and defined as

$$R(G) = \frac{1}{2}\left[\frac{1}{\sqrt{\sum_{i=1}^{n}\left(T_S(u_i), F_S(u_i)\right)d(u_i) \cdot \left(T_S(u_i), F_S(u_i)\right)d(v_i)}}\right] \quad \forall \; i \neq j$$

If G is the IFG in Figure 11.1 and Section 11.3.1, we have

$$R(G) = \frac{1}{2}\left[\frac{1}{\sqrt{0.7 \times 0.42}} + \frac{1}{\sqrt{0.7 \times 0.88}} + \frac{1}{\sqrt{0.32 \times 0.42}} + \frac{1}{\sqrt{0.32 \times 0.88}}\right]$$

$$= \frac{1}{2}\left[\frac{1}{\sqrt{0.294}} + \frac{1}{\sqrt{0.616}} + \frac{1}{\sqrt{0.1344}} + \frac{1}{\sqrt{0.2816}}\right]$$

$$= \frac{1}{2}[7.7306]$$

$$= 3.8653$$

11.3.7 THEOREM

Let $G = (S, T)$ be the strong IFG. Then, $H(G) \leq R(G)$.

Next, we describe incorporating the topological indices in the algorithm methods.

11.4 ALGORITHMS FOR HARMONIC INDEX AND RANDIĆ INDEX IN INTUITIONISTIC FUZZY GRAPH

Let G be an IFG with membership and non-membership values of vertices S and edges T.

Input: vertex set, edge set, and degrees for G
Output: H(G) = Harmonic index

$$H(G) = \frac{1}{2}\left[\sum_{ij \in T(G)}\left[\frac{1}{\left(T_S(u_i), F_S(u_i)\right)d(u_i) + \left(T_S(u_i), F_S(u_i)\right)d(v_i)}\right]\right]$$

Step 1: sum of the product of vertex and the degrees of the adjacent vertices
Step 2: Input step 1 value.
Step 3: Divide step 2 value by 2.

Input: vertex set, edge set and degrees for G
Output: R(G) = Randić index

$$R(G) = \frac{1}{2}\left[\frac{1}{\sqrt{\sum_{i=1}^{n}\left(T_S(u_i), F_S(u_i)\right)d(u_i) \cdot \left(T_S(u_i), F_S(u_i)\right)d(v_i)}}\right]$$

Step 1: First find the product of the vertex and the degrees of each vertex.
Step 2: Sum the products of the adjacent vertex values from step 1.
Step 3: Take the root of the step 2 value.
Step 4: Enter the step 3 value.
Step 5: Divide step 3 value by 2.

These procedures also give the remaining topological indices for IFGs.

11.5 CONCLUSION

Intuitionistic fuzzy graphs have applications in decision-making, pattern recognition, neural networks, computer networks, expert systems, image and signal processing, and various domains where uncertainty and imprecision are prevalent; they provide a more nuanced representation of uncertainty in graph-based models, allowing for a richer description of relationships between vertices and edges.

REFERENCES

Akram Muhammad, Davvaz Bijan, "Strong intuitionistic fuzzy graphs," *Filomat,* 2014, https://doi.org/10.2298/FIL1201177A

Akram, A., Alshehehri, N.O., "Intuitionistic fuzzy cycles and intuitionistic fuzzy trees," *Science World Journal,* 2014, https://doi.org/10.1155/2014/305836

Bhattacharya, P., "Some remarks on fuzzy graphs," *Pattern Recognition Letters,* 1987, https://doi.org/10.1016/0167-8655(87)90012-2

Binu, M., Mathew, S., Mordeson, J.N., "Wiener index of a fuzzy graph and application to illegal immigration networks," *Fuzzy Sets and Systems,* 2019, https://doi.org/10.1016/j.fss.2019.01.022

Javeria, D., Rehman, S.U., "Wiener index for an intuitionistic fuzzy graph and its application in water pipeline network," *Ain Shams Engineering Journal,* 2023, https://doi.org/10.1016/j.asej.2022.101826

Kalaithan, S., Ramalingam, S., Raman, S., Srinivasan, N., "Some topological indices in fuzzy graphs," 2019, https://doi.org/10.1007/978-3-030-23756-1_11

Muhammad, A., Akmal, R., "Intuitionistic fuzzy graph structures," *Kragujevac Journal of Mathematics,* 2017, https://doi.org/10.1016/j.fiae.2017.01.001

Nagoorgani, A., Akram, M., Anupriya, S. "Double domination on Intuitionistic fuzzy graphs," *Journal of Applied mathematics and computing,* 52, 515–528 (2016). https://doi.org/10.1007/s12190-015-0952-0.

Rosenfeld, A., "Fuzzy graphs," *Fuzzy Sets and Their Applications to Cognitive and Decision Process,* 1975. https://doi.org/10.1016/B978-0-12-775260-0.50008-6

Shiram, K., Ramalingam, S., Raman, S., Srinivasan, N., "Some topological indices in fuzzy graphs," *Journal of Intelligent and Fuzzy Systems,* 2020, https://doi.org/10.3233/JIFS-189077

Szmidt, A., Kacprzk, J., "Distance between intuitionistic fuzzy sets and systems," *Fuzzy Sets and Systems,* 2000, https://doi.org/10.1016/S0165-0114(98)00244-9

Wael Ahmad Alzoubi, As'ad Mahmoud As'ad Alnaser, "A study on connectivity concept in intuitionistic fuzzy graphs," *WEAS Transaction on Systems and Control,* 2021, https://doi.org/10.37394/23203.2021.16.5

Zadeh, L.A., "Fuzzy sets," *Information and Control,* 1965, https://doi.org/10.1016/S0019-9958(65)90241-X

12 Advancements in Multicriteria Decision-Making

Exploring Innovative Approaches

Nhut T. M. Vo[#], Cong Truyen Duong[#], and Van Thanh Tien Nguyen[*]

12.1 INTRODUCTION

The multiplicity of applications, approaches, and examples advanced in this work serves as a profound and versatile resource for navigating intricate decision-making processes across various fields. The chapters presented in this part of the book are the outcome of the joint endeavors of top-notch professionals who added their knowledge to complete the puzzles. In theory and practice, they help reveal a more comprehensive perspective about the relationship between multicriteria decision-making (MCDM) and machine learning (ML). In this chapter, we discuss in detail Chapters 13 to 24.

12.2 ADVANCEMENTS IN MULTICRITERIA DECISION-MAKING: EXPLORING INNOVATIVE APPROACHES

Shafik (Chapter 13) combined MCDM with ML techniques with a focus on fuzzy systems and neural networks. The author outlines refinements in the field as well as the difficulties refining ML models and algorithms to operate efficiently in various engineering contexts.

In Chapter 14, Murugesan et al. apply ML combined with neutrosophic TOPSIS to locate power stations for electric vehicles, demonstrating how the algorithms can be used in making rational, sustainable, and environmentally friendly choices. Still, preserving the quality of the models and input data is vital even when integrating ML into the decision-making process.

[#]The authors have made equal contributions to this study.
[*]Corresponding author: nguyenvanthanhtien@iuh.edu.vn

DOI: 10.1201/9781032635170-12

Devi et al. (Chapter 15) did brilliant work on MCDM modeling using ML through spherical neutrosophic similarity and found a high potential for using ML in complex, sophisticated decision-making environments fraught with high uncertainty and ambiguity. Nevertheless, building robust and explainable models, especially for MCDM problems, is one of the critical challenges.

In Chapter 16, Kannan et al. explored machine-learning twig graphs on the hyperWiener index of complete graphs. They clearly show how ML can offer new possibilities to advance graph theory in computer science. The most significant disadvantages of ML algorithms when analyzing graphs at a large scale are scalability and computational performance.

Chapter 17 explores the application of cryptographic methods to enhance the security and integrity of multicriteria decision-making (MCDM) processes. It demonstrates how encryption can protect sensitive decision-related data and improve the reliability of MCDM conclusions. The chapter also suggests future research directions, highlighting the potential of cryptographic technologies in various fields.

Chapter 18 is Yang et al.'s report on AI-powered decision-making applications for sustainable development and intelligence advancements. They also discuss ethical fears and algorithmic bias that should be thoroughly studied.

In Chapter 19, Khekare et al. conducted empirical analysis of actual deployments of AI algorithms for decision-making. They discuss how the issue of AI models' transparency and interpretability needs to be resolved before responsible decision-making systems can be created.

In Chapter 20, Sathvik et al. explore fusion sorting as one of the hybrid sorting algorithms. They highlight that ML improves the efficiency of sorting algorithms. However, the selection and adjustment of the fusion sort parameters remain problematic, thus making it difficult to calibrate the algorithmization across various data collections.

Hai-Ninh Do (Chapter 21) investigates using social network analysis to choose a destination strategies. Adding social network analysis to MCDM contributes new knowledge on making complex decisions in tourism and destination selection. However, the trustworthiness and credibility of social network data remains pressing in the field of destination decision-making.

Sathish et al. analyze students' performance in outcome-based education by applying MCDM and ML algorithms in Chapter 22 to evaluate and optimize educational outcomes; techniques support teachers' decision-making based on their experience, but there are ethical concerns and privacy violation issues regarding the students' data.

Kala Raja Mohan et al. used MCDM algorithms in Chapter 23 to identify best teacher awardees. ML identified the conditions under which exemplary teaching practices could be recognized and rewarded. However, it is crucial to consider fairness and transparency in selecting best teachers.

Finally, in Chapter 24, Nagaram et al. fuse ML with MCDM to detect lumpy skin disease effectively. This iterative process achieved a commendable accuracy rate of 86.667% in classifying instances, gleaned from scrutinizing the confusion matrix. As technological advancements persist, these complementary tools will be pivotal in bolstering the effectiveness of disease surveillance and control endeavors within the livestock sector. Ultimately, such advancements will mitigate economic losses and uphold animal welfare by proactively addressing health-related concerns. The authors shed light on how veterinary diseases can be detected and tackled promptly in their early stages, but data scarcity and challenges with model generalization make designing accurate and robust disease prediction models challenging.

Recent progress in research on machine learning algorithms for multicriteria decision-making and optimization design shows excellent promise to address challenging engineering problems in multiple fields. However, given the varied array of studies that the authors in this next section discuss above, many challenges remain in realizing the full potential of machine learning in practice to support decision-making and optimization.

Given these difficulties, synthesizing findings and delineating future research needs, in conclusion, constitute critical stages in furthering the state of the art in engineering optimization and decision-making. Next, we discuss the advantages and disadvantages of ML approaches and offer optimal practices for removing the existing limitations and enhancing the techniques' associated benefits and the potential of these innovative methods in practical applications.

12.3 CONCLUSIONS

Through forward-thinking methodologies and provoking case studies, this volume's contributors have shed light on the multidimensional interaction between MCDM and machine learning approaches, offering actionable advice and "food for thought" for various seemingly irreconcilable decision-making scenarios. Be it ML twig graphs that feature practical implications for graph theory applications or AI-based algorithms that could be instrumental to sustainable development, both the experimental and the theoretical examples in the book underscore the benefits of interdisciplinary strategies for innovation and sound decision-making practice.

Given the complexities of the contemporary world, the intersection between MCDM and machine learning appears to be a promising avenue for addressing today's most pressing challenges and doing more with less. Machine learning is a highly effective tool and a treasure trove of unexplored opportunities that can help people gain more insights, make even better decisions, and innovate further.

We hope this book inspires further research, experimentation, and application of machine learning in multidimensional decision spaces. Ideally, we aim to drive eventual positive impact and advancements in virtually every field. Therefore, future research should carefully balance both strengths and limitations in question and in so doing provide new insights into breaking existing barriers and reaping ML's full potential in the real world.

12.4 ACKNOWLEDGMENTS

The authors thank the Industrial University of Ho Chi Minh City, Vietnam, and the National Kaohsiung University of Science and Technology, Taiwan, for their assistance. Additionally, we would like to thank the reviewers and editors for their constructive comments and suggestions for improving our work.

12.5 CONFLICTS OF INTEREST

The authors state that they have no known competing financial interests or personal relationships that could have appeared to influence the work reported in this article.

Two authors (Van Thanh Tien Nguyen and Nhut T.M. Vo) are members of this book's editorial board but were not involved in the editorial review or the decision to publish this article. The authors state that the work has no potential conflict of interest. All authors have read and agreed to the published version of the manuscript.

13 Machine Learning Techniques for Multicriteria Decision-Making

Wasswa Shafik

13.1 INTRODUCTION

In an increasingly interconnected and complex world, decision-making often involves more than a single criterion or objective. Decision-making extends beyond mere quantitative analysis, as numerous factors of diverse nature and significance contribute to evaluating alternatives (Sooklall & Fonou-Dombeu, 2023). Multicriteria decision-making (MCDM) emerges as a systematic and structured approach to navigating the intricate choices that require consideration of multiple, often conflicting, criteria. By providing decision-makers with methodologies to effectively weigh and balance these criteria, MCDM offers invaluable insights and aids in making informed decisions that align with desired outcomes (Ahmed et al., 2023).

The essence of MCDM lies in acknowledging that decisions are rarely one-dimensional. Whether it pertains to selecting an investment opportunity, determining the best course of action in project management, or making choices in environmental planning, decisions involve various considerations (Choudhary et al., 2023). For instance, in selecting a location for a new factory, factors such as cost, proximity to suppliers and customers, environmental impact, and regulatory compliance all contribute to the decision-making process.

MCDM methods facilitate structuring these multifaceted scenarios by providing a framework for objectively evaluating, comparing, and ranking choices. These methods can incorporate both quantitative and qualitative criteria to capture the nuances and complexities of real-world decision scenarios (Bardini & DiCarlo, 2023). As technological advancements continue to enable the collection and analysis of vast amounts of data, the significance of MCDM in handling data-driven decisions has only increased.

MCDM techniques aid in tackling the challenges posed by decisions influenced by various criteria, which might have differed levels of importance or uncertainty. By delving into traditional approaches and contemporary integrations with machine learning (ML), we aim here to comprehensively understand MCDM's evolution and adaptability (Mba Kouhoue et al., 2023). As we navigate the intricate web of choices that shape our personal, professional, and societal landscapes, MCDM serves as a guiding compass, offering systematic pathways to decipher complexity and facilitate well-considered decisions. In the subsequent sections of this chapter, we categorize MCDM techniques, highlight their real-world applications through case studies, and explore the implications of integrating ML in this domain (Rathoriya et al., 2023). In an era where informed decisions are paramount, MCDM stands as a crucial tool in the decision-maker's toolkit, enhancing the ability to make choices that align with multiple objectives and criteria beyond traditional approaches (Nguyen & Vo, 2024).

In an era characterized by information abundance and interconnectedness, decision-makers face challenges beyond mere quantitative analysis. The success of an organization, the feasibility of a project, or the sustainability of a policy hinges on holistically evaluating and reconciling various criteria. MCDM emerges as a response to this complexity, offering a structured methodology to navigate the intricate decision-making landscape in scenarios involving multiple objectives, criteria, and stakeholders (Kavya et al., 2023). Traditional decision-making approaches often fall short when confronted with the multifaceted nature of real-world choices. The oversimplification of complex scenarios into single-criterion evaluations can lead to suboptimal outcomes and missed opportunities. MCDM recognizes that decisions rarely conform to a single dimension and, therefore, embraces a broader perspective (Deb et al., 2023). It encompasses objective and subjective factors, accounting for quantitative metrics and qualitative considerations such as social, environmental, and ethical aspects.

The core of MCDM lies in striking a balance between conflicting objectives. The challenge is not just to rank alternatives based on individual criteria but to find a way to harmonize these criteria into a coherent decision framework (Joudar et al., 2023). MCDM provides tools for quantifying and managing trade-offs, ensuring that decisions reflect a comprehensive evaluation. Integrating technology and analytics has brought new dimensions to MCDM in this ever-evolving landscape. ML techniques have shown promise in handling vast data and deriving patterns from intricate decision spaces (Chowdhury et al., 2023). This fusion of MCDM and ML brings about the potential to enhance decision-making with predictive modeling, adaptive strategies, and insights drawn from historical data.

13.1.1 THE CONTRIBUTION OF THE CHAPTER

This chapter makes a number of contributions. For one, we thoroughly examine integrating ML techniques into the seven MCDM processes we graphically present in Figure 13.1. We highlight the importance of MCDM across diverse domains, emphasizing the need for advanced techniques in complex decision scenarios. We delineate how ML techniques enhance MCDM, automating the analysis of large datasets to facilitate more informed and accurate decisions.

Our categorization of ML techniques into supervised, unsupervised, reinforcement, and deep learning also provides a structured framework for understanding their application in MCDM. We explore each technique comprehensively, supported by clear explanations and illustrative examples, aiding readers in grasping their utility, and real-world case studies showcase tangible benefits such as improved

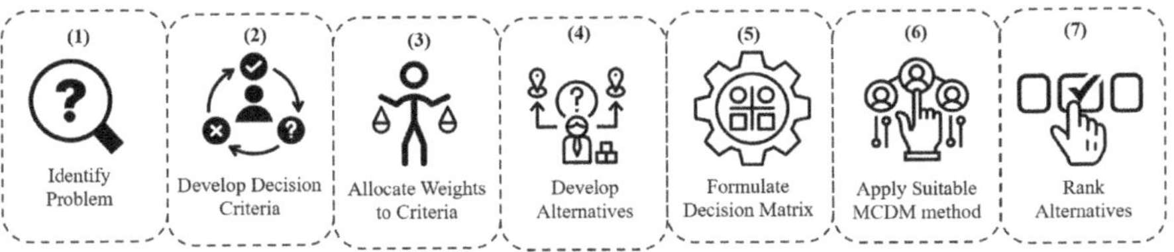

FIGURE 13.1　The seven multicriteria decision-making processes.

accuracy and efficiency. In the process, we briefly compare the ML techniques', outlining their strengths and limitations within diverse MCDM contexts to aid in technique selection. Finally, we propose future research directions, highlighting the potential of hybrid approaches that integrate different ML techniques to address evolving challenges in MCDM.

13.1.2　THE CHAPTER STRUCTURE

In Section 13.2, we define and elaborate on the fundamental concepts of MCDM, discussing the challenges and complexities associated with decision-making involving multiple criteria. Section 13.3 presents the role of ML in MCDM, discussing how ML can help handle large-scale data, automate decision-making processes, and improve accuracy. In Section 13.4, we categorize different ML techniques based on their application in MCDM, and in Section 13.5, we review specific techniques within each category and how they can be applied to MCDM problems, for instance, decision trees for rule-based decision-making, clustering algorithms for grouping similar criteria, neural networks for complex decision patterns, and reinforcement learning for adaptive decision-making. Section 13.6, we present challenges and considerations such as data quality, interpretability, bias, and the need for domain expertise.

In Section 13.7, we give real-world case studies of successfully employing ML in MCDM, highlighting the benefits and outcomes achieved; then in Section 13.8, we compare the strengths, weaknesses, and applicability of different ML techniques for MCDM depending on the specific problem context. Section 13.9 presents the future directions and proposes potential areas of research and development in integrating ML with MCDM; we discuss emerging trends, challenges yet to be addressed, and possibilities for hybrid approaches. Finally, Section 13.10 presents the conclusion, emphasizing the significance of using ML to enhance MCDM and make more informed decisions.

13.2　MCDM

13.2.1　FUNDAMENTAL MCDM CONCEPTS

MCDM is a structured approach used to make decisions when there are multiple criteria or objectives to consider, each with varying degrees of importance. The fundamental concepts of MCDM provide the foundation for understanding its principles and methodologies.

13.2.1.1   Criteria, Objectives, and Trade-Offs

Criteria are the distinct factors that contribute to the decision-making process; these can be quantitative or qualitative and represent the dimensions along which alternatives are evaluated. Conversely, objectives are the desired outcomes or goals that the decision-maker aims to achieve. Each criterion corresponds to a specific objective, and the decision process involves optimizing these objectives simultaneously (Sooklall & Fonou-Dombeu, 2023). Trade-offs occur when improving one criterion leads to a deterioration in another; for example, when choosing a mode of transportation, speed might conflict with cost. Decision-makers must balance these trade-offs to make choices that align with their priorities.

13.2.1.2   Preference, Weights, and Alternatives

Assigning weights to criteria reflects their relative importance. These weights are often based on the preferences of the decision-makers, stakeholders, or experts. Criteria with higher weights are deemed more critical in influencing the decision. Assigning weights can be a subjective process and involves considering the significance of each criterion (Ahmed et al., 2023). Alternatives are the different options under consideration. In a real estate decision, alternatives might be different properties or locations. MCDM evaluates and compares these alternatives based on their performance across the defined criteria.

13.2.1.3   Scoring, Ranking, and Aggregation

Scoring involves assigning numeric values to alternatives based on their performance on each criterion; these scores quantify how well each alternative meets the specific criteria. The ranking process then orders the alternatives based on their aggregated scores, from the most preferred to the least preferred. Aggregation methods combine the individual criterion scores to evaluate each alternative overall. Weighted sum and weighted product methods multiply the criterion scores by their respective weights and sum or multiply the results, respectively (Choudhary et al., 2023). Analytical hierarchy involves pairwise comparisons to derive weights and construct a hierarchical model for aggregation.

13.2.1.4   Dominance, Preference, and Sensitivity Analysis

Dominance occurs when one alternative is better than another across all criteria. When dominance is absent, preference information comes into play. Decision-makers might express their preferences using linguistic terms (for example, "strong preference") or numeric scales (for instance, 1 to 5) to compare alternatives and establish rankings. Sensitivity analysis assesses how changes in criteria weights or scores impact the ranking of alternatives. It helps decision-makers understand which criteria most influence the outcomes and provides insights into the robustness of their decisions (Bardini & DiCarlo, 2023). Sensitivity analysis ensures that small changes in input data do not disproportionately alter the results.

13.2.1.5 Pareto Optimality and Decision Support Systems

Pareto optimality signifies that no alternative can be improved in one criterion without sacrificing performance in another. The set of Pareto optimal alternatives represents the trade-off frontiers that were improving one criterion that necessitated trade-offs in others (Mba Kouhoue et al., 2023). Decision-makers can then explore these alternatives based on their risk tolerance and objectives.

Decision support systems (DSSs) are software tools that facilitate the MCDM process. They integrate criteria weighting, scoring, aggregation, and visualization methods, enabling decision-makers to interactively explore different scenarios, evaluate alternatives, and make informed choices. Understanding these fundamental concepts empowers decision-makers to navigate complex decision scenarios systematically, ensuring that MCDM processes yield outcomes that align with the objectives and criteria (Bardini & DiCarlo, 2023). The appropriate application of these concepts depends on the decision's context, the criteria's nature, and the stakeholders' preferences with different methods, as presented in Figure 13.2.

13.2.2 The Importance of ML for MCDM

MCDM inherently involves the consideration of multiple criteria, often leading to intricate and nonlinear relationships between criteria and alternatives. ML techniques offer several ways to augment MCDM, addressing challenges and limitations associated with traditional methods.

13.2.2.1 Data Handling and Analysis and Pattern Recognition

ML algorithms excel at managing and analyzing large volumes of data. In MCDM, decision-makers must consider many criteria, each with varying data sources and formats; ML techniques can automate data preprocessing tasks, such as cleaning, transformation, and normalization, ensuring that data is ready for analysis. This streamlines the decision-making process and reduces the risk of biased or incomplete information affecting the results. ML models are adept at identifying patterns and relationships within data that may not be immediately apparent (Mba Kouhoue et al., 2023). In MCDM, these patterns can reveal important insights about how

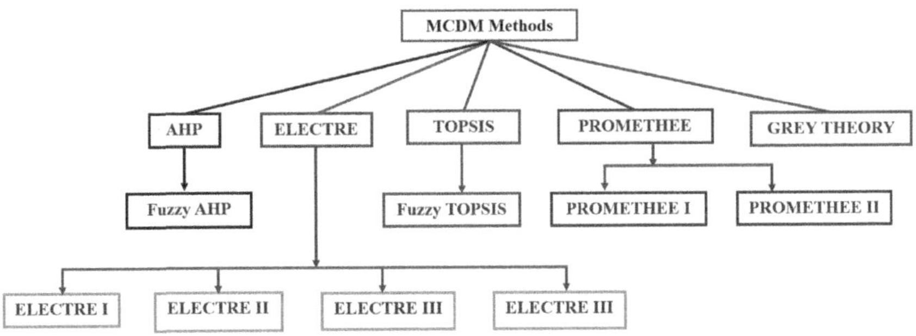

FIGURE 13.2 Multicriteria decision-making methods.

different criteria interact and influence each other. By uncovering hidden relationships, decision-makers can understand how various criteria impact the overall decision and use this knowledge to make more informed choices.

13.2.2.2 Predictive Modeling and Complex Decision Patterns

ML allows for creating models to predict the outcomes of different decision alternatives. For example, in urban planning, ML models can predict the future impact of different zoning policies on factors like traffic congestion or air quality. This predictive capability empowers decision-makers to anticipate the consequences of their choices and select options that align with long-term goals. MCDM problems often involve intricate decision patterns that are difficult to model using traditional methods (Rathoriya et al., 2023). ML techniques like neural networks and deep learning can capture and represent these complex relationships. This enables decision-makers to handle nonlinearity and interactions between criteria, providing a more accurate representation of the decision landscape (Kavya et al., 2023).

13.2.2.3 Optimization and Adaptive Decision-Making

ML algorithms, particularly optimization algorithms like genetic algorithms, can explore a vast solution space to find the best possible compromise solution. This is crucial in MCDM, where trade-offs between criteria need to be carefully balanced. Genetic algorithms, for instance, evolve a set of potential solutions over generations, converging toward an optimal or near-optimal solution that satisfies multiple criteria (Deb et al., 2023). Therefore, ML techniques support adaptive decision-making by continuously learning from new data. As new information becomes available, ML models can update their predictions and recommendations, allowing decision-makers to adapt their strategies based on evolving circumstances. Figure 13.3 presents a comprehensive framework for evaluating hydropower advancement. This is especially valuable in dynamic environments where criteria importance or preferences might change over time.

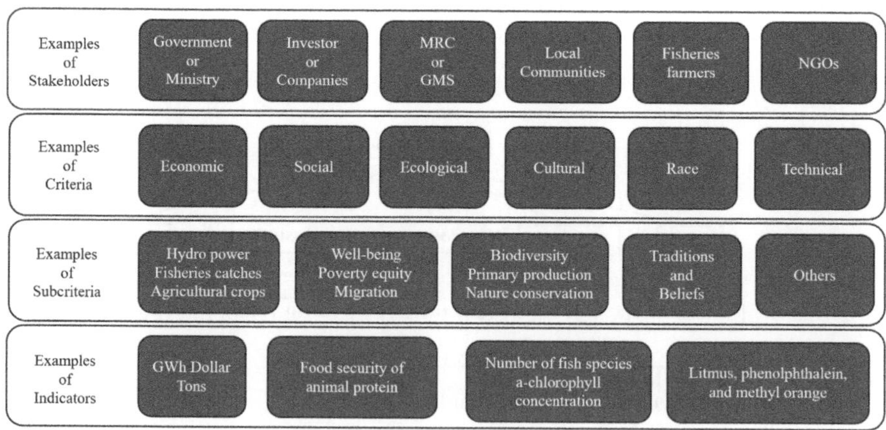

FIGURE 13.3 A comprehensive framework for evaluating hydropower improvement.

13.2.2.4   Reducing Subjectivity and Ensemble Methods

MCDM often involves assigning weights to criteria based on subjective judgments. ML can offer data-driven insights to inform these weight assignments, reducing the potential for bias and improving the transparency of the decision process. For instance, ML algorithms can analyze historical data to determine the relative importance of criteria based on their impact on past decisions. Ensemble methods in ML involve combining predictions from multiple models to achieve better overall performance (Khanorkar & Kane, 2023). In MCDM, combining the results of different ML techniques or perspectives can lead to more robust and reliable decisions. This ensemble approach can mitigate the limitations of individual models and provide decision-makers with a more comprehensive view of the alternatives.

13.2.2.5   Handling Uncertainty and Real-Time Decision Support

ML techniques that incorporate probabilistic modeling or fuzzy logic are well-suited to handling uncertainty in MCDM. Decision-makers often face scenarios with incomplete or imprecise data. ML can quantify the potential range of outcomes by modeling uncertainty and enabling decision-makers to make robust choices. ML-powered DSSs can continuously analyze incoming data and provide real-time recommendations. This is particularly valuable in fast-paced environments where decisions need to be made rapidly. ML algorithms can process new information as it arrives, offering decision-makers up-to-date insights and allowing them to make timely and well-informed choices (Shahiri Tabarestani et al., 2023). Incorporating ML techniques into MCDM processes can significantly enhance the quality and efficiency of decision-making, particularly in scenarios characterized by complexity, data abundance, and evolving conditions.

13.3   THE ROLE OF ML IN MCDM

Integrating ML techniques into MCDM processes introduces a range of capabilities that significantly enhance the decision-making process. ML leverages computational power to handle complex data, uncover patterns, and make predictions, thereby enriching MCDM in various ways.

13.3.1   Data Handling and Analysis, Pattern Recognition, and Insights

ML techniques are well-suited for managing and processing large, diverse data sets commonly encountered in MCDM scenarios. They can handle different data formats, identify missing values, and perform data cleaning, ensuring that the input data is accurate and suitable for analysis. This capacity is significant when dealing with multiple criteria and alternatives, often involving various data sources and types. ML algorithms identify complex patterns, correlations, and trends within datasets (Fan et al., 2023). In MCDM, these insights can help decision-makers understand how different criteria interact and contribute to overall evaluations. For instance, ML might reveal that high cost is often associated with high environmental impact, offering a deeper understanding of the relationships between criteria.

13.3.2 Predictive Modeling and Complex Decision Patterns

ML models such as regression or time-series models can predict outcomes based on historical data. In MCDM, this predictive capability allows decision-makers to estimate the potential consequences of different alternatives. For example, a predictive model might forecast the impacts of different marketing strategies on sales revenue. MCDM problems can involve intricate relationships between criteria and alternatives that aren't easily captured by traditional methods (Hoda & Mondal, 2023). ML techniques like neural networks, which can model complex nonlinear relationships, can capture these patterns. This enhances the accuracy of evaluations and rankings, especially when criteria interactions are intricate.

13.3.3 Optimization and Adaptive Decision-Making

ML optimization tools such as genetic algorithms and particle swarm optimization offer a wide range of potential solutions to find those that best satisfy multiple criteria. In MCDM, these algorithms navigate the trade-offs between criteria, enabling decision-makers to identify compromise solutions that align with their objectives. ML enables decision-making that adapts to changing conditions (Yilmaz et al., 2023). ML models can update their predictions and recommendations by continuously learning from new data. This adaptability is particularly beneficial in dynamic environments where criteria importance or preferences might evolve.

13.3.4 Reducing Subjectivity and Ensemble Methods

ML can provide data-driven insights to inform subjective decisions, such as assigning criteria weights. Instead of relying solely on human judgment, historical data or sensitivity analyses can guide the determination of the importance of criteria, leading to more objective and transparent decisions. Ensemble techniques in ML combine predictions from multiple models to improve overall performance (Liao et al., 2023). In MCDM, combining results from various decision-making methods can mitigate the biases of individual techniques and provide a more robust ranking of alternatives; other benefits are presented in Figure 13.4.

13.3.5 Handling Uncertainty and Real-Time Decision Support

ML methods incorporating probabilistic modeling or fuzzy logic can accommodate uncertainty in MCDM. This is valuable when dealing with criteria values that are uncertain or vague. These techniques quantify uncertainty and give decision-makers a clearer picture of potential outcomes. ML-powered DSSs can analyze streaming data and provide real-time recommendations (Khuc et al., 2023). This is advantageous in time-sensitive situations, where up-to-date information is crucial for making informed decisions. ML algorithms continuously process incoming data, ensuring that decisions are based on the latest insights. The successful integration of ML techniques into MCDM processes requires a thorough understanding of the

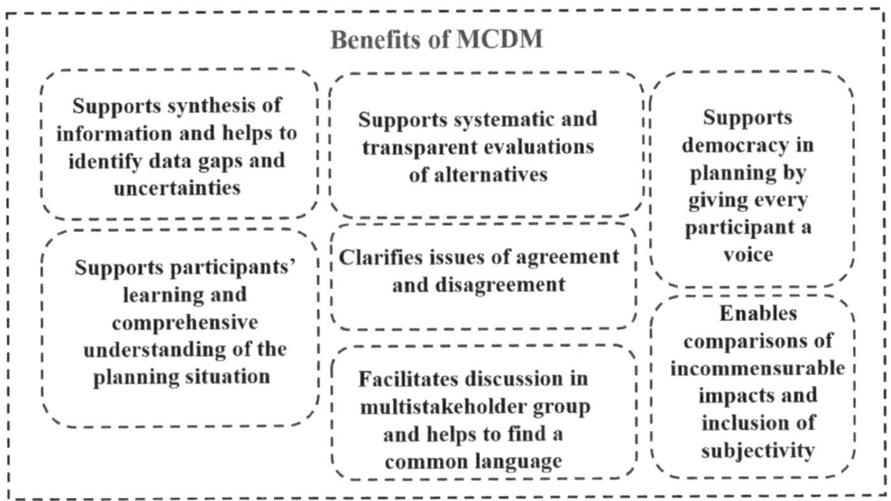

FIGURE 13.4 Multicriteria decision-making benefits.

problem context, careful selection of appropriate algorithms, validation of models, and consideration of ethical and interpretability aspects (Pourkhodabakhsh et al., 2023). ML can significantly enhance MCDM by handling data complexity, offering insights, and supporting more accurate and informed decision-making when leveraged effectively.

13.4 CLASSIFYING ML TECHNIQUES FOR MCDM

13.4.1 Supervised Learning

Supervised learning in MCDM involves using labeled training data to build models that can predict outcomes, evaluate alternatives, or assist in assigning criteria weights. Here are some common supervised learning techniques and their applications in MCDM.

13.4.1.1 Regression Models and Classification Models

Regression models, such as linear, polynomial, and support vector regression, are used to predict numeric values based on input variables. In MCDM, regression can be applied to predict the performance of different alternatives across multiple criteria, for instance, each alternative's expected revenue or cost. Classification models assign labels to instances based on learned patterns, including decision trees, random forests, and support vector machines. In MCDM, supervised learning can be employed to categorize alternatives into predefined groups based on their suitability for specific criteria (Pourkhodabakhsh et al., 2023). For example, based on historical performance, investment options can be categorized as low, moderate, or high risk.

13.4.1.2 Ensemble Methods and Predictive Analytics

Ensemble techniques like bagging and boosting combine multiple models to improve prediction accuracy. In MCDM, ensembles can provide more robust evaluations of alternatives by aggregating predictions from multiple models. This can lead to more reliable rankings and decisions by mitigating the bias of individual models. Supervised learning can be used for predictive analytics, where historical data is used to make predictions about future outcomes (Dutta et al., 2023). For instance, supervised learning models in health care management can predict patient readmission rates based on various medical and demographic factors, aiding in resource allocation and treatment planning.

13.4.1.3 Weight Assignment and Preference Modeling

Supervised learning can assist in assigning weights to criteria based on historical data or expert knowledge. By analyzing past decisions and their outcomes, these models can provide insights into how different criteria have influenced previous decisions. This helps in reducing subjectivity in the weight assignment process. Supervised learning can capture decision-maker preferences by learning from past choices. This allows the model to predict which alternatives a decision-maker will likely prefer, considering their preferences' historical data. Preference modeling can be valuable when dealing with decision-makers who have made similar choices in the past (Henrique Alves Ribeiro & Reynoso-Meza, 2023).

13.4.2 Unsupervised Learning

Unsupervised learning in MCDM is employed to uncover hidden patterns, group similar alternatives, or reduce the dimensionality of data without the need for labeled training examples. The following are some common unsupervised learning techniques and their applications in MCDM.

13.4.2.1 Clustering Algorithms and Dimensionality Reduction

Clustering techniques like k-means, hierarchical clustering, and density-based spatial clustering of applications with noise (DBSCAN) group similar alternatives based on shared characteristics. In MCDM, clustering can help segment alternatives with similar performance across multiple criteria, aiding decision-makers in identifying distinct categories of alternatives. Dimensionality reduction methods, for example, t-SNE and principal component analysis, are used to reduce the number of variables while preserving important information (Shakibaei et al., 2023). In MCDM, these techniques can be employed to handle high-dimensional data with many criteria by transforming it into a lower-dimensional space.

13.4.2.2 Self-Organizing Maps, Association Rule Mining, and Anomaly Detection

Self-organizing mapping is a neural network-based technique that maps high-dimensional data onto a lower-dimensional grid. In MCDM, SOM can help visualize and understand relationships between criteria and alternatives, assisting decision-makers in identifying clusters or patterns within the data. Association rule mining discovers patterns in large data sets by identifying frequent item sets and their associations. In MCDM, this technique can uncover relationships between criteria and alternatives that might not be apparent through manual analysis, aiding in understanding decision factors. Anomaly detection techniques identify instances that deviate significantly from the norm (Rashidi Shikhteymour et al., 2023). In MCDM, these techniques can be applied to identify outliers or unusual alternatives requiring special consideration due to their extreme behavior across criteria.

13.4.2.3 Exploratory Data Analysis and Pattern Recognition

Exploratory data analysis (EDA) involve visualizing and summarizing data to uncover trends and relationships. In MCDM, it can help decision-makers gain initial insights into the distribution and interactions of criteria, guiding further analysis and decision processes (Agarwal et al., 2023). Unsupervised learning techniques can identify intricate patterns in data that might not be immediately apparent and techniques provide valuable tools for exploring and understanding the structure and relationships within MCDM data (Jari et al., 2023). By revealing hidden patterns and grouping similar alternatives, these techniques contribute to more informed decision-making by helping decision-makers identify clusters, outliers, and trends that might impact their choices.

13.4.3 Reinforcement Learning

Reinforcement learning (RL) can be applied to MCDM scenarios where decisions are made in sequential steps to maximize cumulative rewards. Here are some common RL techniques and their applications in MCDM.

13.4.3.1 Dynamic Programming and Markov Decision Processes

Dynamic programming algorithms, such as the Bellman equation and policy iteration, solve decision problems involving sequential actions. In MCDM, dynamic programming can be applied to scenarios where decisions made at each step affect future outcomes, such as resource allocation over time. Markov decision process model decision-making problems with uncertainty and rewards; they can capture the dynamic nature of decisions by considering the probabilistic outcomes of different alternatives and optimizing decisions to maximize expected rewards (Sooklall & Fonou-Dombeu, 2023).

13.4.3.2 Q-Learning and Policy Gradient Methods

Q-learning is a popular RL technique that helps an agent learn a policy to maximize cumulative rewards. In MCDM, Q-learning can guide decisions over multiple steps, considering criteria and objectives to reach optimal or near-optimal solutions. Policy gradient methods optimize policies directly to find actions that maximize cumulative rewards (Ahmed et al., 2023). In MCDM, these methods can be applied to decision processes that involve complex interactions among criteria and alternatives, where traditional optimization methods might be challenging.

13.4.3.3   Monte Carlo Methods and Adaptive Learning

Monte Carlo methods such as Monte Carlo tree search simulate different decision paths to estimate optimal strategies and can help navigate complex decision trees and explore alternative paths to find solutions that satisfy multiple criteria. RL can enable adaptive decision-making, where policies are continuously updated based on new data (Bardini & DiCarlo, 2023). In MCDM, this is useful when criteria or preferences change over time, allowing the decision process to remain relevant and responsive.

13.4.3.4   Stochastic Optimization and Sequential Decision-Making

RL can be used for stochastic optimization, where decisions are made under uncertainty. In MCDM, this can involve selecting alternatives with the highest expected utility or reward based on probabilistic criteria. RL techniques can handle sequential decision-making scenarios in MCDM, where decisions at each step affect the available choices and their outcomes in subsequent steps (Bardini & DiCarlo, 2023). This is particularly relevant in resource allocation, project scheduling, and other dynamic decision scenarios. RL techniques offer a powerful approach to handling MCDM problems that involve sequential decision processes and uncertain outcomes. By optimizing decisions over time to maximize cumulative rewards, these techniques provide a framework for tackling complex decision scenarios and finding strategies that align with multiple criteria and objectives (Mba Kouhoue et al., 2023).

13.4.4   Deep Learning

Deep learning techniques leverage neural networks with multiple layers to capture complex patterns and relationships within data. In MCDM, these techniques offer enhanced capabilities for handling intricate decision landscapes and making accurate evaluations. The following are some common deep learning techniques and their applications in MCDM.

13.4.4.1   Neural Networks and Convolutional Neural Networks

Neural networks can model complex nonlinear relationships between criteria and alternatives. In MCDM, neural networks can predict outcomes, evaluate alternatives, or assist in assigning criteria weights based on historical data. They can capture intricate decision patterns that traditional methods might not easily capture. Convolutional neural networks (CNNs) are particularly effective for scenarios with grid-like structures such as images or spatial data (Rathoriya et al., 2023). In MCDM, CNNs can be applied to spatial criteria or multidimensional data scenarios. For instance, they can analyze satellite images to assess the impact of different criteria on land use decisions.

13.4.4.2   Recurrent Neural and Long Short-Term Memory Networks

Recurrent neural networks (RNNs) are designed for sequential data, making them suitable for scenarios where the order of decisions or criteria matters. In MCDM, RNNs can model decision processes that unfold over time, such as project scheduling or resource allocation. They can consider historical decisions and criteria values to make predictions about future outcomes. Long short-term memory (LSTM) networks are a type of RNN that can capture long-range dependencies in sequential data (Kavya et al., 2023). In MCDM, LSTM networks can be employed to model decision processes that involve complex interactions between criteria and alternatives across multiple time steps.

13.4.4.3 Transformer Networks and Deep RL

Transformer networks are adept at capturing complex patterns and relationships in sequences, making them valuable for MCDM scenarios with intricate decision landscapes. They can handle long-range dependencies and interactions among criteria and alternatives to provide accurate evaluations. Deep RL combines deep learning with reinforcement learning, enabling agents to learn optimal strategies by interacting with an environment (Deb et al., 2023). In MCDM, it can be applied to scenarios where decisions are made in sequence, optimizing actions to maximize cumulative rewards based on criteria and objectives.

13.4.4.4 Transfer Learning

Transfer learning involves training a model on one task and transferring its knowledge to a related task. In MCDM, transfer learning can be used to adapt a pretrained deep learning model to a specific decision problem, leveraging learned patterns from similar domains. Deep learning offers advanced tools for handling complex data, capturing intricate patterns, and making accurate predictions (Joudar et al., 2023).

13.5 SPECIFIC ML TECHNIQUES

So far, we have discussed conceptually how ML techniques enable decision-makers to gain deeper insights into decision landscapes, enhance evaluations of alternatives, and handle scenarios with dynamic, nonlinear, and high-dimensional characteristics. In this section, we discuss representative ML techniques within each category and how they can be applied to MCDM problems.

13.5.1 DECISION TREES FOR RULE-BASED DECISION-MAKING

Decision trees, a supervised learning technique, are a fundamental tool that is widely used in rule-based decision-making. They offer a transparent and intuitive way to model complex decision scenarios, making them valuable tools for MCDM. The following is a review of how decision trees are applied to MCDM (Chowdhury et al., 2023).

Decision trees are algorithms that partition data into subsets based on features and labels. Each internal node represents a decision based on a specific feature, and each leaf node represents an outcome or class label (Sharma et al., 2023). In MCDM, decision trees can evaluate alternatives based on multiple criteria by recursively splitting the data at each node based on different criteria values. This helps decision-makers understand which criteria are most influential in driving decisions. Decision trees can help assign weights to criteria by analyzing historical decision data and identifying how different criteria have affected past choices (Ahasan et al., 2023). This aids in reducing subjectivity in the weight assignment process.

Decision trees provide a visual representation of decision paths and outcomes. In MCDM, this visualization helps decision-makers understand the logical flow of decisions and the criteria contributing to each choice. Decision trees can be converted into rule sets that clearly represent decision-making logic. These rules can provide insights into the decision criteria and how they interact with alternatives.

Decision trees naturally handle nonlinear relationships between criteria and alternatives (Torkjazi & Raz, 2023). They can capture complex interactions that might be difficult to model using linear methods. They are also highly interpretable, making them suitable for scenarios where decision transparency and understanding are crucial. The importance of decision paths and criteria can be easily explained to stakeholders (ForouzeshNejad, 2023).

Decision trees provide transparent decision logic that can be easily understood and communicated to decision-makers and stakeholders. They can capture intricate relationships between criteria and alternatives, accommodating complex decision landscapes. Decision trees can suffer from overfitting, where the model fits the training data too closely and performs poorly on new data (Urgessa & Esfandiari, 2023). Small changes in data can also lead to different decision paths, making decision trees less stable compared to some other algorithms. However, addressing issues like overfitting and instability is crucial to ensure the robustness and reliability of the resulting models.

Decision trees can favor criteria with more categories or higher variability, potentially leading to biased decisions. They are a valuable ML technique for rule-based MCDM (Nayagam et al., 2023), offering transparency, interpretability, and the ability to handle complex relationships. Decision trees can provide actionable insights and guidance for making informed decisions across multiple criteria and alternatives when carefully designed, regularized, and validated.

13.5.2 Clustering Algorithms for Grouping Similar Criteria

Clustering algorithms are unsupervised ML tools that group similar data points based on their features. In the context of MCDM, clustering can be applied to group similar criteria, aiding decision-makers in understanding relationships among criteria and simplifying the decision-making process.

Clustering algorithms work by identifying patterns in data by partitioning them into groups, where data points within the same cluster share similarities and are dissimilar from those in other clusters. These algorithms help identify natural groupings in data without requiring prior knowledge of the group labels. Clustering algorithms can assess the similarity between criteria based on their performance across alternatives (Rodríguez et al., 2023). This helps decision-makers identify which criteria tend to be evaluated similarly or differently, providing insights into criteria interactions. Grouping similar criteria can aid in decision consistency, as decision-makers can focus on evaluating alternatives based on criteria that belong to the same cluster (Mohammadifar et al., 2023). This reduces cognitive load and potential confusion when assessing alternatives.

Clustering can assist in assigning criteria weights by grouping criteria that have similar impacts on decisions. Decision-makers can then assign similar weights to criteria within the same cluster, promoting a more balanced evaluation. Clustering results can be visualized using plots or dendrograms, allowing decision-makers to understand the relationships and distances between criteria (Kaya & Derin, 2023). This visualization aids in identifying clusters and their significance. Clustering can help in performing sensitivity analysis by evaluating how changes in criteria values within a cluster affect the overall ranking of alternatives. This analysis informs decision-makers about the robustness of their choices (Kumar et al., 2023).

K-means partitions data into a predetermined number of clusters, with each data point assigned to the cluster whose centroid is nearest. In MCDM, this algorithm can group criteria based on their performance across alternatives. Hierarchical clustering builds a hierarchy of clusters by iteratively merging or splitting clusters. This results in a dendrogram that displays the relationships between criteria. It helps understand the hierarchy of similarity (Belmecheri et al., 2023). DBSCAN groups data points that are densely packed in feature space. This algorithm can identify dense criteria groups with high correlations in their evaluations.

Agglomerative clustering starts with each data point as a separate cluster and iteratively merges them based on similarity. This approach can yield insights into how criteria naturally form clusters. Clustering algorithms identify patterns and relationships among criteria, helping decision-makers understand similarities and differences (Habib et al., 2023). Grouping similar criteria simplifies the decision process, as decision-makers can focus on fewer clusters instead of individual criteria. Clustering supports decision consistency by evaluating criteria within the same cluster similarly (Arunagiri et al., 2023).

Determining the optimal number of clusters (k) can be challenging and might require domain knowledge or additional analysis. Clustering algorithms can be sensitive to the scale of features, requiring preprocessing like feature scaling. The choice of algorithm should align with the data's characteristics and the decision problem's goals (Albahri et al., 2023).

13.5.3 Neural Networks for Complex Decision Patterns

Neural networks are deep learning techniques that excel at capturing intricate and nonlinear relationships within data. In the context of MCDM, they can be harnessed to model complex decision patterns and interactions among criteria and alternatives (Paz et al., 2023). Neural networks consist of interconnected nodes, or neurons, organized in layers. They use weighted connections to process and transform data, allowing them to learn complex patterns and relationships. This makes neural networks well suited for capturing intricate decision patterns in MCDM. They can capture nonlinear relationships between criteria and alternatives (Siddiqui et al., 2023). This is particularly valuable in MCDM, where criteria might interact in complex and nonlinear ways that cannot be adequately modeled using linear methods.

MCDM often involves multidimensional data with multiple criteria and alternatives. Neural networks can handle high-dimensional data, capturing interactions and dependencies across different criteria. They can automatically learn relevant features from raw data, reducing the need for manual feature engineering (Ankita et al., 2023). This is beneficial when dealing with complex criteria that might not have straightforward representations. Neural networks can adapt to changes in data and decision patterns over time. This adaptability is useful in dynamic MCDM scenarios where criteria are important, or preferences evolve. Neural networks excel at recognizing intricate patterns within data (Albahri et al., 2023). In MCDM, this capability aids in understanding how different criteria contribute to the overall evaluation of alternatives. These are the most common types of neural networks, consisting of input, hidden, and output layers. They are used for various MCDM tasks, such as predicting outcomes, evaluating alternatives, and assigning criteria weights (Salih et al., 2023).

CNNs are well-suited for data with grid-like structures, such as images or spatial data. In MCDM, they can be applied to spatial criteria or multidimensional data scenarios. RNNs are designed for sequential data and are useful for decision processes that unfold over time (Jassim et al., 2023). In MCDM, RNNs can capture the order of decisions and how they impact subsequent evaluations. LSTM networks are a type of RNN that can capture long-range dependencies in sequences. In MCDM, LSTM networks can be employed to model decision processes involving complex interactions. Neural networks can capture intricate and nonlinear relationships between criteria and alternatives, accommodating complex decision patterns (Wang et al., 2023).

Neural networks often require substantial amounts of data for training, which might be a challenge in cases with limited data availability. Deep neural networks can be complex and require significant computational resources for training (Jia et al., 2023). While their complexity and data requirements need consideration, neural networks offer a sophisticated approach to enhancing the understanding and accuracy of MCDM processes. As neural networks can be black-box models, interpreting their decisions can be challenging, especially when decision transparency is crucial (Hezam et al., 2023).

13.5.4 RL for Adaptive Decision-Making

RL is a ML paradigm that involves training agents to make sequential decisions to maximize cumulative rewards. In MCDM, it can be applied to create adaptive decision-making systems that evolve based on feedback and changing conditions. Here is how RL can be used for adaptive decision-making in MCDM (Gandomi et al., 2023).

Reinforcement learning involves an agent interacting with an environment and learning optimal actions through trial and error to maximize cumulative rewards. In MCDM, the environment represents the decision problem, the agent makes decisions, and rewards correspond to how well the decisions align with the objectives and criteria (Rodriguez et al., 2023). RL can adapt to changing criteria, weights, preferences, or objectives. For instance, in a scenario involving resource allocation over time, such as budget allocation, RL can help the agent allocate resources dynamically based on evolving conditions and changing criteria (Jari et al., 2023). Adaptive reinforcement learning can allocate resources optimally over time in scenarios with changing resource availability (Fan et al., 2023).

RL agents can offer real-time decision support by adapting to incoming data and providing recommendations based on the current state of the environment.

Reinforcement learning involves a trade-off between exploring new actions to learn and exploiting the knowledge gained so far. In adaptive MCDM, the agent explores different decision paths while exploiting learned strategies that align with changing criteria (Khanorkar & Kane, 2023). The policy defines the agent's decision-making strategy.

RL involves learning an optimal policy that maximizes cumulative rewards over time, adapting to evolving criteria and objectives. Designing an appropriate reward function is critical. In adaptive MCDM, the reward function should reflect the objectives and criteria, capturing how well the agent's decisions align with the changing priorities (Shahiri Tabarestani et al., 2023; Shafik, 2023a). Balancing exploration and exploitation is crucial to prevent the agent from getting stuck in suboptimal paths or missing out on learning opportunities. Designing a practical reward function that accurately represents the evolving objectives and criteria can be challenging. However, addressing challenges related to exploration, reward function design, and training complexity is essential for harnessing the full potential of adaptive reinforcement learning in MCDM (Hoda & Mondal, 2023). Reinforcement learning offers a dynamic approach to adaptive decision-making in MCDM scenarios. By continuously learning and adjusting decisions based on evolving criteria and objectives, adaptive agents can provide decision-makers with up-to-date recommendations that align with changing priorities (Yilmaz et al., 2023).

13.6 CHALLENGES AND CONSIDERATIONS

Addressing these challenges and considerations is pivotal in successfully integrating machine learning techniques into the MCDM process. ML can provide valuable insights and support for making well-informed decisions across intricate decision landscapes by carefully planning, selecting appropriate techniques, validating models, and involving relevant stakeholders.

13.6.1 Data Quality and Availability and Model Complexity

High-quality data is essential for machine learning success. In the context of MCDM, decision-making relies on accurate and representative data. Poor data quality, such as missing values or inaccuracies, can lead to biased models and unreliable results (Sooklall & Fonou-Dombeu, 2023). A comprehensive data preprocessing strategy is required, involving techniques like data cleaning, imputation, and outlier detection. Additionally, efforts should be made to ensure that the data is relevant and captures the full spectrum of criteria and alternatives (Bardini & DiCarlo, 2023; Shafik, 2023b). While they are capable of capturing intricate relationships, complex ML models might lack transparency. This raises concerns about understanding how the model arrives at its decisions. In MCDM, where decision-makers require insights into the decision process, opting for simpler models or employing techniques like feature importance analysis can enhance transparency. Striking the right balance between model accuracy and interpretability is crucial (ForouzeshNejad, 2023).

13.6.2 Overfitting Generalization and Model Selection

Overfitting occurs when a model learns noise in the training data rather than capturing general patterns. In MCDM, the focus is on making accurate predictions across diverse scenarios. Techniques like L1 or L2 regularization, cross-validation, and early stopping during training can be employed to address overfitting. Cross-validation helps assess how well the model generalizes to unseen data by simulating its performance on different subsets (Fan et al., 2023). The choice of ML technique greatly influences the model's performance in MCDM. Different algorithms have varying strengths and weaknesses, and selecting the most appropriate one demands a deep understanding of the problem (Yilmaz et al., 2023). To make an informed choice, practitioners need to consider factors such as the amount of available data, the nature of criteria interactions, computational complexity, and the desired level of interpretability.

13.6.3 Feature Engineering and Interpretable Models

In MCDM, criteria selection and feature engineering significantly impact model effectiveness. Feature engineering involves selecting relevant criteria, transforming data, and creating informative features that capture decision patterns (Khuc et al., 2023). This step requires domain knowledge to identify criteria with the most impact on decisions and to design features that effectively represent the interactions among criteria and alternatives. The interpretability of ML models is vital in decision-making scenarios where decision transparency is a priority (Dutta et al., 2023; Shafik, 2023c). While complex models like deep neural networks can yield accurate predictions, their inner workings can be challenging to interpret. In MCDM, using inherently interpretable models such as decision trees or linear regression or employing techniques like feature importance analysis can help ensure that decision-makers can understand and trust the model's recommendations (Shakibaei et al., 2023).

13.6.4 Scalability, Changing Preferences, and Criteria

As MCDM problems grow in complexity, scalability becomes a concern. Larger data sets and higher-dimensional decision landscapes demand substantial computational resources and time for model training and evaluation. To address scalability, practitioners can explore algorithms optimized for large data sets, leverage parallel computing techniques, and consider distributed computing frameworks to process and analyze data efficiently (Rodriguez et al., 2023; Shafik, 2023d). MCDM scenarios often involve dynamic criteria and shifting preferences. Traditional models might struggle to adapt to these changes effectively. Reinforcement learning, a dynamic ML paradigm, can be employed to build models that learn and adjust decisions over time based on changing criteria and preferences. This ensures that the decisions remain aligned with evolving goals and objectives (Jassim et al., 2023).

13.6.5 Model Validation and Evaluation and Ethics and Bias Concerns

Ensuring the reliability and accuracy of ML models is paramount in decision support. Rigorous validation techniques are essential for assessing the model's performance and robustness. Cross-validation involves splitting the data into training and testing subsets, preventing models from being overly tailored to specific data points (Albahri et al., 2023). Additionally, benchmarking against existing methods helps establish the model's value relative to existing decision-making approaches. ML models can inadvertently inherit biases in the training data, perpetuating unfair or discriminatory outcomes. Addressing these concerns requires ongoing vigilance and effort (Siddiqui et al., 2023). Practitioners must conduct thorough audits to identify and mitigate biases, apply techniques like fairness-aware learning, and design reward functions in reinforcement learning scenarios that promote fairness across different groups and demographics (Sun et al., 2023).

13.6.6 Decision-Maker Involvement and Transparent Communication

Collaborating with domain experts and decision-makers throughout the modeling process is essential to building effective ML models. These stakeholders possess critical domain knowledge, which helps shape the problem definition, data selection, feature engineering, and model evaluation (Arunagiri et al., 2023). Involvement from the beginning ensures that the models capture relevant insights and criteria accurately, enhancing the model's practicality and real-world applicability. Transparent communication of model results, limitations, and uncertainties is crucial to fostering trust and understanding among decision-makers and stakeholders (Habib et al., 2023). Clarifying how the model generates predictions, the rationale behind its decisions, and any potential biases or uncertainties helps decision-makers make informed choices. Transparent communication also highlights the model's value while acknowledging its limitations and constraints (Belmecheri et al., 2023).

13.7 CASE STUDIES

This section presents some real-world scenarios where ML techniques have been successfully employed in MCDM processes.

13.7.1 Health Care Resource Allocation

Hospitals face the challenge of efficiently allocating health care resources such as staff, equipment, and beds to ensure high-quality patient care. ML has been utilized to optimize these allocations by analyzing historical patient data, staff availability, equipment usage, and patient acuity levels (Choudhary et al., 2023; Shafik, 2023e). Models trained on this data help hospitals predict patient admission patterns, staff requirements, and equipment needs; an example of algorithm is presented in Figure 13.5. Hospitals can make data-driven decisions that enhance patient care and operational efficiency by considering multiple criteria

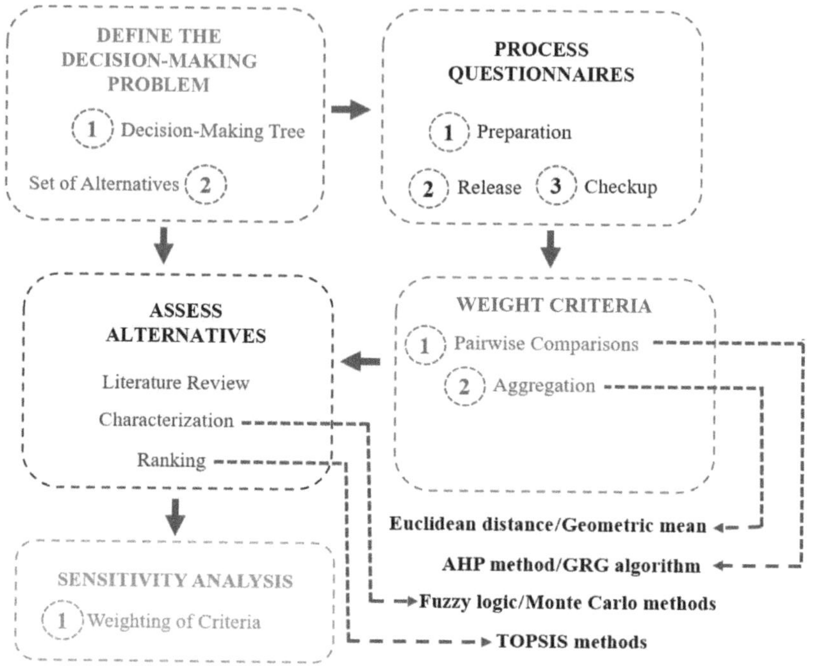

FIGURE 13.5 Algorithms of the multicriteria decision-making methodology.

simultaneously, such as patient safety, staff workload, and resource utilization (Bardini & DiCarlo, 2023).

13.7.2 Energy Efficiency in Buildings

Optimizing energy consumption in commercial buildings involves balancing occupant comfort, energy savings, and operational costs. ML techniques have been applied to develop predictive models that learn patterns from historical data on temperature, occupancy, and HVAC (heating, ventilation, and air conditioning) usage. These models adapt HVAC settings in real time, considering multiple criteria like energy usage, indoor air quality, and occupant preferences (Rathoriya et al., 2023). By dynamically adjusting HVAC systems based on current conditions and historical trends, energy consumption is minimized while maintaining comfortable indoor environments.

13.7.3 Environmental Impact Assessment

Urban development projects require comprehensive environmental impact assessments to ensure sustainable growth. ML has been employed to predict and evaluate the environmental impacts of such projects (Chowdhury et al., 2023; Shafik, 2024a; Deb et al., 2023). These models analyze data from various sources, including satellite imagery, sensor networks, and historical data on pollution levels. By considering criteria such as air quality, noise pollution, traffic congestion, and green space availability, these models provide insights into the potential effects of development on the environment, helping city planners make informed decisions (ForouzeshNejad, 2023).

13.7.4　FINANCIAL PORTFOLIO MANAGEMENT

Constructing investment portfolios involves balancing risk and return while considering individual investor preferences. ML algorithms have optimized portfolio allocation by analyzing historical financial data, market trends, and risk factors. These models consider expected return, volatility, and correlation among assets (Urgessa & Esfandiari, 2023; Shafik, 2024e). By incorporating these criteria, the models generate diversified portfolios that align with an investor's risk appetite and financial goals, leading to more effective investment strategies.

13.7.5　SUPPLY CHAIN OPTIMIZATION

Supply chain management involves inventory levels, transportation routes, and order fulfillment decisions. ML techniques have been applied to optimize these decisions by considering criteria like cost, delivery time, and demand variability. Models trained on historical data predict demand patterns and identify optimal inventory levels (Nayagam et al., 2023). Real-time data on orders, inventory levels, and transportation options are then used to adjust supply chain decisions dynamically. Simultaneously considering multiple criteria, organizations can streamline supply chain operations and reduce costs.

13.7.6　SMART CITY TRAFFIC MANAGEMENT

Traffic congestion is a pressing issue affecting road efficiency and environmental impact in urban areas. ML algorithms have optimized traffic signal timings, considering criteria such as vehicle flow, pedestrian safety, and emergency response times. These models utilize real-time sensor data, historical traffic patterns, and predictive analytics to adjust signal timings dynamically (Rodríguez et al., 2023; Shafik, 2024b). Traffic management systems can simultaneously optimize these criteria, reduce congestion, improve traffic flow, and enhance overall urban mobility. These detailed case studies showcase the diverse applications of ML in MCDM. By considering multiple criteria, historical data, and real-time information, ML techniques enable organizations to make more informed, efficient, and effective decisions across various domains, ultimately leading to improved outcomes and resource utilization (Shahiri Tabarestani et al., 2023).

13.8　COMPARISON AND EVALUATION

The choice of ML technique depends on the problem context, including data availability, the complexity of criteria interactions, interpretability requirements, and the dynamic nature of the decision landscape.

13.8.1 SUPERVISED LEARNING

Supervised learning methods such as decision trees, random forests, and support vector machines excel in predicting outcomes based on labeled training data. They are suitable when historical data is available and past patterns guide decisions. These techniques can handle linear and nonlinear relationships among criteria and alternatives, making them adaptable to various MCDM scenarios (Fan et al., 2023; Kalinki et al., 2024). However, they might struggle with capturing intricate decision patterns and interactions, especially in cases where criteria exhibit complex nonlinear relationships.

13.8.2 UNSUPERVISED LEARNING

Unsupervised learning methods like clustering and dimensionality reduction are valuable for exploring patterns within MCDM data. Clustering algorithms group similar criteria or alternatives together, aiding decision-makers in identifying criterion interactions. Dimensionality reduction techniques such as principal component analysis reduce the complexity of high-dimensional data by retaining its most informative components (Yilmaz et al., 2023; Shafik, 2024c). Unsupervised learning techniques, however, might not directly provide predictive insights for decision-making and might not capture subtle relationships among criteria.

13.8.3 REINFORCEMENT LEARNING

RL stands out in dynamic MCDM scenarios where criteria or preferences evolve. RL agents learn optimal decisions by interacting with the environment and accommodating changing priorities. This adaptability is advantageous for problems with dynamic criteria weights or shifting objectives. RL, however, demands substantial computational resources for training and might require careful reward function design to ensure that the agent's learning aligns with the desired outcomes (Pourkhodabakhsh et al., 2023). Additionally, explaining RL decisions to decision-makers can be challenging due to their complex learning process.

13.8.4 DEEP LEARNING

Deep learning, particularly neural networks, captures complex and nonlinear relationships among criteria and alternatives. They can learn intricate decision patterns and adapt to high-dimensional data. Convolutional neural networks are suitable for image or spatial data, while recurrent neural networks handle sequential data. Deep learning models often require substantial data for training and can be computationally intensive (Dutta et al., 2023; Shafik, 2024d). Their black-box nature can hinder interpretability, which might be crucial in MCDM contexts that require transparency.

The choice of ML technique for MCDM should align with the specific problem's characteristics. For example, if the goal is to predict outcomes based on historical data and interpretability is essential, supervised techniques like decision trees might be appropriate. In cases of complex criteria interactions, clustering techniques can reveal underlying relationships (Shakibaei et al., 2023). In dynamic environments with changing criteria, RL techniques could provide adaptability. Deep learning techniques might capture intricate patterns for highly complex and nonlinear decision landscapes. Striking a balance between accuracy, interpretability, data availability, and the nature of criteria interactions is key to selecting the most suitable ML technique for a given MCDM problem (Shikhteymour et al., 2023).

13.9 FUTURE DIRECTIONS

Applying ML techniques to MCDM holds tremendous potential to revolutionize decision-making processes across diverse domains. By addressing emerging challenges, embracing interdisciplinary collaborations, and tailoring ML methods to the unique demands of MCDM, researchers and practitioners can unlock innovative solutions that enable more informed, efficient, and impactful decisions.

13.9.1 HYBRID APPROACHES AND EXPLAINABLE AI FOR MCDM

The integration of machine learning with traditional MCDM methods presents an exciting avenue. Future research will likely focus on developing effective ways to combine the strengths of different techniques. Hybrid models can harness the interpretability of traditional methods and ML's predictive power, leading to more accurate and understandable decision support systems (Khanorkar & Kane, 2023). Understanding how ML models arrive at their conclusions becomes paramount in complex decision scenarios. The research will delve deeper into creating explainable AI methods tailored to MCDM. These techniques will provide insights into model features, criteria interactions, and decision pathways, enhancing decision-maker trust and facilitating informed choices (Fan et al., 2023).

13.9.2 HANDLING UNCERTAINTY AND DYNAMIC CRITERIA LEARNING

Real-world decisions are often plagued by uncertainty stemming from various sources. Future directions will involve developing ML techniques that can explicitly handle uncertainty. Probabilistic models, Bayesian approaches, and techniques for uncertainty propagation will become essential to make decisions robust in the face of unpredictable factors (Hoda & Mondal, 2023; Shafik & Matinkhah, 2021). As decision contexts change over time, ML models for MCDM will need to adapt dynamically. Future research will explore RL and adaptive algorithms that can continuously learn from evolving criteria weights and shifting preferences. These models will ensure that decisions remain aligned with changing objectives (Henrique Alves Ribeiro & Reynoso-Meza, 2023).

13.9.3 HUMAN–AI COLLABORATION, ETHICAL CONSIDERATIONS, AND BIAS MITIGATION

The future of MCDM will revolve around collaboration between AI and human decision-makers. Researchers will focus on designing AI systems that assist rather than replace humans. These systems will be capable of providing insights, suggestions, and alternate scenarios, allowing decision-makers to leverage AI's computational power while retaining their domain expertise (Hezam et al., 2023). The ethical implications of AI-driven decisions will be a central concern. Future researchers will develop methods for identifying and rectifying biases in machine learning models. Fairness-aware learning and bias detection techniques will ensure equitable outcomes across diverse user groups (Salih et al., 2023).

13.9.4 CUSTOMIZATION, PERSONALIZATION, AND RL IN REAL-WORLD SCENARIOS

The future of MCDM will witness the rise of personalized decision support. ML models will be tailored to individual preferences, constraints, and decision contexts. By accounting for specific needs, these customized models will enhance user engagement and decision outcomes (Paz et al., 2023; Shafik et al., 2022). Expanding the application of RL from simulated environments to real-world scenarios will be a significant advancement. Researchers will explore the challenges of safely deploying reinforcement learning agents in physical environments, overcoming safety, ethics, and interaction dynamics obstacles (Albahri et al., 2023).

13.9.5 BIG DATA, SCALABILITY, AND HUMAN-CENTRIC DESIGN

With the proliferation of data, ML models for MCDM will need to scale effectively. Future research will focus on developing algorithms that can handle big data efficiently, leveraging techniques such as distributed computing, parallel processing, and data compression. The human aspect of MCDM will gain prominence (ForouzeshNejad, 2023). Researchers will delve into incorporating psychological and behavioral insights into machine learning models. These models will align with the cognitive processes of decision-makers, ensuring AI tools that are intuitive, user-friendly, and aligned with human decision-making patterns. The intricate interplay between ML and MCDM holds the potential to revolutionize decision support systems across various sectors (Ahasan et al., 2023; Shafik et al., 2023). As researchers and practitioners navigate these detailed future directions, they will shape innovative methodologies, tools, and frameworks that enhance decision-making effectiveness, transparency, and adaptability (Joudar et al., 2023).

13.10 CONCLUSION

The fusion of machine learning and MCDM represents a significant leap forward in tackling complex decisions across diverse domains. This synthesis leverages the computational power of ML to address the intricate interplay of multiple criteria, enabling more informed, accurate, and efficient decisions. Throughout this exploration, we have delved into the foundational concepts of MCDM, elucidating its principles, methods, and real-world significance. The role of ML in enhancing MCDM processes is undeniable.

ML offers predictive capabilities, revealing patterns hidden within vast data sets that might elude traditional methods. Their adaptability enables dynamic decision-making in the face of shifting preferences and evolving criteria weights. The comprehensive overview of ML categories from supervised to deep learning underscores their applicability and showcases their diverse strengths. In navigating the challenges of data quality, model complexity, and interpretability, the symbiotic relationship between ML and MCDM becomes evident. The success of this synergy hinges on meticulous model selection, feature engineering, and validation processes.

The future unfolds with promising horizons. Hybrid approaches, explainable AI, and personalized models are poised to reshape decision-making paradigms. As ML algorithms evolve to handle uncertainty, adapt to dynamic criteria, and collaborate harmoniously with human decision-makers, the potential for effective, equitable, and impactful decisions soars. By embracing this dynamic partnership, we stand on the precipice of a new era of informed and transformative decision-making across industries, ushering in a realm where multidimensional considerations converge to shape a better future.

ABBREVIATIONS

AI	Artificial Intelligence
CNNs	Convolutional neural networks
DSS	Decision support systems
LSTM	Long Short-Term Memory
MCDM	Multicriteria decision-making
ML	Machine learning
RL	Reinforcement learning
RNNs	Recurrent neural networks

REFERENCES

Agarwal, R., Suthar, J., Panda, S. K., & Mohanty, S. N., "Fuzzy and Machine Learning based Multi-Criteria Decision Making for Selecting Electronics Product," *EAI Endorsed Transactions on Scalable Information Systems,* 10(5), 2023, https://doi.org/10.4108/eetsis.3353

Ahasan, M. A. Al, Hu, M., & Shahriar, N., "OFMCDM/IRF: A Phishing Website Detection Model based on Optimized Fuzzy Multi-Criteria Decision-Making and Improved Random Forest," *2023 Silicon Valley Cybersecurity Conference* (SVCC 2023), Burlingame, California, 2023, https://doi.org/10.1109/SVCC56964.2023.10165344

Ahmed, M. A., Al-Qaysi, Z. T., Albahri, A. S., Alqaysi, M. E., Kou, G., Albahri, O. S., Alamoodi, A. H., Albahri, S. A., Alnoor, A., Al-Samarraay, M. S., Hamid, R. A., Garfan, S., & Alotaibi, F. S., "Intelligent Decision-Making Framework for Evaluating and Benchmarking Hybridized Multi-Deep Transfer Learning Models: Managing COVID-19 and Beyond," *International Journal of Information Technology & Decision Making,* 2023, https://doi.org/10.1142/s0219622023500463

Albahri, A. S., Zaidan, A. A., AlSattar, H. A., Hamid, R. A., Albahri, O. S., Qahtan, S., & Alamoodi, A. H., "Towards Physician's Experience: Development of Machine Learning Model for the Diagnosis of Autism Spectrum Disorders Based on Complex T-Spherical Fuzzy-Weighted Zero-Inconsistency Method," *Computational Intelligence,* 39(2), 2023, https://doi.org/10.1111/coin.12562

Albahri, O. S., Al-Samarraay, M. S., AlSattar, H. A., Alamoodi, A. H., Zaidan, A. A., Albahri, A. S., Zaidan, B. B., & Jasim, A. N., "Rough Fermatean Fuzzy Decision-Based Approach for Modelling IDS Classifiers in the Federated Learning of IoMT Applications," *Neural Computing and Applications,* 35(30), 2023. https://doi.org/10.1007/s00521-023-08933-y

Ankita, Rani S., Bashir, A. K., Krichen, M., & Alshammari, A., "A Low-Rank Learning Based Multi-Label Security Solution for Industry 5.0 Consumers using Machine Learning Classifiers," *IEEE Transactions on Consumer Electronics,* 2023, https://doi.org/10.1109/TCE.2023.3282964

Arunagiri, R., Pandian, P., Krishnasamy, V., Ramasamy, R., & Sivaprakasam, R., "Selection of Browsers for Smartphones: A Fuzzy Hybrid Approach and Machine Learning Technique," *Knowledge and Information Systems,* 65(5), 2023, https://doi.org/10.1007/s10115-022-01778-2

Bardini, R., & DiCarlo, S., "Computational Modeling and Optimization of Biofabrication in Tissue Engineering and Regenerative Medicine—a Literature Review," *Computational and Structural Biotechnology Journal,* 2023, https://doi.org/10.1016/j.csbj.2023.12.035

Belmecheri, N., Aribi, N., Lazaar, N., Lebbah, Y., & Loudni, S., "Boosting the Learning for Ranking Patterns," Algorithms, 16(5), 2023, https://doi.org/10.3390/a16050218

Choudhary, S., Jain, J., Pingale, S. M., & Khare, D. "A Comprehensive Review on Mapping of Groundwater Potential Zones: Past, Present and Future Recommendations," *In Springer Water:* Vol. Part F1186, 2023, https://doi.org/10.1007/978-3-031-35279-9_6

Chowdhury, S. J., Mahi, M. I., Saimon, S. A., Urme, A. N., & Nabil, R. H., "An Integrated Approach of MCDM Methods and Machine Learning Algorithms for Employees' Churn Prediction," *International Conference on Robotics, Electrical and Signal Processing Techniques,* Dhaka, Bangladesh, 2023, https://doi.org/10.1109/ICREST57604.2023.10070079

Deb, K., Gondkar, A., & Anirudh, S., "Learning to Predict Pareto-Optimal Solutions from Pseudo-weights," *Lecture Notes in Computer Science* (Including Subseries Lecture Notes in Artificial Intelligence and Lecture Notes in Bioinformatics), 13970 LNCS, 2023, https://doi.org/10.1007/978-3-031-27250-9_14

Dutta, R., Das, S., & De, S., "Multi Criteria Decision Making with Machine-Learning Based Load Forecasting Methods for Techno-Economic and Environmentally Sustainable Distributed Hybrid Energy Solution," *Energy Conversion and Management,* 291, 2023, https://doi.org/10.1016/j.enconman.2023.117316

Fan, S., Liu, G., Tu, Y., Zhu, J., Zhang, P., & Tian, Z., "Improved Multi-Criteria Decision Making Method Integrating Machine Learning for Patent Competitive Potential Evaluation: A Case Study in Water Pollution Abatement Technology," *Journal of Cleaner Production,* 403, 2023, https://doi.org/10.1016/j.jclepro.2023.136896

ForouzeshNejad, A. A., "A Hybrid Data-Driven Model for Project Portfolio Selection Problem Based on Sustainability and Strategic Dimensions: A Case Study of the Telecommunication Industry," *Soft Computing,* 2023, https://doi.org/10.1007/s00500-023-08445-w

Gandomi, M., Pirooz, M. D., Nematollahi, B., Nikoo, M. R., Varjavand, I., Etri, T., & Gandomi, A. H., "Multi-Criteria Decision-Making Optimization Model for Permeable Breakwaters Characterization," *Ocean Engineering,* 280, 2023, https://doi.org/10.1016/j.oceaneng.2023.114447

Habib, A., Khan, Z. A., Riaz, M., & Marinkovic, D., "Performance Evaluation of Healthcare Supply Chain in Industry 4.0 with Linear Diophantine Fuzzy Sine-Trigonometric Aggregation Operations," *Mathematics,* 11(12), 2023, https://doi.org/10.3390/math11122611

Henrique Alves Ribeiro, V., & Reynoso-Meza, G., "Multi-Criteria Decision-Making Techniques for the Selection of Pareto-optimal Machine Learning Models in a Drinking-Water Quality Monitoring Problem," *International Journal of Information Technology and Decision Making,* 2023, https://doi.org/10.1142/S0219622023500104

Hezam, I. M., Mishra, A. R., Rani, P., & Alshamrani, A., "Assessing the Barriers of Digitally Sustainable Transportation System for Persons with Disabilities using Fermatean Fuzzy Double Normalization-Based Multiple Aggregation Method," *Applied Soft Computing*, 133, 2023, https://doi.org/10.1016/j.asoc.2022.109910

Hoda, S. A., & Mondal, A. C., "Studies on Multi-Criteria Decision-Making-Based Healthcare Systems Using the Machine Learning," *Journal of Artificial Intelligence and Technology*, 3(2), 2023, https://doi.org/10.37965/jait.2023.0167

Jari, A., Khaddari, A., Hajaj, S., Bachaoui, E. M., Mohammedi, S., Jellouli, A., Mosaid, H., El Harti, A., & Barakat, A., "Landslide Susceptibility Mapping Using Multi-Criteria Decision-Making (MCDM), Statistical, and Machine Learning Models in the Aube Department, France," *Earth (Switzerland)*, 4(3), 2023, https://doi.org/10.3390/earth4030037

Jassim, M. A., Abd, D. H., & Omri, M. N., "Machine Learning-Based New Approach to Films Review," *Social Network Analysis and Mining*, 13(1), 2023, https://doi.org/10.1007/s13278-023-01042-7

Jia, X., Xie, M., Hu, B., Li, H., He, X., Zhao, W., Deng, W., & Wang, J., "Potential Risk Evaluation for Soil Environmental Quality Assessment in China Based on Spatial Multi-Criteria Decision-Making Theory," *Eurasian Soil Science*, 56(7), 2023, https://doi.org/10.1134/S1064229322601809

Joudar, S. S., Albahri, A. S., Hamid, R. A., Zahid, I. A., Alqaysi, M. E., Albahri, O. S., & Alamoodi, A. H., "Artificial Intelligence-Based Approaches for Improving the Diagnosis, Triage, and Prioritization of Autism Spectrum Disorder: A Systematic Review of Current Trends and Open Issues," *Artificial Intelligence Review*, 56, 2023, https://doi.org/10.1007/s10462-023-10536-x

Kalinaki, K., Shafik, W., Namuwaya, S., & Namuwaya, S., "Perspectives, Applications, Challenges, and Future Trends of IoT-Based Logistics," *Navigating Cyber Threats and Cybersecurity in the Logistics Industry*, 148–171, 2024. IGI Global. https://doi.org.10.4018/979-8-3693-3816-2.ch005

Kavya, R., Gupta, S., Christopher, J., & Panda, S., "Explainable Decision Making Model By Interpreting Classification Algorithms," *Lecture Notes in Networks and Systems*, 715 LNNS, 2023, https://doi.org/10.1007/978-3-031-35507-3_31

Kaya, C. M., & Derin, L., "Parameters and Methods Used in Flood Susceptibility Mapping: A Review," *Journal of Water and Climate Change* 14(6), 2023, https://doi.org/10.2166/wcc.2023.035

Khanorkar, Y., & Kane, P. V., "Selective Inventory Classification using ABC Classification, Multi-Criteria Decision Making Techniques, and Machine Learning Techniques," *Materials Today: Proceedings*, 72, 2023, https://doi.org/10.1016/j.matpr.2022.09.298

Khuc, T. D., Truong, X. Q., Tran, V. A., Bui, D. Q., Bui, D. P., Ha, H., Tran, T. H. M., Pham, T. T. T., & Yordanov, V., "Comparison of Multi-Criteria Decision Making, Statistics, and Machine Learning Models for Landslide Susceptibility Mapping in Van Yen District, Yen Bai Province, Vietnam," *International Journal of Geoinformatics*, 19(7), 2023, https://doi.org/10.52939/ijg.v19i7.2743

Kumar, S., Sarangi, A., & Mohanty, R. P., "Development of Automobile Recommender System Using Machine Learning and AHP Algorithms," *Lecture Notes in Mechanical Engineering*, 2023, https://doi.org/10.1007/978-981-19-9493-7_32

Liao, H., He, Y., Wu, X., Wu, Z., & Bausys, R., "Reimagining Multi-Criterion Decision Making by Data-Driven Methods Based on Machine Learning: A Literature Review," *Information Fusion*, 100, 2023, https://doi.org/10.1016/j.inffus.2023.101970

Mba Kouhoue, J., Lonlac, J., Lesage, A., Doniec, A., & Lecoeuche, S., "A Comprehensive Analysis on Associative Classification in Building Maintenance Datasets," *Lecture Notes in Computer Science (Including Subseries Lecture Notes in Artificial Intelligence and Lecture Notes in Bioinformatics)*, 13926 LNAI, 2023, https://doi.org/10.1007/978-3-031-36822-6_4

Mohammadifar, A., Gholami, H., & Golzari, S., "Novel Integrated Modelling Based on Multiplicative Long Short-Term Memory (mLSTM) Deep Learning Model and Ensemble Multi-Criteria Decision Making (MCDM) Models for Mapping Flood Risk," *Journal of Environmental Management,* 345, 2023, https://doi.org/10.1016/j.jenvman.2023.118838

Nayagam, V. L. G., Suriyapriya, K., & Jagadeeswari, M., "A Novel Similarity Measure Based on Accuracy Score of Conventional Type of Trapezoidal-Valued Intuitionistic Fuzzy Sets and Its Applications in Multi-criteria Decision-Making Problems," *International Journal of Computational Intelligence Systems,* 16(1), 2023, https://doi.org/10.1007/s44196-023-00274-x

Nguyen, T. V., & Vo, N. (Eds.), "Using Traditional Design Methods to Enhance AI-Driven Decision Making," *IGI Global,* 2024. https://doi.org/10.4018/979-8-3693-0639-0

Paz, T. da S. R., Rocha Junior, V. G. da, Campos, P. C. de O., Paz, I., Caiado, R. G. G., Rocha, A. de A., & Lima, G. B. A., "Hybrid Method to Guide Sustainable Initiatives in Higher Education: A Critical Analysis of Brazilian Municipalities," *International Journal of Sustainability in Higher Education,* 24(2), 2023, https://doi.org/10.1108/IJSHE-07-2021-0281

Pourkhodabakhsh, N., Mamoudan, M. M., & Bozorgi-Amiri, A., "Effective Machine Learning, Meta-Heuristic Algorithms and Multi-Criteria Decision Making to Minimizing Human Resource Turnover," *Applied Intelligence,* 53(12), 2023, https://doi.org/10.1007/s10489-022-04294-6

Rashidi Shikhteymour, S., Borji, M., Bagheri-Gavkosh, M., Azimi, E., & Collins, T. W., "A Novel Approach for Assessing Flood Risk with Machine Learning and Multi-Criteria Decision-Making Methods," *Applied Geography,* 158, 2023, https://doi.org/10.1016/j.apgeog.2023.103035

Rathoriya, P., Panda, R. R., & Nagwani, N. K., "A Novel Approach for Bug Triaging Using TOPSIS. Smart Innovation," *Systems and Technologies,* 327, 2023, https://doi.org/10.1007/978-981-19-7524-0_12

Rodríguez, J. V., Martínez, J. R., & Jumbo Salazar, F. F., "Colorectal Cancer Prediction Using Machine Learning and Neutrosophic MCDM Methodology: A Case Study," *International Journal of Neutrosophic Science,* 21(2), 2023, https://doi.org/10.54216/IJNS.210211

Rodriguez, P. A. C., Canon, A. F. G., & Orjuela-Castro, J. A., "Methodologies for Characterization, Evaluation, and Improvement of Logistics in the Food Supply Chain," *Acta Logistica,* 10(2), 2023, https://doi.org/10.22306/al.v10i2.369

Salih, M. M., Ahmed, M. A., Al-Bander, B., Hasan, K. F., Shuwandy, M. L., & Al-Qaysi, Z. T., "Benchmarking Framework for COVID-19 Classification Machine Learning Method Based on Fuzzy Decision by Opinion Score Method," *Iraqi Journal of Science,* 64(2), 2023, https://doi.org/10.24996/ijs.2023.64.2.36

Shafik, W., "A Comprehensive Cybersecurity Framework for Present and Future Global Information Technology Organizations. *Effective Cybersecurity Operations for Enterprise-Wide Systems,* 56–79, 2023a. IGI Global. https://doi.org/10.4018/978-1-6684-9018-1.ch002

Shafik, W., "Artificial Intelligence and Blockchain Technology Enabling Cybersecurity in Telehealth Systems," *Artificial Intelligence and Blockchain Technology in Modern Telehealth Systems,* 285–326, 2023b. IET. https://doi.org/10.1049/PBHE061E_ch11

Shafik, W., "Cyber Security Perspectives in Public Spaces: Drone Case Study," *Handbook of Research on Cybersecurity Risk in Contemporary Business Systems,* 79–97, 2023c. IGI Global. https://doi.org/10.4018/978-1-6684-7207-1.ch004

Shafik, W., "IoT-Based Energy Harvesting and Future Research Trends in Wireless Sensor Networks," *Handbook of Research on Network-Enabled IoT Applications for Smart City Services,* 282–306, 2023d, https://doi.org/10.4018/979-8-3693-0744-1.ch016

Shafik, W., "Making Cities Smarter: IoT and SDN Applications, Challenges, and Future Trends," *Opportunities and Challenges of Industrial IoT in 5G and 6G Networks*, 73–94, 2023e. IGI Global. https://doi.org/10.4018/978-1-7998-9266-3.ch004

Shafik, W., "Introduction to ChatGPT," *Advanced Applications of Generative AI and Natural Language Processing Models*, 1–25. 2024a. IGI Global. https://doi.org/10.4018/979-8-3693-0502-7.ch001

Shafik, W., "Navigating Emerging Challenges in Robotics and Artificial Intelligence in Africa," *Examining the Rapid Advance of Digital Technology in Africa*. 124–144, 2024b, IGI Global. https://doi.org/10.4018/978-1-6684-9962-7.ch007

Shafik, W., "Predicting Future Cybercrime Trends in the Metaverse Era," *Forecasting Cyber Crimes in the Age of the Metaverse*, 78–113, 2024c, IGI Global. https://doi.org/10.4018/979-8-3693-0220-0.ch005

Shafik, W., "Wearable Medical Electronics in Artificial Intelligence of Medical Things," *Handbook of Security and Privacy of AI-Enabled Healthcare Systems and Internet of Medical Things*, 21–40, 2024d. CRC Press. https://doi.org/10.1201/9781003370321-2

Shafik, W. "Toward a More Ethical Future of Artificial Intelligence and Data Science," *The Ethical Frontier of AI and Data Analysis*, 362–388, 2024e. IGI Global. https://doi.org/10.4018/979-8-3693-2964-1.ch022

Shafik, W., & Matinkhah, S. M., "Unmanned Aerial Vehicles Analysis to Social Networks Performance," *The CSI Journal on Computer Science and Engineering*, 18(2), 24–31, 2021. https://jcse.ir/article/263

Shafik, W., Matinkhah, S. M., & Shokoor, F., "Recommendation System Comparative Analysis: Internet of Things Aided Networks," *EAI Endorsed Transactions on Internet of Things*, 8(29), 2022. http://dx.doi.org/10.4108/eetiot.v8i29.1108

Shafik, W., Matinkhah, S. M., & Shokoor, F., "Cybersecurity in Unmanned Aerial Vehicles: A Review," *International Journal on Smart Sensing and Intelligent Systems*, 16(1), 2023, https://doi.org/10.2478/ijssis-2023-0012

Shahiri Tabarestani, E., Hadian, S., Pham, Q. B., Ali, S. A., & Phung, D. T., "Flood Potential Mapping By Integrating the Bivariate Statistics, Multi-Criteria Decision-Making, and Machine Learning Techniques," *Stochastic Environmental Research and Risk Assessment*, 37(4), 2023, https://doi.org/10.1007/s00477-022-02342-8

Shakibaei, H., Farhadi-Ramin, M. R., Alipour-Vaezi, M., Aghsami, A., & Rabbani, M. (2023). "Designing a Post-Disaster Humanitarian Supply Chain Using Machine Learning and Multi-Criteria Decision-Making Techniques," *Kybernetes*, 2023, https://doi.org/10.1108/K-10-2022-1404

Sharma, D., Jamwal, A., Agrawal, R., Jain, J. K., & Machado, J., "Decision Making Models for Sustainable Supply Chain in Industry 4.0: Opportunities and Future Research Agenda," *Lecture Notes in Mechanical Engineering*, 2023, https://doi.org/10.1007/978-3-031-09360-9_15

Siddiqui, S., Mallick, M. U., & Varshney, A., "Breast Tumor Classification using Machine Learning," *EAI Endorsed Transactions on Context-Aware Systems and Applications*, 9(1), 2023, https://doi.org/10.4108/eetcasa.v9i1.3600

Sooklall, A., & Fonou-Dombeu, J. V., "Application of Genetic Algorithm for Complexity Metrics-Based Classification of Ontologies with ELECTRE Tri," *Lecture Notes of the Institute for Computer Sciences, Social-Informatics and Telecommunications Engineering*, LNICST, 459 LNICST, 2023, https://doi.org/10.1007/978-3-031-25271-6_9

Sun, Y., Zhu, D., Li, Y., Wang, R., & Ma, R., "Spatial Modelling the Location Choice of Large-Scale Solar Photovoltaic Power Plants: Application of Interpretable Machine Learning Techniques and the National Inventory," *Energy Conversion and Management*, 289, 2023, https://doi.org/10.1016/j.enconman.2023.117198

Torkjazi, M., & Raz, A. K., "Data-Driven Approach with Machine Learning to Reduce Subjectivity in Multi-Attribute Decision Making Methods," *SysCon 2023-17th Annual IEEE International Systems Conference, Proceedings*, Vancouver, Canada, 2023, https://doi.org/10.1109/SysCon53073.2023.10131094

Urgessa, G., & Esfandiari, J., "A Machine Learning Algorithm for Optimization of Reinforced Concrete Structures Subjected to Blast," *Proceedings of International Structural Engineering and Construction*, 10(1), 2023, https://doi.org/10.14455/10.14455/isec.2023.10(1).str-36

Wang, Z., Tan, W. G. Y., Rangaiah, G. P., & Wu, Z., "Machine Learning Aided Model Predictive Control with Multi-Objective Optimization and Multi-Criteria Decision Making," *Computers and Chemical Engineering*, 179, 2023, https://doi.org/10.1016/j.compchemeng.2023.108414

Yilmaz, I., Adem, A., & Dağdeviren, M., "A Machine Learning-Integrated Multi-Criteria Decision-Making Approach Based on Consensus for Selection of Energy Storage Locations," *Journal of Energy Storage*, 69, 2023, https://doi.org/10.1016/j.est.2023.107941

14 Locating Electric Vehicle Power Stations Using Neutrosophic TOPSIS

Regan Murugesan, Sathish Kumar Kumaravel, Kala Raja Mohan, Nagadevi Bala Nagaram, and R. Narmada Devi

14.1 INTRODUCTION

Global economic development and the depletion of natural resources have led to severe energy crises and ecological deterioration, posing significant challenges to sustainable development. In response, many countries have formulated strategies for energy utilization. China, as the world's largest CO_2 emitter, sees about 6% of carbon emissions stemming from the transportation sector. The sector's energy consumption and carbon emissions have outpaced national averages due to urbanization and growing automobile demand.

Electric vehicles (EVs) are crucial for reducing atmospheric pollution and fostering low-carbon transportation. With advancing battery technology and economies of scale, EVs have become the forefront of new energy automobiles, offering the potential to shift power peak loads, provide spinning reserves, and enhance renewable energy penetration, thus contributing to the safe, stable, and economical operation of electric power grids. Governments and the public increasingly recognize the importance of EVs in addressing fossil resource depletion and worsening environmental pollution, promoting urban sustainable development.

Electric vehicle charging stations (EVCSs) serve as the backbone of EV industry development, influencing consumer willingness to purchase and overall industry growth. The site selection for EVCSs, being a critical aspect in the entire life cycle, significantly impacts service quality and operational efficiency. For this paper, we approach EVCS site selection from a sustainability perspective considering economic, social, and environmental dimensions. Economic factors such as construction costs, annual operation and maintenance costs, and investment payback periods are integrated into the decision-making process. Social considerations align EVCS with urban road network and power grid development, traffic convenience, service capability, and societal impacts. Environmental aspects evaluate the impacts of EVCS on vegetation, water, waste discharge, greenhouse gas emissions, and fine particulate matter emissions.

DOI: 10.1201/9781032635170-14

Hwang and Yoon (2023) pioneered the TOPSIS method (technique for order preference by similarity to the ideal solution), an MCDM strategy. Classical TOPSIS, an advanced MCDM, is meant to tackle problems requiring crisp numbers using complex calculation procedures that are difficult to understand and implement. In real-world MCDM scenarios, attribute values are frequently expressed with incomplete information. Decision-makers might prefer a straightforward method that produces the same outcomes as a complicated algorithm. This work seeks to fill that void by making two crucial contributions. To begin, we introduce simplified TOPSIS, a novel MCDM method that simplifies calculations while producing the same results as classic TOPSIS, is introduced. Second, we combine neutrosophic and simplified TOPSIS methods to tackle genuine MCDM issues with poor information.

M. Bahramloo et al. (2013) developed the concept of a complicated neutrosophic set. For ranking options, the authors used MCDM in intuitionist fuzzy sets, which Smarandache (2006) extended to neutrosophic sets. Biswas et al. (2016) summed up their work and looked into aggregation operators, like weighted neutrosophic and aggregated single-valued neutrosophic TOPSIS to solve MCDM problems. Broumi and Smarandache (2013) researched multiple attribute decision-making employing uncertain linguistic variables with intervals, and created an MCDM system in a multi-hesitant, fuzzy environment using aggregation operators and TOPSIS.

14.2　PRELIMINARIES

Neutrosophic TOPSIS operates as a compensatory method and presumes total aggregation among criteria, constructing a linear additive model. The determination of weights and scores primarily involves pairwise comparisons conducted between all available options.

14.2.1　Fuzzy TOPSIS

A fuzzy set $\overline{a}$ in universe of discourse X is defined by a membership function $\mu_{\overline{a}}(x)$ that associates each element x in X with a real number in the interval [0, 1], where $\mu_{\overline{a}}(x)$ represents the degree of membership of x in $\mu_{\overline{a}}(x)$, with a higher value indicating a stronger association. The membership function quantifies the degree to which elements in the universe X belong to the $\overline{a}$. When $\mu_{\overline{a}}(x)$ is closer to unity, it signifies a higher degree of membership for x in $\overline{a}$.

14.2.2　Neutrosophic TOPSIS

Neutrosophic TOPSIS is an extension of the classical TOPSIS method that incorporates neutrosophic set theory. TOPSIS is a decision-making technique used for ranking alternatives based on their proximity to an ideal solution. Neutrosophic set theory is an extension of classical set theory that allows for the representation of indeterminacy, uncertainty, and inconsistency. In neutrosophic TOPSIS, the decision matrix and the criteria weights are expressed in the form of neutrosophic numbers. Neutrosophic numbers consist of three components: a truth value (T), an indeterminacy value (I), and a falsity value (F). These values help capture the uncertainty and indeterminacy associated with decision-making.

14.2.3 CRITERIA

To ensure an equitable selection of decisions, elements related to the decision problem are identified as criteria. Criteria encompass a range of rules, values, and conditions, acting as representations that contribute to improving the impartiality and fairness of the decision-making process. These criteria serve as essential benchmarks, guiding the evaluation and comparison of alternatives and ultimately promoting a balanced and just decision-making outcome.

14.2.4 DECISION MATRIX

A decision matrix is a structured matrix with dimensions m × n where the values are determined by the relationships between attributes and criteria. In this context, m denotes the number of attributes, and n represents the number of criteria. The decision matrix provides a systematic representation of how each attribute is associated with each criterion in the decision-making process. It serves as a comprehensive tool for organizing and evaluating the relationships between different aspects of the decision problem, aiding decision-makers in making informed and rational choices based on the specified criteria.

14.2.5 NORMALIZED DECISION MATRIX

The matrix obtained by dividing the elements of the decision matrix by its Euclidean norm is termed a normalized matrix. The Euclidean norm is computed by taking the square root of the sum of squares of the elements within the decision matrix. This normalization process is valuable as it standardizes the values in the matrix, facilitating a more straightforward comparison and analysis of different elements. Normalizing the matrix mitigates the influence of the magnitude of the original values, providing a relative scale that aids in meaningful comparisons across attributes and criteria in the decision-making context.

14.2.6 WEIGHTED NORMALIZED MATRIX

Weights in a decision-making process are measures of the preference assigned to each attribute. The matrix obtained by multiplying the normalized decision matrix by the weighted matrix is referred to as the weighted normalized matrix. This matrix integrates the preferences assigned to each attribute (via the weights) with the normalized values derived from the decision matrix. This comprehensive representation takes into account both the significance of each attribute, as indicated by the weights, and the standardized relationships between attributes and criteria. The weighted normalized matrix serves as a crucial intermediary step in the decision-making process, capturing the combined impacts of attribute preferences and their normalized influences on the overall evaluation of alternatives.

14.2.7 RANKING

Now, a set of alternatives can be ranked based on the descending order of C_i^*. Higher C_i^* indicates closer proximity of an alternative to the neutrosophic ideal solution, considering both benefit and cost attributes. By organizing alternatives in descending order of C_i^*, the ranking reflects the relative strengths of each alternative in meeting the decision criteria, offering a clear and prioritized understanding of their performance in the context of the neutrosophic TOPSIS approach

14.3 DEFINING THE NEUTROSOPHIC TOPSIS ALGORITHM

Location Selection: Evaluate potential locations for new electric vehicle charging stations based on criteria such as accessibility, demand, cost, and environmental impact. Consider factors like population density, proximity to highways, and existing charging infrastructure.

Charging Station Design: Assess different designs for charging stations considering factors like the number of charging points, charging speed, and user experience. Consider the preferences and feedback of potential users, as well as technical and economic considerations.

Power Infrastructure Planning: Optimize the selection of power sources and grid connectivity for charging stations based on criteria such as reliability, cost, and environmental sustainability. Consider the integration of renewable energy sources and energy storage solutions to enhance sustainability.

Operational Efficiency: Rank charging station operational strategies, such as pricing models and maintenance plans, to maximize efficiency and user satisfaction. Consider factors like service reliability, payment convenience, and station maintenance requirements.

Integration with Smart Grids: Evaluate the integration of charging stations with smart grid technologies, considering factors like demand response, grid stability, and load balancing. Assess the impact on overall energy consumption and environmental sustainability.

User Experience and Accessibility: Rank charging stations based on user-centric criteria, including ease of use, payment options, and accessibility for individuals with disabilities. Consider user reviews and feedback to enhance the overall charging experience.

Cost-Benefit Analysis: Assess the cost-effectiveness of different charging station implementations, considering initial investment, operational costs, and potential revenue streams. Consider factors like the economic viability and return on investment over time.

Environmental Impact: Evaluate the environmental impacts of charging stations based on factors such as energy source, carbon footprint, and potential for emissions reduction. Consider the overall sustainability and alignment with environmental goals.

Scalability and Future Expansion: Rank charging station designs based on their scalability and adaptability for future expansion. Consider criteria such as growth potential, technological advancements, and changing user needs.

Regulatory Compliance: Assess charging station designs for compliance with local and national regulations, safety standards, and environmental regulations. Consider factors like permits, zoning requirements, and adherence to industry standards.

Neutrosophic TOPSIS can provide a structured and flexible approach to decision-making in the electric vehicle charging station domain, accommodating the inherent uncertainties and complexities associated with multiple criteria and decision factors.

14.4 APPLYING THE ALGORITHM

MCDM neutrosophic TOPSIS follows these steps:

Step 1: The aggregated single-valued neutrosophic decision matrix is constructed based on assessments from decision-makers (Equation (14.1)):

$$R = (d_{ij})_{1 \leq i \leq n} = (A_{ij}, B_{ij}, C_{ij})_{1 \leq i \leq n} \; 1 \leq j \leq m, 1 \leq j \leq m \tag{14.1}$$

where A_{ij} denotes truth, B_{ij} represents indeterminacy, and C_{ij} signifies falsity in the membership scores of preferences i concerning criterion j in the single-valued neutrosophic context. Furthermore, a weight vector $w = (\omega_1, \omega_2, \omega_n)$ is introduced, with w_i being a single-valued neutrosophic weight assigned to criterion i, specifically, $\omega_i = (a_i, b_i, c_i)$.

Step 2: Aggregate the weighted neutrosophic decision matrix by combining the information provided by decision-makers with the assigned neutrosophic weights. The aggregated score for each criterion is determined by considering the truth, indeterminacy, and falsity, as represented in Equation (14.2):

$$D^w = R \otimes W = (d_{ij}^w)_{1 \leq i \leq n} = (A^w, B^w, C^w)_{1 \leq i \leq n} \; 1 \leq j \leq m \, 1 \leq j \leq m \tag{14.2}$$

Step 3: Determine the relative neutrosophic positive ideal solution (RNPIS) and the relative negative ideal solution. These solutions serve as benchmarks for evaluating the preferences. The RNPIS S^+ and the relative negative ideal solution S^- are computed by considering the maximum and minimum values, respectively, across all aggregated scores obtained in Step 2.

$$A_j^{w+} = \{(\max_i \{A_{ij}^{w_i} \mid j \in S\}), (\min_i \{A_{ij}^{w_i} \mid j \in U$$

$$Q_N^+ = \left(d_1^{w+}, d_2^{w+}, \cdots, d_n^{w+} \right)$$

$$A_j^{w+} = \{(\max_i \{A_{ij}^{w_j} \mid j \in S\}), (\min_i \{A_{ij}^{w_j} \mid j \in U$$

$$B_j^{w+} = \{(\min_i \{B_{ij}^{w_j} \mid j \in S\}), (\max_i \{B_{ij}^{w_j} \mid j \in U$$

$$C_j^{w+} = \{(\min_i \{C_{ij}^{w_j} \mid j \in S\}), (\max_i \{C_{ij}^{w_j} \mid j \in U$$

$$Q_N^- = \left(d_1^{w-}, d_2^{w-}, \cdots, d_n^{w-} \right)$$

$$A_j^{w-} = \{(\min_i \{A_{ij}^{w_j} \mid j \in S\}), (\max_i \{A_{ij}^{w_j} \mid j \in U$$

$$B_j^{w-} = \{(\max_i\{B_{ij}^{w_j} \mid j \in S\}), (\min_i\{A_{ij}^{w_j} \mid j \in U\}$$

$$C_j^{w-} = \{(\max_i\{C_{ij}^{w_j} \mid j \in S\}), (\min_i\{C_{ij}^{w_j} \mid j \in U\}$$

Step 4: Now determine the relative closeness coefficients for each alternative concerning S^+ and S^- by considering the benefit and cost attribute sets, denoted as sets S and U, respectively:

$$D_{Eu}^{i+}\left(d_{ij}^{wj}, d_{ij}^{w+}\right) = \sqrt{\frac{1}{3n}\sum_{j=1}^{n}\left\{\left(A_{ij}^{wj}(x) - A_{ij}^{w+}(x)\right)^2 + \left(B_{ij}^{wj}(x) - B_{ij}^{w+}(x)\right)^2 + \left(C_{ij}^{wj}(x) - C_{ij}^{w+}(x)\right)^2\right\}}$$

$$D_{Eu}^{i+}\left(d_{ij}^{wj}, d_{ij}^{w+}\right) = \sqrt{\frac{1}{3n}\sum_{j=1}^{n}\left\{\left(A_{ij}^{wj}(x) - A_{ij}^{w-}(x)\right)^2 + \left(B_{ij}^{wj}(x) - B_{ij}^{w-}(x)\right)^2 + \left(C_{ij}^{wj}(x) - C_{ij}^{w-}(x)\right)^2\right\}}$$

Step 5: Calculate the relative closeness coefficients for single-valued neutrosophic sets, that is, the distances from each alternative to S^+ and S^-. This coefficient is a measure of the proximity of each alternative to the neutrosophic ideal solution considering both benefit and cost (Equation (14.3)):

$$C_i^* = \frac{MS_i^-}{\left(MS_i^+ + MS_i^-\right)}; i = 1, 2\cdots, m \tag{14.3}$$

Step 6: With C_j^*, rank the alternatives in descending order. Higher C_j^* indicates the greater proximity of an alternative to the neutrosophic ideal solution considering both benefit and cost, making alternatives with better performance those that are positioned at the top.

14.5 NUMERIC ILLUSTRATION

Our selected criteria were environment, economy, energy, and transportation, and we assigned weights to each criterion to reflect their relative importance, ensuring that the weights summed to 1. The next step was to utilize GIS to integrate the determined weights and perform spatial analysis for selecting suitable locations for EV charging stations based on the specified criteria. We applied neutrosophic TOPSIS to rank the potential charging station locations, L1, L2, L3, and L4, compute C_j^* as in Equation (14.1), and build the aggregated single-valued neutrosophic decision matrix in Table 14.1 based on assessments from decision-makers.

TABLE 14.1

Neutrosophic Decision Matrix

	Environment	Economic	Energy	Transportation
L1	(0.386, 0.254, 0.486)	(0.458, 0.456, 0.324)	(0.521, 0.315, 0.428)	(0.546, 0.423, 0.304)
L2	(0.543, 0.498, 0.752)	(0.621, 0.587, 0.642)	(0.573, 0.572, 0.467)	(0.487, 0.542, 0.354)
L3	(0.643, 0.545, 0.684)	(0.592, 0.423, 0.364)	(0.751, 0.467, 0.576)	(0.756, 0.532, 0.467)
L4	(0.621, 0.598, 0.425)	(0.723, 0.645, 0.568)	(0.821, 0.578, 0.632)	(0.647, 0.458, 0.687)

The weight of each criterion is calculated in the form of $w = \left(\omega_1, \omega_2, \ldots, \omega_n\right)$ with ω_i being the single-valued weights in Table 14.2 calculated from Equation (14.2).

The next stage involves aggregating the weighted neutrosophic decision matrix by combining the information provided by decision-makers depending on the assigned neutrosophic weights shown in Table 14.3. The aggregated score for each criterion is determined by considering the truth, indeterminacy, and falsity components, as represented in Equation (14.4):

$$D^w = R \otimes W = (d_{ij}^w)_{1 \leq i \leq n} = (A^w, B^w, C^w, C^w)_{1 \leq i \leq n} 1 \leq j \leq m 1 \leq j \leq m \quad (14.4)$$

From identifying the maximum and minimum neutrosophic ideal solutions for each criterion or each alternative, subtract the neutrosophic value of the alternative from the corresponding value in the maximum and minimum ideal solutions. Use the absolute values to ensure that the separation measures are positive or negative, shown in Table 14.4 and Table 14.5, respectively.

Table 14.6 indicates that alternative L1 had the highest C_1, making L1 the optimal location for the EV power station, the same ranking yielded by neutrosophic TOPSIS.

TABLE 14.2

Neutrosophic Single-Valued Weight Matrix

	Environment	Economic	Energy	Transportation
Weight	(0.225, 0.354, 0.421)	(0.356, 0.351, 0.293)	(0.213, 0.458, 0.329)	(0.356, 0.202, 0.442)

TABLE 14.3

Weighted Neutrosophic Decision Matrix

d_{ij}^w	Environment	Economic	Energy	Transportation
L1	(0.08685, 0.089916, 0.204606)	(0.163048, 0.160056, 0.094932)	(0.110973, 0.14427, 0.140812)	(0.194376, 0.085446, 0.134368)
L2	(0.122175, 0.176292, 0.316592)	(0.221076, 0.206037, 0.188106)	(0.122049, 0.261976, 0.153643)	(0.173372, 0.109484, 0.156468)
L3	(0.144675, 0.19293, 0.287964)	(0.210752, 0.148473, 0.106652)	(0.159963, 0.213886, 0.189504)	(0.269136, 0.107464, 0.206414)
L4	(0.139725, 0.211692, 0.178925)	(0.257388, 0.226395, 0.166424)	(0.174873, 0.264724, 0.207928)	(0.230332, 0.092516, 0.303654)

TABLE 14.4

The Neutrosophic Positive Ideal Solutions

	Environment	Economic	Energy	Transportation
d_{ij}^{w+}	(0.144675, 0.089916, 0.178925)	(0.163048, 0.226395, 0.188106)	(0.174873, 0.14427, 0.140812)	(0.269136, 0.085446, 0.134368)

TABLE 14.5

The Neutrosophic Negative Ideal Solutions

	Environment	Economic	Energy	Transportation
d_{ij}^{w-}	(0.08685, 0.211692, 0.316592)	(0.257388, 0.148473, 0.094932)	(0.110973, 0.264724, 0.207928)	(0.173372, 0.109484, 0.303654)

TABLE 14.6

The Ranks of the Alternatives

Alternatives	MS_i^+	MS_i^-	Ci
L1	0.182621	0.329464	**0.643377**
L2	0.263501	0.212495	**0.446422**
L3	0.250567	0.187489	**0.428003**
L4	0.306219	0.217404	**0.415192**

14.6 CONCLUSION

In this chapter, we introduced neutrosophic TOPSIS and described its use in selecting an optimum site for an electric vehicle power station. We defined the neutrosophic TOPSIS algorithm's individual components and then computed a weighted decision matrix. Integrating the maximum and minimum neutrosophic ideal solutions to build the matrix, calculate ranks, and make the final decision.

REFERENCES

Bahramloo, M., & Hoseini, M. A multiple criteria decision making for raking alternatives using preference relation matrix based on intuitionistic fuzzy sets. Decision Science Letters, 2(4), 281–286, 2013.

Biswas, P., Pramanik, S., & Giri, B. C. TOPSIS method for multi-attribute group decision-making under single-valued neutrosophic environment. Neural Computing and Applications, 27, 727–737, 2016.

Broumi, Said, & Smarandache, Florentin. Several similarity measures of neutrosophic sets. Infinite Study, 410, 2013.

Huang, Y., Li, X., Liu, S., & Wu, Z. A hybrid model combining traditional techniques and AI for decision-making in financial markets. Expert Systems with Applications, 195, 115501, 2023. DOI: 10.1016/j.eswa.2022.115501

Smarandache, Florentin. "Neutrosophic set-a generalization of the intuitionistic fuzzy set." In IEEE international conference on granular computing, pp. 38–42. IEEE, 2006.

15 Multicriteria Decision-Making Modeling Using Spherical Neutrosophic Similarity Measures

R. Narmada Devi, Regan Murugesan, Nagadevi Bala Nagaram, Kala Raja Mohan, and Sathish Kumar Kumaravel

15.1 INTRODUCTION

Computers can learn on their own, without being taught, to carry out certain tasks, according to the core idea of machine learning. Given its capacity to automatically identify patterns in data and predict future outcomes, machine learning has become increasingly in demand for making judgments in a variety of areas, including financial trading and medical diagnosis.

In order to resolve ambiguous circumstances, fuzzy theory [11] is essential. After the fuzzy set was extended, a new set known as intuitionistic fuzzy sets [1] was developed. Smarandache developed the general idea of neutrosophic sets [9], and later scholars described variations such as Pythagorean, bipolar [2, 5], and Fermatean [7] neutrosophic sets. Broumi measured the Hausdorff distance between two neutrosophic sets [2, 3], and digital neutrosophic topological spaces were first suggested by Narmada Devi [4, 5, 6]. Incomplete and ambiguous data in MCDM [10] challenges have exposed the variable of approximate reasoning and pushed the decision-making process to new advancements and innovation.

The theory of neutrosophic logics extends to include number theory, operations research, algebra, topology, graphs, probability, and numeric measures. Field neutrosophic set theory is applied in a wide range of fields including logic in image processing, large-scale image and multimedia processing, machine learning, big data analytics, deep learning, data visualization, feature learning, classification, regression, and clustering, virtual reality, heterogeneous data mining multimodal sensor data, wireless sensor networks, astronomy, space sciences, bioinformatics and medical analytics, retrieval of medical images, brain–machine interfaces, medical signal analysis, and large-scale health care data.

In this chapter, we introduce the basic concepts of spherical neutrosophic sets (SNSs; mathematically, $\mathcal{SPhN\_S'}$) and their characteristics and introduce SNS similarity measure; in more detail, we discuss tangent as an SNS similarity measures.

DOI: 10.1201/9781032635170-15

We used SNS to represent uncertain MCDM issues using Python programming to analyze a real-world decision-making problem.

15.2 BASIC CONCEPTS

15.2.1 Spherical Neutrosophic Set

First, an SNS, calculated as $A = \left\{ \left\langle A_{\mathbb{T}}\left(\mathfrak{p}\right), A_{\mathbb{I}}\left(\mathfrak{p}\right), A_{\mathbb{F}}\left(\mathfrak{p}\right) \right\rangle : \mathfrak{p} \in X \right\}$ on $\mathbb{X}$, satisfies the following properties:

(i) $A_{\mathbb{T}}\left(\mathfrak{p}\right), A_{\mathbb{I}}\left(\mathfrak{p}\right), A_{\mathbb{F}}\left(\mathfrak{p}\right) \in \left[0,1\right]$

(ii) $0 \leq A_{\mathbb{T}}\left(\mathfrak{p}\right) + A_{\mathbb{I}}\left(\mathfrak{p}\right) + A_{\mathbb{F}}\left(\mathfrak{p}\right) \leq 1$

(iii) $0 \leq A_{\mathbb{T}}\left(\mathfrak{p}\right)^{2} + A_{\mathbb{I}}\left(\mathfrak{p}\right)^{2} + A_{\mathbb{F}}\left(\mathfrak{p}\right)^{2} \leq 1$, for all $\mathfrak{p} \in \mathbb{X}$

The family of SNSs on $\mathbb{X}$ is denoted $\mathcal{SPhN\_S}(\mathbb{X})$.

15.2.2 Operation of SNS

Let $\mathcal{M}$ and $\mathcal{L}$ be any two $\mathcal{SPhN\_S}$ on $\mathbb{X}$. Then

(i) $\mathcal{M}^{c} = \left\{ \left\langle \mathcal{M}_{\mathbb{F}}\left(\mathfrak{p}\right), 1 - \mathcal{M}_{\mathbb{I}}\left(\mathfrak{p}\right), \mathcal{M}_{\mathbb{T}}\left(\mathfrak{p}\right) \right\rangle : \mathfrak{p} \in X \right\}$ is the complement of $\mathcal{M}$

(ii) $\mathcal{M} \sqcap \mathcal{L} = \left\{ \left\langle \mathcal{M}_{\mathbb{T}}\left(\mathfrak{p}\right) \wedge \mathcal{L}_{\mathbb{T}}\left(\mathfrak{p}\right), \mathcal{M}_{\mathbb{I}}\left(\mathfrak{p}\right) \wedge \mathcal{L}_{\mathbb{I}}\left(\mathfrak{p}\right), \mathcal{M}_{\mathbb{F}}\left(\mathfrak{p}\right) \vee \mathcal{L}_{\mathbb{F}}\left(\mathfrak{p}\right) \right\rangle \right\}$

(iii) $\mathcal{M} \sqcup \mathcal{L} = \left\{ \left\langle \mathcal{M}_{\mathbb{T}}\left(\mathfrak{p}\right) \vee \mathcal{L}_{\mathbb{T}}\left(\mathfrak{p}\right), \mathcal{M}_{\mathbb{I}}\left(\mathfrak{p}\right) \vee \mathcal{L}_{\mathbb{I}}\left(\mathfrak{p}\right), \mathcal{M}_{\mathbb{F}}\left(\mathfrak{p}\right) \wedge \mathcal{L}_{\mathbb{F}}\left(\mathfrak{p}\right) \right\rangle \right\}$

(iv) $0_{\mathcal{SPhN}} = \left\{ \left\langle 0,0,1 \right\rangle : \mathfrak{p} \in \mathbb{X} \right\}$ is a Null $\mathcal{SPhN\_S}$ on $\mathbb{X}$

(v) $1_{\mathcal{SPhN}} = \left\{ \left\langle 1,0,0 \right\rangle : \mathfrak{p} \in \mathbb{X} \right\}$ is a Full $\mathcal{SPhN\_S}$ on $\mathbb{X}$

15.2.3 Similarity Measures between SNSs

The similarity measure between the two $\mathcal{SPhN_S}$s $\mathcal{G}$ and $\mathcal{H}$ is defined as $\left\langle \varpi\left(\mathcal{G}, \mathcal{H}\right) = \varpi_{\mathbb{T}}\left(\mathcal{G}, \mathcal{H}\right), \varpi_{\mathbb{I}}\left(\mathcal{G}, \mathcal{H}\right), \varpi_{\mathbb{F}}\left(\mathcal{G}, \mathcal{H}\right) \right\rangle$ where $\varpi : \mathcal{SPhN\_S}(\mathbb{X}) \times \mathcal{SPhN\_S}(\mathbb{X}) \to [0,1]$ such that ϖ satisfies the following properties of conditions:

(i) $\varpi\left(\mathcal{G}, \mathcal{H}\right) = \varpi\left(\mathcal{B}, \mathcal{A}\right)$

(ii) $\varpi\left(\mathcal{G}, \mathcal{H}\right) = \left(1,0,0\right) = 1_{\mathcal{SPhN}}$ if $\mathcal{A} = \mathcal{B}$ for all $\mathcal{G}, \mathcal{H} \in \mathcal{SPhN\_S}(\mathbb{X})$

(iii) $\varpi_{\mathbb{T}}\left(\mathcal{G}, \mathcal{H}\right) \geq 0 \, \varpi_{\mathbb{I}}\left(\mathcal{G}, \mathcal{H}\right) \geq 0 \text{ and } \varpi_{\mathbb{F}}\left(\mathcal{G}, \mathcal{H}\right) \geq 0$

(iv) If $\mathcal{G} \subseteq \mathcal{H} \subseteq \mathcal{D}$ for all $\mathcal{G}, \mathcal{H}, \mathcal{D} \in \mathcal{SPhN\_S}(\mathbb{X})$, then $\varpi\left(\mathcal{G}, \mathcal{H}\right) \geq \varpi\left(\mathcal{G}, \mathcal{D}\right)$ and $\varpi\left(\mathcal{H}, \mathcal{D}\right) \geq \varpi\left(\mathcal{G}, \mathcal{D}\right)$

where $\varpi_{\mathbb{T}}\left(\mathcal{G}, \mathcal{H}\right)$, $\varpi_{\mathbb{I}}\left(\mathcal{G}, \mathcal{H}\right)$, and $\varpi_{\mathbb{F}}\left(\mathcal{G}, \mathcal{H}\right)$ represented as the degrees of $\mathbb{T}$, $\mathbb{I}$, and $\mathbb{F}$ similarity between $\mathcal{G}$ and $\mathcal{H}$, respectively.

15.2.4 TRIGONOMETRIC SIMILARITY MEASURES

Let $\mathcal{M}$ and $\mathcal{L}$ be any two $\mathcal{SPhN_S}$s on $\mathbb{X}$. Then the three trigonometric similarity measures tangent, cosine, and cotangent are defined as follows:

(i) $\mathcal{TAN\_S}(\mathcal{M}, \mathcal{L}) =$

$$\frac{1}{n}\left[\sum_{i=1}^{n} 1 - \tan\left[\left(\frac{\pi}{12}\right)\left[\left|\mathcal{M}_{\mathbb{T}}(\mathfrak{p}_i) - \mathcal{L}_{\mathbb{T}}(\mathfrak{p}_i)\right| + \left|\mathcal{M}_{\mathbb{I}}(\mathfrak{p}_i) - \mathcal{L}_{\mathbb{I}}(\mathfrak{p}_i)\right|\right.\right.\right.$$

$$\left.\left.\left. + \left|\mathcal{M}_{\mathbb{F}}(\mathfrak{p}_i) - \mathcal{L}_{\mathbb{F}}(\mathfrak{p}_i)\right|\right]\right]\right]$$

(ii) $\mathcal{COT\_S}(\mathcal{M}, \mathcal{L}) =$

$$\frac{1}{n}\left[\sum_{i=1}^{n} \cot\left[\frac{\pi}{4} + \left(\frac{\pi}{12}\right)\left[\left|\mathcal{M}_{\mathbb{T}}(\mathfrak{p}_i) - \mathcal{L}_{\mathbb{T}}(\mathfrak{p}_i)\right| + \left|\mathcal{M}_{\mathbb{I}}(\mathfrak{p}_i) - \mathcal{L}_{\mathbb{I}}(\mathfrak{p}_i)\right|\right.\right.\right.$$

$$\left.\left.\left. + \left|\mathcal{M}_{\mathbb{F}}(\mathfrak{p}_i) - \mathcal{L}_{\mathbb{F}}(\mathfrak{p}_i)\right|\right]\right]\right]$$

(iii) $\mathcal{COS\_S}(\mathcal{M}, \mathcal{L}) =$

$$\frac{1}{n}\left[\sum_{i=1}^{n} \frac{\mathcal{M}_{\mathbb{T}}(\mathfrak{p}_i)\cdot\mathcal{L}_{\mathbb{T}}(\mathfrak{p}_i) + \mathcal{M}_{\mathbb{I}}(\mathfrak{p}_i)\cdot\mathcal{L}_{\mathbb{I}}(\mathfrak{p}_i) + \mathcal{M}_{\mathbb{F}}(\mathfrak{p}_i)\cdot\mathcal{L}_{\mathbb{F}}(\mathfrak{p}_i)}{\sqrt{\mathcal{M}_{\mathbb{T}}(\mathfrak{p}_i)^2 + \mathcal{M}_{\mathbb{I}}(\mathfrak{p}_i)^2 + \mathcal{M}_{\mathbb{F}}(\mathfrak{p}_i)^2}\cdot\sqrt{\mathcal{L}_{\mathbb{T}}(\mathfrak{p}_i)^2 + \mathcal{L}_{\mathbb{I}}(\mathfrak{p}_i)^2 + \mathcal{L}_{\mathbb{F}}(\mathfrak{p}_i)^2}}\right]$$

15.2.5 PROPERTIES OF TRIGONOMETRIC SIMILARITY MEASURES

Theorem 2.5.1 Let $\mathcal{M}$ and $\mathcal{L}$ be any two $\mathcal{SPhN_S}$s on $\mathbb{X}$. Then

  (i) $0 \leq \mathcal{TAN\_S}(\mathcal{M}, \mathcal{L}) \leq 1$

  (ii) $\mathcal{TAN\_S}(\mathcal{M}, \mathcal{L}) = 1$ if and only if $\mathcal{M} = \mathcal{L}$

  (iii) $\mathcal{TAN\_S}(\mathcal{M}, \mathcal{L}) = \mathcal{TAN\_S}(\mathcal{L}, \mathcal{M})$

  (iv) If $\mathcal{Q}$ is a $\mathcal{SPhN\_S}$ in X and $\mathcal{M} \subseteq \mathcal{Q} \subseteq \mathcal{L}$, then $\mathcal{TAN\_S}(\mathcal{M}, \mathcal{Q}) \leq \mathcal{TAN\_S}(\mathcal{M}, \mathcal{L})$ and $\mathcal{TAN\_S}(\mathcal{M}, \mathcal{Q}) \leq \mathcal{TAN\_S}(\mathcal{L}, \mathcal{Q})$

Proof:

  (i) $\mathbb{T}$, $\mathbb{I}$, and $\mathbb{F}$ lie between 0 and 1. Hence, $0 \leq \mathcal{TAN\_S}(\mathcal{M}, \mathcal{L}) \leq 1$.

  (ii) For any two $\mathcal{SPhN_S}$s $\mathcal{M}$ and $\mathcal{L}$, if
$\mathcal{M} = \mathcal{L} \Rightarrow \mathcal{M}_{\mathbb{T}}(\mathfrak{x}) = \mathcal{L}_{\mathbb{T}}(\mathfrak{x}), \mathcal{M}_{\mathbb{I}}(\mathfrak{p}) = \mathcal{L}_{\mathbb{I}}(\mathfrak{x})$ and $\mathcal{M}_{\mathbb{F}}(\mathfrak{p}) = \mathcal{L}_{\mathbb{F}}(\mathfrak{p})$.
Therefore, $\mathcal{TAN\_S}(\mathcal{M}, \mathcal{L}) = 1$.
Conversely, if $\mathcal{TAN\_S}(\mathcal{M}, \mathcal{L}) = 1$
$\Rightarrow \left|\mathcal{M}_{\mathbb{T}}(\mathfrak{p}_i) - \mathcal{L}_{\mathbb{T}}(\mathfrak{p}_i)\right| = \left|\mathcal{M}_{\mathbb{I}}(\mathfrak{p}_i) - \mathcal{L}_{\mathbb{I}}(\mathfrak{p}_i)\right| = \left|\mathcal{M}_{\mathbb{F}}(\mathfrak{p}_i) - \mathcal{L}_{\mathbb{F}}(\mathfrak{p}_i)\right| = 0 \cdot$
Therefore, $\mathcal{M} = \mathcal{L}$.

(iii) The proof is obvious.

(iv) Since $\mathcal{M}_T(\mathfrak{p}) \le \mathcal{Q}_T(\mathfrak{p}) \le \mathcal{L}_T(\mathfrak{p})$, $\mathcal{M}_I(\mathfrak{p}) \le \mathcal{Q}_I(\mathfrak{p}) \le \mathcal{L}_I(\mathfrak{p})$ and
$\mathcal{M}_F(\mathfrak{p}) \ge \mathcal{Q}_F(\mathfrak{p}) \ge \mathcal{L}_F(\mathfrak{p})$,
$\Rightarrow \left| \mathcal{M}_T(\mathfrak{p}) - \mathcal{L}_T(\mathfrak{p}) \right| \le \left| \mathcal{M}_T(\mathfrak{p}) - \mathcal{Q}_T(\mathfrak{p}) \right|$
$\left| \mathcal{M}_I(\mathfrak{p}) - \mathcal{L}_I(\mathfrak{p}) \right| \le \left| \mathcal{M}_I(\mathfrak{p}) - \mathcal{Q}_I(\mathfrak{p}) \right|$ and
$\left| \mathcal{M}_F(\mathfrak{p}) - \mathcal{L}_F(\mathfrak{p}) \right| \ge \left| \mathcal{M}_F(\mathfrak{p}) - \mathcal{Q}_F(\mathfrak{p}) \right|$. Hence
$\left| \mathcal{L}_T(\mathfrak{p}) - \mathcal{Q}_T(\mathfrak{p}) \right| \le \left| \mathcal{M}_T(\mathfrak{p}) - \mathcal{Q}_T(\mathfrak{p}) \right|$, $\left| \mathcal{L}_I(\mathfrak{p}) - \mathcal{Q}_I(\mathfrak{p}) \right| \le \left| \mathcal{M}_I(\mathfrak{p}) - \mathcal{Q}_I(\mathfrak{p}) \right|$
and $\left| \mathcal{L}_F(\mathfrak{p}) - \mathcal{Q}_F(\mathfrak{p}) \right| \ge \left| \mathcal{M}_F(\mathfrak{p}) - \mathcal{Q}_F(\mathfrak{p}) \right|$.
Therefore, $TAN\_\mathcal{S}(\mathcal{M},\mathcal{Q}) \le TAN\_\mathcal{S}(\mathcal{M},\mathcal{L})$ and
$TAN\_\mathcal{S}(\mathcal{M},\mathcal{Q}) \le TAN\_\mathcal{S}(\mathcal{L},\mathcal{Q})$.

Theorem 2.5.2 Let $\mathcal{M}$ and $\mathcal{L}$ be any two $\mathcal{SPhN_S}$s on $\mathbb{X}$. Then

(i) $0 \le COS\_\mathcal{S}(\mathcal{M},\mathcal{L}) \le 1$

(ii) $COS\_\mathcal{S}(\mathcal{M},\mathcal{L}) = 1$ if and only if $\mathcal{M} = \mathcal{L}$

(iii) $COS\_\mathcal{S}(\mathcal{M},\mathcal{L}) = COS\_\mathcal{S}(\mathcal{L},\mathcal{M})$

(iv) If $\mathcal{Q}$ is a $\mathcal{SPhN\_S}$ in X and $\mathcal{M} \subseteq \mathcal{Q} \subseteq \mathcal{L}$, then $COS\_\mathcal{S}(\mathcal{M},\mathcal{Q}) \le COS\_\mathcal{S}(\mathcal{M},\mathcal{L})$ and $COS\_\mathcal{S}(\mathcal{M},\mathcal{Q}) \le COS\_\mathcal{S}(\mathcal{L},\mathcal{Q})$

Theorem 2.5.3 Let $\mathcal{M}$ and $\mathcal{L}$ be any two $\mathcal{SPhN_S}$s on $\mathbb{X}$. Then

(i) $0 \le COT\_\mathcal{S}(\mathcal{M},\mathcal{L}) \le 1$

(ii) $COT\_\mathcal{S}(\mathcal{M},\mathcal{L}) = 1$ if and only if $\mathcal{M} = \mathcal{L}$

(iii) $COT\_\mathcal{S}(\mathcal{M},\mathcal{L}) = COT\_\mathcal{S}(\mathcal{L},\mathcal{M})$

(iv) If $\mathcal{Q}$ is a $\mathcal{SPhN\_S}$ in X and $\mathcal{M} \subseteq \mathcal{Q} \subseteq \mathcal{L}$, then $COT\_\mathcal{S}(\mathcal{M},\mathcal{Q}) \le COT\_\mathcal{S}(\mathcal{M},\mathcal{L})$ and $COT\_\mathcal{S}(\mathcal{M},\mathcal{Q}) \le COT\_\mathcal{S}(\mathcal{L},\mathcal{Q})$.

15.3 THE DECISION-MAKING ALGORITHM

Let $Å = \{Å_1, Å_2, ..., + Å_m\}$ be the collection of persons, $\mathbb{C} = \{\mathbb{C}_1, \mathbb{C}_2, ..., \mathbb{C}_n\}$ be the set of criteria for each person, and $\mathcal{B} = \{\mathcal{B}_1, \mathcal{B}_2, ..., \mathcal{B}_l\}$ be the family of alternatives of persons.

Step 1: Estimate the connection between persons and attributes $(\mathfrak{R}_1)$

The connections between person $Å_m$ ($j = 1, 2, ...,$) and attribute $\mathbb{C}_j$ ($k = 1, 2 ... n$) are given in Table 15.1 in terms of SNSs.

Step 2: Find the association between attributes and alternatives $(\mathfrak{R}_2)$

$\mathfrak{R}_2$ can be represented in the form of decision Table 15.2 based on the SNSs.

Step 3: Calculate the trigonometric similarity measures

Use the code in Python to calculate the different trigonometric similarity measures between $\mathfrak{R}_1$ and $\mathfrak{R}_2$ based on Section 15.2.4; the measures are presented in Table 15.3.

TABLE 15.1

Connections between Pupils and Attributes

$\mathfrak{R}_1$	$\mathbb{C}_1$	$\mathbb{C}_2$	$\cdots$	$\mathbb{C}_n$
$\mathring{A}_1$	$\langle \mathbb{T}_{11}, \mathbb{I}_{11}, \mathbb{F}_{11} \rangle$	$\langle \mathbb{T}_{12}, \mathbb{I}_{12}, \mathbb{F}_{12} \rangle$	$\cdots$	$\langle \mathbb{T}_{1n}, \mathbb{I}_{1n}, \mathbb{F}_{1n} \rangle$
$\mathring{A}_2$	$\langle \mathbb{T}_{21}, \mathbb{I}_{21}, \mathbb{F}_{21} \rangle$	$\langle \mathbb{T}_{22}, \mathbb{I}_{22}, \mathbb{F}_{22} \rangle$	$\cdots$	$\langle \mathbb{T}_{2n}, \mathbb{I}_{2n}, \mathbb{F}_{2n} \rangle$
$\cdots$	$\cdots$	$\cdots$	$\cdots$	$\cdots$
$\mathring{A}_m$	$\langle \mathbb{T}_{m1}, \mathbb{I}_{m1}, \mathbb{F}_{m1} \rangle$	$\langle \mathbb{T}_{m2}, \mathbb{I}_{m2}, \mathbb{F}_{m2} \rangle$	$\cdots$	$\langle \mathbb{T}_{mn}, \mathbb{I}_{mn}, \mathbb{F}_{mn} \rangle$

TABLE 15.2

Connections between Attributes and Alternatives

$\mathfrak{R}_2$	$\mathcal{B}_1$	$\mathcal{B}_2$	$\cdots$	$\mathcal{B}_l$
$\mathbb{C}_1$	$\langle \mathbb{T}_{11}, \mathbb{I}_{11}, \mathbb{F}_{11} \rangle$	$\langle \mathbb{T}_{12}, \mathbb{I}_{12}, \mathbb{F}_{12} \rangle$	$\cdots$	$\langle \mathbb{T}_{1l}, \mathbb{I}_{1l}, \mathbb{F}_{1l} \rangle$
$\mathbb{C}_2$	$\langle \mathbb{T}_{21}, \mathbb{I}_{21}, \mathbb{F}_{21} \rangle$	$\langle \mathbb{T}_{22}, \mathbb{I}_{22}, \mathbb{F}_{22} \rangle$	$\cdots$	$\langle \mathbb{T}_{2l}, \mathbb{I}_{2l}, \mathbb{F}_{2l} \rangle$
$\cdots$	$\cdots$	$\cdots$	$\cdots$	$\cdots$
$\mathbb{C}_n$	$\langle \mathbb{T}_{n1}, \mathbb{I}_{n1}, \mathbb{F}_{n1} \rangle$	$\langle \mathbb{T}_{n2}, \mathbb{I}_{n2}, \mathbb{F}_{n2} \rangle$	$\cdots$	$\langle \mathbb{T}_{nl}, \mathbb{I}_{nl}, \mathbb{F}_{nl} \rangle$

TABLE 15.3

Trigonometric Similarity Measures ($\mathcal{TSM}$)

$\mathfrak{R}_1 \times \mathfrak{R}_2$	$\mathcal{B}_1$	$\mathcal{B}_2$	$\cdots$	$\mathcal{B}_l$
$\mathring{A}_1$	$\mathcal{TSM}_{11}$	$\mathcal{TSM}_{12}$	$\cdots$	$\mathcal{TSM}_{1l}$
$\mathring{A}_2$	$\mathcal{TSM}_{21}$	$\mathcal{TSM}_{22}$	$\cdots$	$\mathcal{TSM}_{2l}$
$\cdots$	$\cdots$	$\cdots$	$\cdots$	$\cdots$
$\mathring{A}_m$	$\mathcal{TSM}_{m1}$	$\mathcal{TSM}_{m2}$	$\cdots$	$\mathcal{TSM}_{ml}$

Step 4: Order the alternatives

Arrange the scores in Table 15.3 in descending order based on the individual pupils' ordering of the alternatives.

Step 5: Make the decision

The highest score reflects the best alternative.

15.4 APPLYING THE TRIGONOMETRIC SIMILARITY MEASURES

After students in India pass their secondary exams, they must consider a higher secondary education program; after grade 12, students pursue coursework aimed at employment. Most students in this critical stage of their studies make hasty, confused decisions that they subsequently regret. Here, we describe how to apply the spherical neutrosophic set trigonometric similarity measures to solve the real-world multicriteria problem of how students should choose their higher education and career paths.

$Å = \{Å_1 = Raj, Å_2 = Naruva, Å_3 = Devi\}$ is the set of three pupils. $B = \{B_1, B_2, B_3, B_4\}$ is the set of instruction streams the students are considering: B_1 = science, B_2 = humanities or arts, B_3 = business, and B_4 = career training. $\mathbb{C} = \{\mathbb{C}_1, \mathbb{C}_2, \mathbb{C}_3, \mathbb{C}_4, \mathbb{C}_5\}$ is a set of qualities that the students could possess: $\mathbb{C}_1$ = basic scientific and mathematical understanding, $\mathbb{C}_2$ = depth of language, $\mathbb{C}_3$ = successful performance on secondary exams, $\mathbb{C}_4$ is concentration and $\mathbb{C}_5$ is working diligently respectively. Our approach is helps to assess the pupils and determine which educational track is best for them based on $SP\hbar N\_S$'s and given the following Table 15.1 and Table 15.2. Using the steps below, the decision-making process is demonstrated.

Step 1: The connection between pupils' qualities and their own in the form $SP\hbar N\_S$ sets is presented in Table 15.4.

Step 2: The association between a student's characteristics and academic pathways in the form $SP\hbar N\_S$ sets is presented in Table 15.5.

Step 3: Calculate all three trigonometric similarity measures for the data in Tables 15.4 and 15.5 using Python. The obtained scores are presented in Table 15.6.

Step 4: The highest trigonometric measures for each student, with their respective education streams, are presented in Table 15.7. Figures 15.1 through 15.3 graph the tangents, cotangents, and cosines for each student to their instruction stream.

TABLE 15.4

Connections between Students and Attributes

$\Re_1$	∂_1	∂_2	∂_3	∂_4	∂_5
$Å_1 = Raj$	(0.8,0.2,0.2)	(0.7,0.2,0.1)	(0.6,0.3,0.3)	(0.8,0.3,0.2)	(0.5,0.4,0.2)
$Å_2 = Naruva$	(0.7,0.1,0.3)	(0.6,0.3,0.2)	(0.7,0.2,0.1)	(0.7,0.4,0.1)	(0.8,0.0,0.2)
$Å_3 = Devi$	(0.5,0.3,0.3)	(0.6,0.3,0.2)	(0.7,0.3,0.1)	(0.8,0.0,0.2)	(0.8,0.3,0.1)

TABLE 15.5

Associations between Attributes and Education Streams

$\Re_2$	B_1	B_2	B_3	B_4
∂_1	(0.8,0.3,0.1)	(0.9,0.1,0.2)	(0.7,0.3,0.2)	(0.8,0.2,0.2)
∂_2	(0.5,0.4,0.1)	(0.6,0.3,0.2)	(0.7,0.4,0.0)	(0.9,0.1,0.1)
∂_3	(0.6,0.4,0.1)	(0.7,0.3,0.3)	(0.7,0.3,0.2)	(0.9,0.1,0.2)
∂_4	(0.6,0.2,0.2)	(0.7,0.2,0.2)	(0.8,0.2,0.1)	(0.5,0.4,0.2)
∂_5	(0.7,0.3,0.2)	(0.6,0.3,0.2)	(0.7,0.4,0.1)	(0.6,0.4,0.1)

TABLE 15.6 $\mathcal{TSM}$

	$\mathcal{B}_1$			$\mathcal{B}_2$			$\mathcal{B}_3$			$\mathcal{B}_4$		
	TAN_S	COT_S	COS_S	TAN_S	COT_S	COS_S	TAN_S	COT_S	COS_S	TAN_S	COT_S	COS_S
$\mathring{A}_1 = Raj$	0.921277	0.854486	0.965121	0.947578	0.90085	0.987323	0.937075	0.881875	0.979068	0.92108	0.858162	0.968906
$\mathring{A}_2 = Naruva$	0.900147	0.819011	0.95011	0.926448	0.865914	0.961948	0.90534	0.828978	0.945977	0.878754	0.786354	0.939132
$\mathring{A}_3 = Devi$	0.915945	0.8464	0.963021	0.92097	0.859169	0.960349	0.93178	0.873463	0.973621	0.862854	0.760512	0.916757

TABLE 15.7
The Highest Trigonometric Similarity Measures for Each Student

Student/Measure	TAN_S	COT_S	COS_S
$\mathring{A}_1 = Raj$	0.94757 for $\mathcal{B}_2$	0.90085 for $\mathcal{B}_2$	0.987323 for $\mathcal{B}_2$
$\mathring{A}_2 = Naruva$	0.92644 for $\mathcal{B}_2$	0.865914 for $\mathcal{B}_2$	0.961948 for $\mathcal{B}_2$
$\mathring{A}_3 = Devi$	0.93177 for $\mathcal{B}_3$	0.873463 for $\mathcal{B}_3$	0.973621 for $\mathcal{B}_3$

Based on all three trigonometric measures in the measures and the three figures, the three students should choose the following instruction paths for their higher education: Raj and Naruva should select arts, $\mathcal{B}_2$, and Devi should select business, $\mathcal{B}_3$.

15.5   CONCLUSION

In this chapter, we have presented the two fundamentals of data mining: similarity measures and machine learning in uncertainty contexts, with a focus on spherical

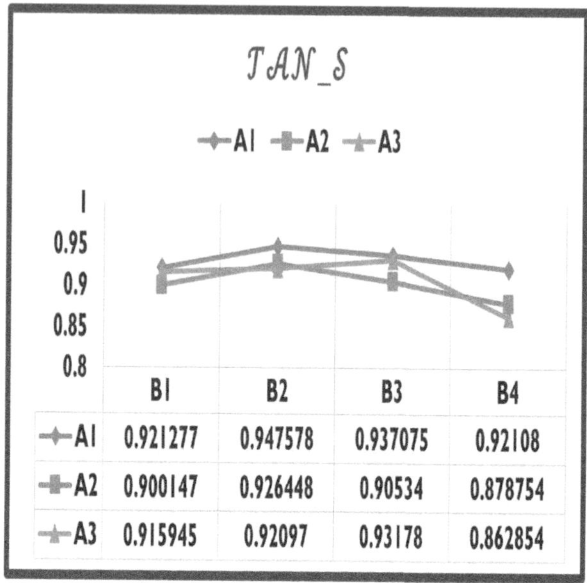

FIGURE 15.1 Line graph of TAN_S.

Notes: Å1 = Raj, Å2 = Naruva, Å3 = Devi; B_1= science, B_2 = humanities or arts, B_3 = business, and B_4 = career training.

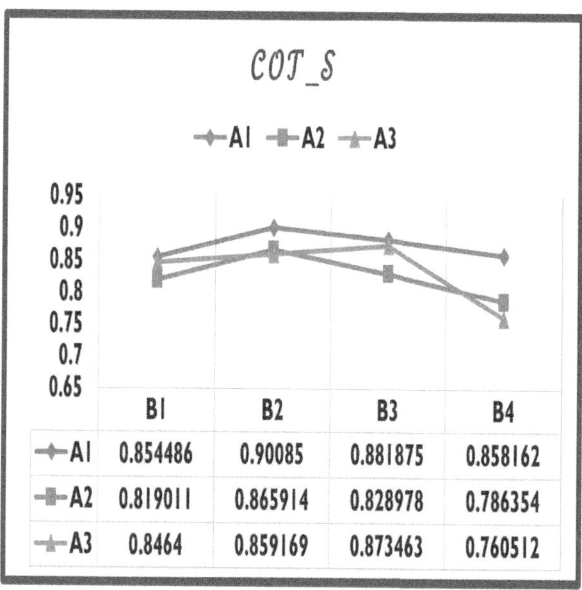

FIGURE 15.2 Line graph of COT_S.

Notes: Å1 = Raj, Å2 = Naruva, Å3 = Devi; B_1 = science, B_2 = humanities or arts, B_3 = business, and B_4 = career training.

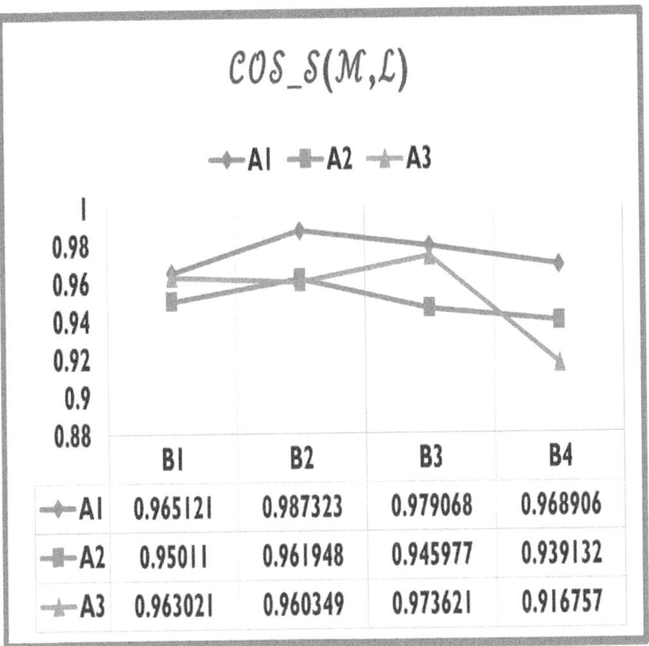

	BI	B2	B3	B4
AI	0.965121	0.987323	0.979068	0.968906
A2	0.95011	0.961948	0.945977	0.939132
A3	0.963021	0.960349	0.973621	0.916757

FIGURE 15.3 Line graph of COS_S.

Notes: $Å^1$ = Raj, $Å^2$ = Naruva, $Å^3$ = Devi; B_1 = science, B_2 = humanities or arts, B_3 = business, and B_4 = career training.

neutrosophic sets. Specifically, we introduced trigonometric similarity measures—tangent, cotangent, and cosine—and demonstrated how to apply SNSs and similarity measures to a real-world multicriteria decision task. This approach will be valuable for mining large amounts of indeterminate data. In future developments in machine learning, this process will be applicable in the interval SNS environment.

REFERENCES

[1] K. Atanassov, Intuitionistic Fuzzy Sets, Fuzzy Sets and Systems, Vol. 20, PP. 87–96, 1986. DOI: 10.1016/S0165-0114(86)80034-3

[2] R. Jansi, K. Mohana, F. Smarandache, Correlation Measure for Pythagorean Neutrosophic Sets with T and F as Dependent Neutrosophic Components, Neutrosophic Sets and System, Vol. 30, PP. 202–212, 2019. DOI: 10.5281/zenodo.3569785

[3] R. Jansi, K. Mohana, Tangent and Cotangent Similarity Measures of Pythagorea Neutrosophic Sets with T and F are Dependent Neutrosophic Components, Infokara Research, Vol. 8, PP. 657–671, 2019. DOI: 16.10089.IR.2019.V8I8.285311.2769

[4] N. Devi Rathinam, S. Rasappan, A. J. Obaid, Novel on Digital Neutrosophic Topological Spaces, Proceedings of 2nd International Conference on Mathematical Modeling and Computational Science, Vol. 29, PP. 213–219, 2022. DOI: 10.1007/978-981-19-0182-9_22.

[5] R. Narmada Devi, Y. Parthiban, Decision Making Process over Neutrosophic Pythagorean Soft Sets Using Measure of Correlation, Southeast Europe Journal of Soft Computing, Vol. 12, No. 2, September 2023, PP. 6–9. DOI: 10.21533/scjournal.v12i2.276

[6] R. Murugesan, Y. Parthiban, R. Narmada Devi, K. Raja Mohan, S. Kumar Kumaravel, A Comparative Study of Fuzzy Cognitive Maps and Neutrosophic Cognitive Maps on Covid Variants, Neutrosophic Sets and Systems, Vol. 55, PP. 329–343, 2023. https://doi.org 10.5281/zenodo.7832763

[7] T. Sanapati, R. Yager, Fermatean Fuzzy Sets, Journal of Ambient Intelligence and Humanized Computing, Vol. 11, PP. 663–674, 2019. DOI: 10.1007/s12652-019-01377-0

[8] S. Kumar Sahool, S. Shubhra Goswami, A Comprehensive Review of Multiple Criteria Decision-Making (MCDM) Methods: Advancements, Applications, and Future Directions, Decision Making Advances, Vol. 1, No. 1, PP. 25–48, 2023. https://doi.org/10.31181/dma1120237

[9] Smarandache (editor), Proceedings of the First International Conference on Neutrosophy, NeutrosophicLogic, Set, Probability and Statistics, University of New Mexico, Gallup, NM 87301, USA (2002).

[10] R. R. Yager, Pythagorean Membership Grades in Multicriteria Decision Making. IEEE Trans Fuzzy System, Vol. 22, PP. 958–965, 2014. DOI: 10.1109/TFUZZ.2013.2278989

[11] L. A. Zadeh, Fuzzy Sets, Information and Control, Vol. 8, PP. 338–353, 1965. DOI: 10.1016/S0019-9958(65)90241-X.

16 A Study on Machine Learning Twig Graphs on the Hyper Wiener Index of Complete Graph

Kannan A, Meenakshi Annamalai,
Rajkumar. P, and Kujani T

16.1 INTRODUCTION

In a chemical graph, atoms and bonds represent vertices and edges, respectively, and molecules' physical properties can be described with graph theory descriptors or indices. For instance, $W_{TG}(G)$ is a graph-based measure of the boiling point of a molecule by adding up the lengths of all shortest pathways. In this chapter, we introduce the concept of power of three acyclic graphs and proves that they have a graceful labeling property. We also show that every binomial and power of the three acyclic graphs has a cube sum labeling (Meenakshi and Kannan, 2022).

The study of the impact of pure structural variation on the boiling point of paraffins was of theoretical interest early on (Wiener, 1947). The Wiener index was introduced in the mathematical literature in 1976 (Entringer et al., 1976) and was first used to represent the properties of acyclic compounds called alkanes. This led to the development of the theory of Wiener indices of trees (Dobrynin et al., 2002), and later researchers investigated the link between a tree's hyper-Wiener index and the Wiener indices of T and certain of its pieces (Gutman and Furtula, 2003).

The hyper-Wiener indices of the Cartesian product, composition, join, and disjunction of graphs are calculated (Khalifeh et al., 2008), and trees with the highest and lowest Wiener indices are identified based on the number of vertices and degree sequence (Su et al., 2023). Machine learning has gained much interest as artificial intelligence technology has advanced, with numerous methods being suggested and effectively used in a variety of real-world applications (Xia et al., 2020). Let G_{TG} be an undirected connected graph with vertex set $V_{TG}(G)$ and edge set $E_{TG}(G)$ such that every edge of G_{TG} is related to a pair of vertices in G_{TG}. Since distance matrix D_{TG} $\left(D_{TG} = D_{TG}(G)\right)$ and graph G_{TG} have n vertices, and the ith row's sum is $W_{TG}i$, the Wiener index of G is defined by $W_{TG}(G) = \frac{1}{2}\sum_{i} W_{TG}i$.

DOI: 10.1201/9781032635170-16

16.2 PRELIMINARY DEFINITIONS

16.2.1 DEFINITION

A simple graph has edges that connect every pair of unique vertices. It is represented by K_n, where n > 2 vertices. Figure 16.1 shows an example of a complete graph with only five vertices.

16.2.2 DEFINITION

A path graph consists of vertices and edges with the property that every vertex in the sequence is adjacent to the vertex next to it. It is denoted by P_n. Figure 16.2 shows a path graph P_4.

16.2.3 DEFINITION

A graph $P_m \odot K_1$ is obtained from a path graph, $m \geq 3$ by appending m pendent vertices to all the vertices of path graph Pn.

Let $G_{TG} = P_m \odot K_1$ be a graph obtained from P_n with m vertices where $m \geq 3$, and let $V_{TG}(G) = \{u_1, u_2, ..., u_m, v_1, v_2, ...v_m\}$ and $E_{TG}(G) = \{u_i u_i + 1; 1 \ i \ m - 1\} \cup \{u_1 v_1, u_2 v_2, ..., u_m v_m\}$ be, respectively, the vertex set and edge set of $P_m \odot K_1$ (Figure 16.3).

16.2.4 DEFINITION

A twig graph $P_m \odot 2K_1$ is obtained from a Pm where $m \geq 3$ by appending $2m$ pendent vertices to all the vertices of the graph.

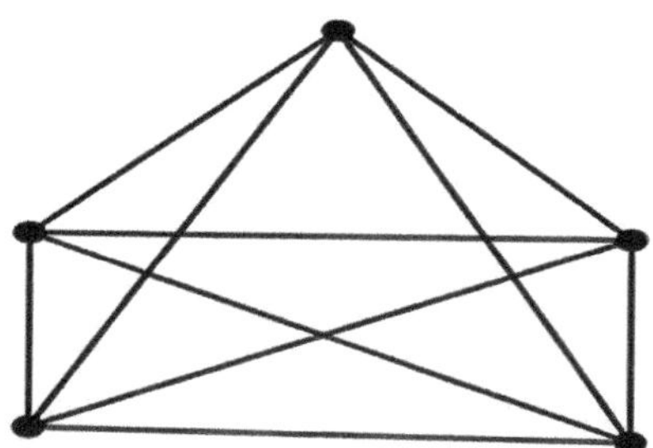

FIGURE 16.1 A complete graph with K_5 vertices.

FIGURE 16.2 A path graph with four vertices, P_4.

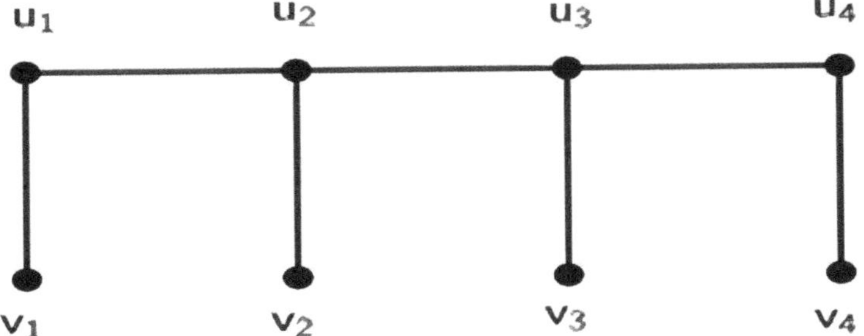

FIGURE 16.3   Graph of $P_4 \odot K_1$.

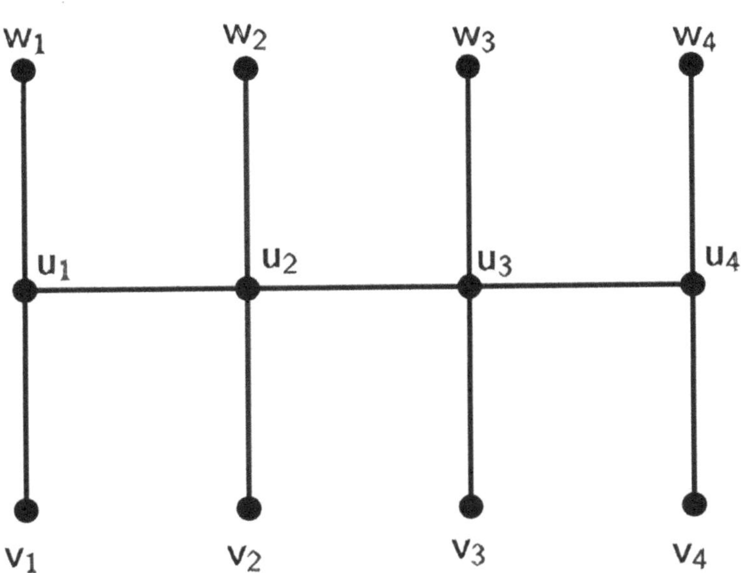

FIGURE 16.4   A twig graph T_n.

Let $T_m = P_m \odot 2K_1$ be graph P_m with $2m$ vertices where $m \geq 3$, and let $V_{TG}(G) = \{u_1, u_2, \ldots, u_n, v_1, v_2, \ldots v_n, w_1, w_2, \ldots, w_n\}$ & $E_{TG}(G) = \{u_i u_i + 1; 1 \leq i \leq m-1\}$ $\cup \{u_1 v_1, u_1 w_1, u_2 v_2, u_2 w_2, \ldots u_m v_m, u_m w_m\}$ be, respectively, the vertex set and edge set of T_m Figure 16.4).

16.2.5   DEFINITION

A graph obtained connecting complete graph K_m and path graph $P_m \odot K_1$ is denoted by $K_m \odot P_m \odot K_1$ (Figure 16.5).

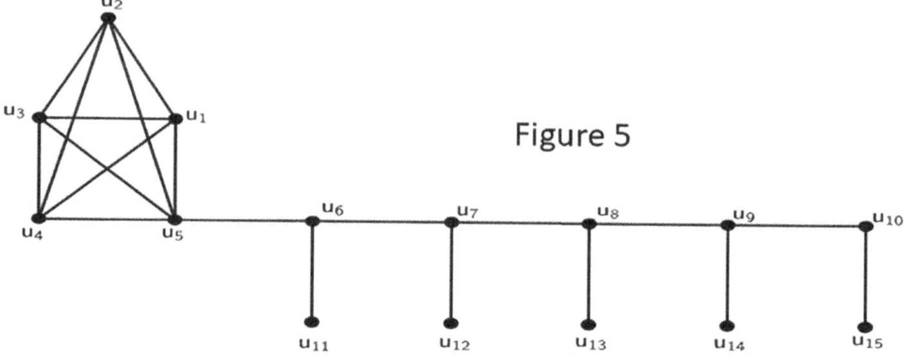

FIGURE 16.5 Graph of $K_m \odot P_m \odot K_1$.

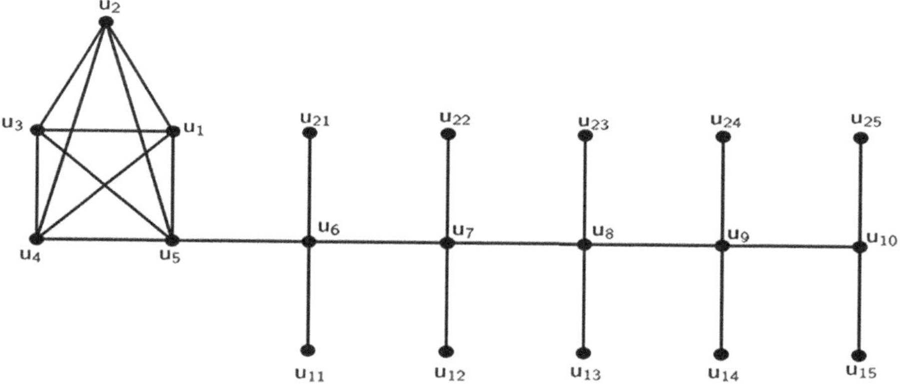

FIGURE 16.6 Graph of $K_5 \odot T_5$.

16.2.6 DEFINITION

A graph obtained by connecting K_m with $m \geq 3$ and T_m is denoted by $K_m \odot T_m$ (Figure 16.6).

16.2.7 DEFINITION

A graph's $W_{TG}(G)$ is defined as the total of the distances between all vertex pairs.

$$W_{TG}(G) = \frac{1}{2} \sum_{\{u,v\} \in V(G)} d(u,v)$$

16.2.8 Definition

A graph's $HW_{TG}(G)$ can be calculated by adding the graph's wiener index and half of the total squared distances between all of the graph's vertex pairs:

$$HW_{TG}(G) = \frac{1}{2} \sum_{\{u,v\} \in V(G)} d(u,v) + \frac{1}{2} \sum_{\{u,v\} \in V(G)} d(u,v)^2$$

16.3 RESULTS AND DISCUSSION

16.3.1 Hyper-Wiener Index of a Graph

16.3.1.1 Theorem 1

Let $K_m @ (P_m \odot K_1)$ be a graph with $m \geq 3$ vertices. Then the hyper-Wiener index of the graph $K_m @ (P_m \odot K_1)$ is

$$HW_{TG}\left(K_m @ (P_m \odot K_1)\right) = \frac{m\left(m^3 + 21m^2 + 144m - 378\right)}{81}$$

Proof

Theorem is proved by mathematical induction on the $m \geq 3$ vertices. If $m = 3$, then the total number of vertices of $K_3 \odot (P_3 \odot K_1)$ is $n = 3m$, that is, 9. Then, the hyper-Wiener index of $K_3 \odot (P_3 \odot K_1)$ is 372 as follows:

$$HW_{TG}\left(K_m \odot (P_m \odot K_1)\right) = \frac{m\left(m^3 + 21m^2 + 144m - 378\right)}{81}$$

$$= \frac{9\left[(9)^3 + 21(9)^2 + 144(9) - 378\right]}{81} = 372$$

Consequently, for $m = 3$ vertices, the theorem holds valid. Assume that for $m = s$, the theorem holds valid.

$$HW_{TG}\left(K_m \odot (P_m \odot K_1)\right) = \frac{m\left(m^3 + 21m^2 + 144m - 378\right)}{81}$$

To Prove

For $m = s + 1$, the theorem holds valid. That is, to prove

$$HW_{TG}\left(K_{s+1} \odot (P_{s+1} \odot K_1)\right) = \frac{(s+1)\left((s+1)^3 + 21(s+1)^2 + 144(s+1) - 378\right)}{81}$$

Now, let $m = s + 1$. Then

$$\mathrm{HW}_{\mathrm{TG}}\left(K_{s+1}\Theta\left(P_{s+1}\Theta K_1\right)\right) = \mathrm{HW}\left(K_s\Theta\left(P_s\Theta K_1\right)\right) + \sum_{k=1}^{m} d\left(U_{m+s}, V_i\right) + \sum_{k=1}^{m} d\left(U_{m+s}, V_i\right)^2,$$

$$S = 1, 2, 3$$

$$= \frac{s\left(s^3 + 21s^2 + 144s - 378\right)}{81} + \sum_{k=1}^{m} d\left(U_{m+s}, V_i\right) + \sum_{s=1}^{m} d\left(U_{m+s}, V_i\right)^2$$

$$= \frac{(s+1)\left((s+1)^3 + 21(s+1)^2 + 144(s+1) - 378\right)}{81}$$

Hence, for $m = s + 1$, the theorem holds valid for every value of m. Hence, the hyper-Wiener index of $K_m \odot \left(P_m \odot K_1\right)$ is $HW_{TG}\left(K_m \odot \left(P_m \odot K_1\right)\right) = \frac{m\left(m^3 + 21m^2 + 144m - 378\right)}{81}$

16.3.1.2 Theorem 2

Let $K_m \odot T_m$ be a graph with n vertices where $m \geq 3$. Then the Hyper Wiener index of the graph $K_n @ T_n$ is

$$\mathrm{HW}_{\mathrm{TG}}\left(K_m\Theta T_m\right) = \frac{m\left(7m^3 + 216m^2 + 2192m - 6528\right)}{1024}$$

Proof

Theorem is proved by mathematical induction on m ≥ 3 vertices. If m = 3, then the total number of vertices of $K_m \odot T_m$ is m = 4n, or 12. The hyper-Wiener index of $K_n @ T_n$ is 738.

$$\mathrm{HW}_{\mathrm{TG}}\left(K_m\Theta T_m\right) = \frac{m\left(7m^3 + 216m^2 + 2192m - 6528\right)}{1024}$$

$$\mathrm{HW}_{\mathrm{TG}}\left(K_m\Theta T_m\right) = \frac{12\left(7(12)^3 + 216(12)^2 + 2192(12) - 6528\right)}{1024} = 738$$

Consequently, for $m = 3$ vertices, the theorem holds valid. Assume that for $m = s$, the theorem holds valid.

$$\mathrm{HW}_{\mathrm{TG}}\left(K_m\Theta T_m\right) = \frac{s\left(7s^3 + 216s^2 + 2192s - 6528\right)}{1024}$$

To Prove

For $m = s + 1$, the theorem holds valid. That is.

to prove $\mathrm{HW_{TG}}\left(K_m\Theta T_m\right) = \dfrac{(s+1)\left(7(s+1)^3 + 216(s+1)^2 + 2192(s+1) - 6528\right)}{1024}$

Now, let $m = s + 1$. Then

$$\mathrm{HW_{TG}}\left(K_{s+1}\Theta T_{s+1}\right) = \mathrm{HW_{TG}}\left(K_s\Theta T_s\right) + \sum_{s=1}^{m} d\left(U_{m+s}, V_i\right) + \sum_{k=1}^{m} d\left(U_{m+s}, V_i\right)^2$$

$$= \frac{s\left(7s^3 + 216s^2 + 2192s - 6528\right)}{1024} + \sum_{s=1}^{m} d\left(U_{m+s}, V_i\right) + \sum_{s=1}^{m} d\left(U_{m+s}, V_i\right)^2 \quad s = 1,2,3$$

$$= \frac{(s+1)\left(7(s+1)^3 + 216(s+1)^2 + 2192(s+1) - 6528\right)}{1024}$$

Hence, for $m = s + 1$, the theorem holds valid for every value of m. Hence, the hyper-Wiener index of $K_m \odot T_s$ is

$$\mathrm{HW_{TG}}\left(K_m\Theta T_m\right) = \frac{m\left(7m^3 + 216m^2 + 2192m - 6528\right)}{1024}$$

16.4   PERFORMANCE ANALYSIS

16.4.1   The C++ Coding for $\mathrm{HW_{TG}}\left(K_m\Theta\left(P_m\Theta K_1\right)\right)$

```cpp
// Graph (Km⊖(Pm⊖K1))
#include<iostream.h>
#include<math.h>
#include<conio.h>
main(void)
  {
  long int i,n,j,z,s=0,t,b,h=0,a=0;
  clrscr();
  cout<<"\n Enter No. of vertices in a Graph:\t";
  cin>>n;
  for(i=1;i<3*n;i++,s=0,t=1,b=2)
  {
  cout<<"\n"<<i<<"\t";
    for(z=2;z<=i;z++)
      cout<<" ";
    for(j=i+1;j<=3*n;j++)
      {
      if(j<=n)
```

```cpp
        {
          s=1;
          cout<<" "<<s;
          a=a+s;
          h=h+s*s;
        }
        if(j>n && j<=2*n+1)
        {
          s=s+1;
          cout<<" "<<s;
          a=a+s;
          h=h+s*s;
          }
        if(j>2*n+1)
           {
           if(i>2*n)
           {
             b=b+1;
             cout<<" "<<b;
             a=a+b;
             h=h+b*b;
           }
           else
           {
           if(s==1)
               {
               t=t+1;
               cout<<" "<<t;
               a=a+t;
               h=h+t*t;
               }
           else
           {
             s=s-1;
             cout<<" "<<s;
             a=a+s;
             h=h+s*s;
           }
           }
           }
        }
     return 0;
   }
cout<<"\n\nWiemer Index of G \t:\t"<<a;
cout<<"\n\nHyper Wiener Index of G is:\t"<<a+h;
  getch();
}
```

16.4.2 THE C++ CODING FOR $HW_{TG}\left(K_m \Theta T_m\right)$

```cpp
// Graph (Km⊖Tm)
#include<iostream.h>
#include<conio.h>
void main(void)
  {
  int n,i,j,s=0,a=0,t,b,h=0;
  clrscr();
  cout<<"\n Enter No. of vertices in a Graph:\t";
  cin>>n;
  for(i=1;i<4*n;i++,s=0,t=1,b=2)
    {
    for(j=i+1;j<=4*n;j++)
    {
      if(j<=n)
      {
        s=1;
        cout<<" "<<s;
        a=a+s;
        h=h+s*s;
      }
      if(j>n && j<=2*n+1)
      {
        s=s+1;
        cout<<" "<<s;
        a=a+s;
        h=h+s*s;
      }
      if(j>2*n+1)
      {
        if(i>2*n)
        {
          if(j%2==1) b=b+1;
          cout<<" "<<b;
          a=a+b;
          h=h+b*b;
        }
        else
        {
        if(s==1)
          {
        if(j%2==1) t=t+1;
          cout<<" "<<t;
          a=a+t;
          h=h+t*t;
          }
```

```
    else
      {
      if(j%2==1)  s=s-1;
      cout<<" "<<s;
      a=a+s;
      h=h+s*s;
      }
    }
    }
  }
  cout<<"\n";
  }
cout<<"\n\nWiemer Index of G \t:\t"<<a;
cout<<"\n\nHyper Wiemer Index of G:\t"<<h+a;
  getch();
}
```

16.5 CONCLUSION

Once a model is trained and evaluated satisfactorily, deploy it for predicting the hyper-Wiener index of new, unseen graphs. This predictive capability can be useful in various applications, including chemical graph theory, network analysis, computational biology, radar tracking, remote control, radio-astronomy and communication networks, etc. Estimation of Wiener indices for other trees is under investigation. This will entail preparing data sets in which each data point represents a graph, and the target variable will be the hyper-Wiener index of that graph; next will be generating a diverse set of graphs with varying structural properties to effectively train and evaluate models.

REFERENCES

Annamalai, M. and K. Adhimoolam, A Note on Graceful Trees. AIP Conference Proceedings. 2516(1) (2022). AIP Publishing. https://doi.org/10.1063/5.0108571

Dobrynin, A. A., I. Gutman, S. Klavzar and P. Zigert, Wiener Index of Hexagonal Systems, Acta appl. Math. 72 (2002), 247–294. https://doi.org/10.1023/A:1010767517079

Entringer, R. C., D. E. Jackson and D. A. Snyder, Distance in Graphs, Czechoslovak Math. J. 26 (1976), 283–296. http://dx.doi.org/10.21136/CMJ.1976.101401

Gutman, I. and B. Furtula, Hyper-Wiener index vs. Wiener index. Two Highly Correlated Structure-Descriptors, Monatshefte für Chemie/Chemical Monthly 134 (2003), 975–981. https://doi.org/10.1007/s00706-003-0003-7

Khalifeh, M. H., Hassan Yousefi-Azari, and Ali Reza Ashrafi. The hyper-Wiener Index of Graph Operations, Computers & Mathematics with Applications, 56(5) (2008), 1402–1407. http://dx.doi.org/10.1016/j.camwa.2008.03.003

Meenakshi, A., J. Senbagamalar and A. Kannan. Application of Intuitionistic Fuzzy Network Using Efficient Domination, Fuzzy Logic Applications in Computer Science and Mathematics. (2023a), 213–232. https://doi.org/10.1002/9781394175130.ch14

Meenakshi, A., et al. A Comparative Study of Fuzzy Domination and Fuzzy Coloring in an Optimal Approach. Mathematics. 11.18 (2023b), 4019. https://doi.org/10.3390/math11184019

Meenakshi, A., et al. Efficient Graph Network Using Total Magic Labeling and Its Applications. Mathematics. 11.19 (2023c): 4132. https://doi.org/10.3390/math11194132

Su, Z., Z. Tang and H. Deng. Extremal Wiener Index of Graphs with Given Number of Vertices of Odd Degree. MATCH Communications in Mathematical and in Computer Chemistry. 89.2 (2023), 503–516. https://doi.org/10.46793/match.89-2.503S

Wiener, H. Structural Determination of Paraffin Boiling Points, J. Amer. Chem. Soc. 69 (1947), 17–20. http://dx.doi.org/10.1021/ja01193a005

Xia, X. Y., H. Zhang, W. H. Li, and G. Y. Wang, A Novel Rough Set Algorithm for Fast Adaptive Attribute Reduction in Classification, IEEE Trans. Knowl. Data Eng. 34 (2020), 1231–1242. https://doi.org/10.1109/TKDE.2020.2997039

17 Enhancing Multicriteria Decision-Making through Cryptographic Security Systems

Meenakshi Annamalai, S. Dhanushiya,
O. Mythreyi, and Shivangi Mishra J

17.1 INTRODUCTION

The Iranian scientist Lotfi A. Zadeh developed the mathematical framework known as fuzzy set theory in 1965 as a way to address ambiguity and uncertainty in data and information. Mathematicians such as George Cantor created traditional set theory, which works with crisp or classical sets, in which a member either belongs to a set or does not. But many notions in the real world are neither precisely defined nor binary. In order to overcome this, fuzzy set theory allows objects to have varying degrees of membership in a set as opposed to a strict yes or no membership. Meenakshi et al. [1,2] combined cubic fuzzy graphs and fuzzy graph structures and used hypergraphs to model social networks, providing a rich representation and intuitive modeling language for exploring polyadic interactions between entities.

The RSA (Rivest-Shamir-Adleman) cryptosystem [3–5] is one of the most widely used public key cryptography algorithms; it was introduced in 1977 by Ron Rivest, Adi Shamir, and Leonard Adleman and is based on the mathematical properties of large prime numbers and their difficulty in factorization. It involves the use of a pair of mathematically related keys: a public and a private key. The private key is required for decryption; the public key is utilized for encryption. The difficulty of factoring the product of two huge prime numbers, which serve as the foundation of the keys, is the basis for RSA's security. RSA has been a well-studied and widely used encryption algorithm for several decades. Researchers have conducted extensive studies on various aspects of RSA cryptography, focusing on its security, efficiency, and potential vulnerabilities [6,7].

Pius et al. [8–10] proposed an enhanced version of RSA to increase security for online transactions by replacing the number n in the original RSA algorithm with the new number f. This to the original algorithm eliminates the need to transfer the product of two random prime numbers in the public key, making it difficult for intruders to guess n and ensuring that the encrypted message remains safe from attackers. The

DOI: 10.1201/9781032635170-17

proposed algorithm follows asymmetric key cryptography, which further enhances the security of the system. The authors also present a comparative analysis of the proposed algorithm with the traditional RSA algorithm, highlighting the improvements in security achieved by the enhanced version.

Meenakshi et al. [11] explored the use of effective domination and fuzzy vertex coloring in the context of cryptography and improved the security and effectiveness of cryptographic protocols, and Henry Rowland [12] performed integer factorization and primality testing to establish the use of RSA cryptosystems for data security. He determined that continued study into integer factorization methods like Shor's factoring algorithm and technological advancements, particularly quantum computing, could pose a significant danger to the security of the RSA cryptosystem.

Ivy et al. presented a novel method that speeds up the operations of key generation and decryption by using three prime integers and the Chinese reminder theorem (CRT) for the RSA cryptosystem, and the third integer made RSA more sophisticated and secure. The RSA algorithm includes message encryption, message decryption, and key generation, and Ivy et al. [13] proposed a modified RSA cryptosystem based on n prime numbers to provide maximum security for data over the network. By using three prime integers rather than two, the suggested technique seeks to minimize the time required for key generation and raise the complexity of n analysis [14–16].

Meenakshi et al. [17] presented a new combinatorial technique to encrypt and decrypt secret numbers using labeled graphs and domination theory. The technique involves splitting the secret number into partitions and assigning them to efficient dominating vertex sets and edges in a graph. Tan et al. [18] sought strategies for improving the synergy between multicriteria decision making (MCDM) and building information modelling (BIM) in the architecture, engineering, and construction industry. Combining MCDM with BIM, they identified the major application domains as sustainability, retrofit, supplier selection, safety, and constructability. Tan et al. also discuss five strategies for improving the synergy between MCDM and BIM.

Aliamis et al. [19] proposed a new RSA cryptosystem based on fuzzy set theory, using triangular fuzzy numbers for encryption and decryption and showed that fuzzy set theory is suitable for securing other cryptosystems and provides difficulty for attackers in communication. Kamardan et al. [20] proposed a modified multiple prime RSA consisting of three phases: key generation, encryption, and decryption. The decryption phase utilized the CRT to expedite computation and found that decryption using multiple prime RSA was nearly four times faster than the standard RSA. Additionally, using multiple prime numbers and additional secret keys increases the difficulty of breaking the security of the algorithm.

It is an intriguing idea to combine conventional RSA encryption with fuzzy logic, more particularly, with trapezoidal fuzzy numbers (TFNs). Using the CRT allows for efficient computation by breaking down the RSA problem into smaller modular equations, reducing computational complexity, and enabling faster encryption and decryption. Furthermore, incorporating TFNs enables a more comprehensive representation of uncertainty in the encryption process, enhancing security against attacks that exploit vulnerabilities in standard RSA methods [12,21,22].

In this chapter, we present a detailed theoretical framework that integrates the CRT and TFNs into the RSA encryption algorithm. The main aim of this study is to maintain secure encryption while adding fuzziness or unpredictability to key generation considering uncertainties in prime number selection or exponent choices. Through mathematical formulations and algorithmic procedures, integrating these methodologies improved the computational efficiency and security in encrypting sensitive data.

17.2   KEY GENERATION

1. Choose two distinct prime numbers.

    Select two large prime numbers, often denoted as p and q. These numbers should be randomly chosen and kept secret. Their multiplication will be part of the RSA algorithm.
2. Calculate the modulus.

    Compute modulus n by multiplying the two prime numbers: $n = p \times q$. The modulus is a part of both the public and private keys.
3. Compute Euler's totient function.

    Calculate Euler's totient function $\phi(n)$, which is the number of positive integers less than n that are coprime with n. For two distinct primes p and q,
    $\phi(n) = (p-1) \times (q-1)$
4. Choose an exponent.

    Select an integer e that is relatively prime to $\phi(n)$. This e will be part of the public key.
5. Calculate the private exponent.

    Compute private exponent d as the modular multiplicative inverse of e modulo $\phi(n)$. In other words, find a value for d such that $(d \times e) \bmod \phi(n) = 1$. The value of d is a crucial part of the private key.
6. Create the public key.

    The public key consists of the modulus (n) and the exponent (e). This is the key that will be distributed to others for encryption.
7. Create the private key.

    The private key consists of the modulus (n) and the private exponent (d). This key must be kept secret and should not be shared.
8. Transform the message into cipher text.

    To enable encryption, the message must be transformed into a numerical representation. This is commonly referred to as "padding" and entails numeric value conversion from the plaintext message.

17.2.1   ENCRYPTION

The sender uses the recipient's public key to encrypt the message (m). The following is how the encryption works:

a. Acquire the public key of the receiver, which consists of the encryption exponent e and modulus n.
b. Using the selected encoding scheme, represent the message as a number m.
c. Utilize modular exponentiation to determine the cipher text C by applying the formula.

$$c = m^e \bmod m$$

This calculation takes the modulo n and enhances the plaintext message m to the power of the encryption exponent e.

d. Convert the cipher text integer to TFNs $\langle a,b,c,d \rangle$. For x, choose any number between $\langle a,d \rangle$, giving the fuzzy number

$$
\mu_m(x) = \begin{cases} 0 & \text{if } x < a \\ \dfrac{x-a}{b-a} & \text{if } a \le x \le b \\ \dfrac{d-x}{d-c} & \text{if } c \le x \le b \\ 0 & \text{if } x > d \end{cases}
$$

Now the cipher text is given as $\langle a,b,c^*,d \rangle$ where c^* is the fuzzy number found above. Here the secret key is the value of x and the range $(\yen)$ chosen for finding TFNs $\langle a,b,c^*,d \rangle$.

17.2.2 Decryption

$$
\mu_m(x) = \frac{c^* \times \gamma}{d-c}
$$

$$
\text{where } d-c = Y
$$

$$
\Rightarrow d - Y = c
$$

The RSA algorithm itself does not naturally contain a two-level encryption scheme. RSA encryption uses a public key for encryption and a private key for decryption, resulting in a single level of encryption and decryption. On the other hand, TFNs are combined with RSA encryption for better security. The second level of decryption is as follows:

$$
m_e = \frac{c^{*d}\,(mod\,p) - b}{a}
$$

$$
m_f = \frac{c^{*d}\,(mod\,q) - b}{a}
$$

$$
m_g = \frac{c^{*d}\,(mod\,r) - b}{a}
$$

Therefore, the original message m is now ($m = m_e = m_f = m_g$)

17.3 ILLUSTRATION

1. Choose six distinct prime numbers, say $p = 17$, $q = 19$, $r = 23$, $s = 29$, $t = 31$, $u = 37$

2. Calculate the modulus: $n = p \times q \times r \times s \times t \times u$

$$n = 247110827$$

3. Compute Euler's totient function $\phi(n) = (p-1) \times (q-1) \times (r-1) \times (s-1) \times (t-1) \times (u-1)$

$$\phi(n) = 191600640$$

4. Choose an exponent

$$e \ni 1 < e < \phi(n) \text{ and } gcd(e, \phi(n)) = 1$$

5. Calculate the private exponent d, such that

$$d \times e = 1 \left(mod\, \phi(n) \right)$$

$$d \equiv e^{-1} mod\, \phi(n)$$

$$\equiv 1^{-1} mod\, 191600640$$

$$d = 1$$

6. Create the public key:

$$(n, e) = (247110827, 1)$$

Therefore the secret key $(a, b) = (3, -15)$

7. (a) First-level encryption

For example, the message is 15. The sender encrypts it to a cipher text using

$$c^* = (am + b)^e \,(mod\, n)$$

$$c^* = (3(15) - 15)^1 \,mod\, 247110827$$

$$= 30 \,mod\, 247110827$$

$$c^* = 30$$

(b) Second-level encryption

Converting the cipher text $c^* = 30$ into a trapezoidal fuzzy number $< 10, 20, 30, 40 >$. For x, choose any between $\langle 10, 40 \rangle$, Let $x = 38$.

$$\mu_m(38) = \frac{d-x}{d-c} = \frac{40-38}{40-30} = 0.2$$

Now the cipher text is $\langle 10, 20, 0.2, 40 \rangle$; Key = 38; Range = 10

8. (a) First-level decryption

$$\mu_m(38) = \frac{Encrypted\ fuzzy\ number \times Range}{d-c^*}$$

$$\mu_m(38) = \frac{0.2 \times 10}{d-c} \quad \text{where } d-c^* = range$$

$$40 - c^* = 10$$

$$c^* = 30$$

(b) Second-level decryption

$$m_{17} = \frac{30^1(mod\ 17)+15}{3} = 9.3,$$

$$m_{19} = \frac{30^1(mod\ 19)+15}{3} = 8.6, m_{23} = \frac{30^1(mod\ 23)+15}{3} = 7.3$$

$$m_{29} = \frac{30^1(mod\ 29)+15}{3} = 5.3, m_{31} = \frac{30^1(mod\ 31)+15}{3} = 15,$$

$$m_{37} = \frac{30^1(mod\ 37)+15}{3} = 15$$

$$\therefore m = m_{31} = m_{37} = \ldots = 15$$

Hence the original message is found as 15

17.4 APPLICATIONS

In the realm of MCDM, the RSA cryptosystem finds application primarily in securing sensitive data and communication channels within decision-making processes that involve multiple stakeholders, criteria, and sensitive information. MCDM often involves the exchange of critical and confidential information among multiple parties. RSA encryption can be employed to secure these communications, ensuring that the criteria, evaluations, and decisions remain confidential and protected from unauthorized access or interception.

In scenarios where various decision criteria are assessed and weighted for decision-making, RSA encryption can secure the criteria values and weights. This ensures that any manipulation or unauthorized access to these criteria is prevented, maintaining the integrity of the decision-making process. MCDM processes often involve collaboration among distributed parties. RSA encryption ensures secure collaboration by safeguarding shared documents, databases, or communication channels and maintaining confidentiality and integrity throughout the collaborative decision-making process. In all these applications, the RSA cryptosystem plays a crucial role in ensuring confidentiality, integrity, and authenticity within MCDM processes. By employing RSA encryption, sensitive information remains secure, fostering trust among stakeholders and ensuring the reliability of decision-making outcomes.

Here we give an example of combining RSA with TFNs to select a candidate for a job position based on multiple criteria: education, experience, skills, and personality traits. Criteria weights (the importance assigned to each criterion) for each candidate are encrypted using RSA to maintain confidentiality and integrity. Then, TFNs are employed to represent the uncertainty and imprecision associated with evaluating each candidate across criteria. Expert opinions or historical data are used to assign TFNs representing the education level, years of experience, skill proficiency, and personality traits of each candidate, as in the following example TFNs for Candidate A:

Education: [3.5, 4, 4.5, 5] on a scale of 1 to 5
Experience: [5, 6, 7, 8]
Skills: [0.7, 0.75, 0.8, 0.85]
Personality Traits: [2, 3, 4, 5]

The candidates are evaluated across the criteria using fuzzy weighted sums, and the aggregated evaluations, encrypted with RSA, are decrypted using the RSA private key, allowing access to the final decision values for each candidate. Based on the decrypted fuzzy evaluations, the hiring decision-makers select the candidate with the highest overall performance, considering all criteria for the job position. By integrating RSA encryption for securing criteria weights and candidate information and employing TFNs for fuzzy evaluations, this approach ensures confidentiality, prevents unauthorized access to sensitive data, incorporates uncertainty in evaluations, and facilitates reliable decision-making in selecting the most suitable candidate for the job.

17.5　CONCLUSION

The combination of the TFNs, CRT, and RSA cryptosystems provides an effective approach for improving security and effectiveness in cryptographic processes. Because it relies on the computational complexity of factoring big prime numbers for asymmetric encryption, the RSA method offers a solid foundation for secure communication. By dividing the modular exponentiation into smaller, separate computations, integrating CRT with RSA offers useful optimization. Furthermore, by adding an extra degree of uncertainty and imprecision to the cryptographic system, the use of TFNs improves privacy protection and provides even another layer of security by making it more difficult for adversaries to decode information. The synergy between these mathematical ideas creates opportunities for additional research and development in the field of cryptography, offering more robust and effective cryptographic systems for protecting sensitive data in an increasingly digital world, even as technology advances and security threats grow more complex.

Network security systems use cryptographic techniques in various fields to protect data, ensure privacy, authenticate users, and secure communications. Military agencies employ cryptanalysts to decrypt intercepted enemy communications, providing valuable intelligence for military decision-making. The effectiveness of the approach we discuss in this chapter heavily relies on the availability and quality of the data used to define fuzzy numbers and criteria weights; inaccurate or incomplete data can lead to unreliable results, and assigning membership functions to trapezoidal fuzzy numbers involves subjective judgements, which can introduce bias and variability in the decision-making process.

REFERENCES

[1] Meenakshi, A., & Shivangi Mishra, J. (2023, October). The Modular Product of Two Cubic Fuzzy Graph Structures. In 2023 First International Conference on Advances in Electrical, Electronics and Computational Intelligence (ICAEECI) (pp. 1–7). IEEE. DOI: 10.1109/ICAEECI58247.2023.10370796

[2] Meenakshi, A., & Mythreyi, O. (2023, October). Mathematical Modeling of Social Networks using Hypergraphs. In 2023 First International Conference on Advances in Electrical, Electronics and Computational Intelligence (ICAEECI) (pp. 1–6). IEEE. DOI: 10.1109/ICAEECI58247.2023.10370980

[3] Rivest, R. L., Shamir, A., & Adleman, L. (1978). A Method for Obtaining Digital Signatures and Public-Key Cryptosystems. Communications of the ACM, 21(2), 120–126. https://dl.acm.org/doi/10.1145/359340.359342

[4] Sahu, J., Singh, V., Sahu, V., & Chopra, A. (2017). An Enhanced Version of RSA to Increase the Security. Journal of Network Communications and Emerging Technologies, 7(4), 1–4. https://www.jncet.org/Manuscripts/Volume-7/Issue-4/Vol-7-issue-4-M-01.pdf

[5] Somani, N., & Mangal, D. (2014). An Improved RSA Cryptographic System. International Journal of Computer Applications, 105(16). https://www.ijcaonline.org/archives/volume105/number16/18461-9820

[6] Meenakshi. A, & Mythreyi, O. (2023). Applications of Neutrosophic Social Network Using Max Product Networks. Journal of Intelligent & Fuzzy Systems. https://doi.org/10.3233/JIFS-223484

[7] Meenakshi, A., Senbaga Malar, J., & Kannan, A. (2023). Application of Intuitionistic Fuzzy Network Using Efficient Domination. DOI: 10.1002/9781394175130.ch14.

[8] Ning, L., Ali, Y., Ke, H., Nazir, S., & Huanli, Z. (2020). A Hybrid MCDM Approach of Selecting Lightweight Cryptographic Cipher Based on ISO and NIST Lightweight Cryptography Security Requirements for the Internet of Health Things. IEEE Access, 8, 220165–220187.

[9] Padmaja, C. J., Bhagavan, V. S., & Srinivas, B. (2016). RSA Encryption Using Three Mersenne Primes. International Journal of Chemical Science, 14(4), 2273–2278. https://www.tsijournals.com/articles/rsa-encryption-using-three-mersenne-primes.pdf

[10] Pius, A., & Kirubaharan, D. R. (2021). Application of Cryptography in Data Privacy using Fuzzy Graph Theory. Journal of Discrete Mathematical Sciences and Cryptography, 24(8), 2389–2401. DOI: 10.1080/09720529.2021.2014146

[11] Meenakshi, A., & Dhanushiya, S. (2023, October). Fuzzy Network using Vertex Order Coloring and Efficient Domination. In 2023 First International Conference on Advances in Electrical, Electronics and Computational Intelligence (ICAEECI) (pp. 1–11). IEEE. DOI: 10.1109/ICAEECI58247.2023.10370790

[12] Al-Hamami, A. H., & Aldariseh, I. A. (2012, November). Enhanced Method for RSA Cryptosystem Algorithm. In 2012 International Conference on Advanced Computer Science Applications and Technologies (ACSAT) (pp. 402–408). IEEE. https://doi.org/10.1109/ACSAT.2012.102

[13] Ivy, B. P. U., Mandiwa, P., & Kumar, M. (2012). A modified RSA cryptosystem based on prime numbers. International Journal of Engineering and Computer Science, 1(2), 63–66. https://www.hgsitebuilder.com/files/writeable/uploads/hostgator581320/file/63-66.pdf

[14] Amin, R., & Biswas, G. P. (2015). An Improved RSA-Based User Authentication and Session Key Agreement Protocol Usable in Tmis. Journal of Medical Systems, 39(8), 79. https://pubmed.ncbi.nlm.nih.gov/26123833/

[15] Jamaludin, J., & Romindo, R. (2020). Implementation of Combination Vigenere Cipher and RSA in Hybrid Cryptosystem for Text Security. IJISTECH (International Journal of Information System and Technology), 4(1), 471–481. https://ijistech.org/ijistech/index.php/ijistech/article/view/85

[16] Mfungo, D. E., Fu, X., Xian, Y., & Wang, X. (2023). A Novel Image Encryption Scheme Using Chaotic Maps and Fuzzy Numbers for Secure Transmission of Information. Applied Sciences, 13(12), 7113. https://doi.org/10.3390/app13127113

[17] Meenakshi, A., & Baskar Babujee, J., (2016). Encryption through Labeling using Efficient Domination. Asian Journal of Research in Social Sciences and Humanities, 6, 1967. DOI: 10.5958/2249-7315.2016.00918.7

[18] Tan, T., Mills, G., Papadonikolaki, E., & Liu, Z. (2021). Combining Multi-Criteria Decision Making (MCDM) Methods with Building Information Modeling (BIM): A Review. Automation in Construction. DOI: 10.1016/j.autcon.2020.103451

[19] Aliamis, Hardi. RSA Cryptosystem with Fuzzy Set Theory for Encryption and Decryption. AIP Conference Proceedings. Vol. 30001. No. 2017. 1905. DOI: 10.1063/1.5012147

[20] Kamardan, M. G., Aminudin, N., Che-Him, N., Sufahani, S., Khalid, K., & Roslan, R. (2018, April). Modified Multi Prime RSA Cryptosystem. In Journal of Physics: Conference Series (Vol. 995, No. 1, p. 012030). IOP Publishing. DOI: 10.1088/1742-6596/995/1/012030

[21] Abdaoui, A., Erbad, A., Al-Ali, A. K., Mohamed, A., & Guizani, M. (2021). Fuzzy Elliptic Curve Cryptography for Authentication in Internet of Things. IEEE Internet of Things Journal, 9(12), 9987–9998. DOI: 10.1109/JIOT.2021.3121350

[22] Kumar, A., Sah, B., Singh, A. R., Deng, Y., He, X., Kumar, P., & Bansal, R. C. (2017). A review of Multi-Criteria Decision Making (MCDM) towards sustainable renewable energy development. Renewable and Sustainable Energy Reviews, 69, 596–609. DOI: 10.1016/j.rser.2016.11.191

18 AI-Powered Decision-Making Applications for Sustainable Development

Yang Minghai, Fluturim Saliu,
and Waqar Akbar Khan

18.1 INTRODUCTION

Artificial intelligence (AI) refers to the field of computer science that focuses on developing intelligent machines capable of performing tasks that typically require human intelligence, for instance, speech recognition, decision-making, problem-solving, learning, and visual perception. AI systems are designed to analyze vast amounts of data, make predictions, and automate complex processes, thereby mimicking human cognitive abilities (Copeland, 1993).

The concept of AI dates to ancient times, with myths and legends often depicting artificial beings capable of human-like actions. However, it was in the mid-20th century that AI emerged as a formal discipline. The term "artificial intelligence" was coined by John McCarthy, who organized the Dartmouth Conference in the 1950s, widely regarded as the birth of AI as a field of study. Over the years, AI has witnessed significant advancements driven by breakthroughs in computing power, data availability, and algorithmic innovations. One crucial aspect of AI is machine learning (ML), a subset of AI that focuses on creating algorithms and models that enable machines to learn and improve from data without explicit programming. Deep learning, a subset of ML, utilizes artificial neural networks to simulate the functioning of the human brain and has played a pivotal role in revolutionizing AI applications such as image and speech recognition (Howard, 2019 and McCarthy et al. 2006).

18.1.1 EVOLUTION AND IMPACT OF AI

Goodfellow et al. (2016) describe AI as the ability of machines to execute an array of tasks with support from people. Machines are going to acquire knowledge, resolve complicated tasks, make plans, consider those plans, etc. with the support of AI; however, to make intelligent judgments and carry out activities that would ordinarily need human intellect, AI systems often rely on massive amounts of information as well as techniques. These systems are capable of data analysis, understanding patterns, forecasting language comprehension, and environment interaction. It is expected that in the not-too-distant future, artificial intelligence will significantly change the way people live and, possibly, put an end to all of the world's issues by resolving the key dilemmas (Sarker, 2021a).

AI has found applications across various domains, transforming industries and impacting society in numerous ways. Minghai et al. (2023) described using artificial neural networks in deep learning modeled after the structure and operation of the human brain. In health care, AI is being utilized for early disease detection, medical image analysis, and drug discovery. In finance, AI algorithms are employed for fraud detection, risk assessment, and algorithmic trading (Dhanabalan and Sathish, 2018). These applications can also help to identify fraudulent activities in areas such as financial transactions, insurance claims, and cybersecurity.

AI-powered decision-making applications can optimize supply chain operations by analyzing various factors such as demand, inventory levels, transportation costs, and production schedules (Khan et al., 2023). AI-powered decision-making applications have been acknowledged for their potential to assist in sustainable development (Topol, 2019). AI-powered virtual assistants like Siri and Alexa have become integral parts of our daily lives, and self-driving cars are on the verge of transforming transportation. AI-powered applications can be used for vehicles' navigation in their environments and to optimize energy consumption by analyzing data from smart grids, weather patterns, and consumer behavior (Shoham and Leyton-Brown, 2010).

18.1.2 Leveraging AI for Sustainable Development: Opportunities, Challenges, and Ethical Considerations

Today, sustainable development is a complex concept that aims to strike a balance between economic development, social progress, and environmental preservation to fulfill the demands of the current generation without jeopardizing the ability of future generations to meet their own needs; sustainable development is a comprehensive approach to meeting the requirements of both the present and the future. However, accomplishing sustainable development objectives calls for a significant amount of information processing and analysis competence. By making it easier to anticipate mistakes and more efficiently prepare for achieving sustainable development, AI can greatly support this effort (Iqbal et al., 2023).

Owing to global issues including environmental degradation, declining diversity of life, disparities, and exhaustion of natural resources, there has been a growing awareness of the significance of sustainable development in recent years. It is crucial to include sustainable development ideas in decision-making processes, company operations, and personal behavior (Khan et al., 2023). The United Nations Sustainable Development Goals (UN SDGs), established in 2015, provide a comprehensive framework for solving global sustainability concerns and attaining sustainable development in the 21st century (United Nations, 2015), and AI-powered decision-making can be useful in achieving these goals.

However, even though technology can aid in sustainable development, AI should be used with honesty and transparency. Effort should be made to promote fairness, transparency, and responsibility in AI systems to prevent prejudices or unintended adverse consequences (Khan et al., 2023). Cooperation between industry, academia, and community stakeholders is necessary to fully fulfill AI's potential for sustainable development. However, as AI continues to advance, it also raises important ethical and societal concerns. Issues surrounding data privacy, algorithmic bias, and the impact of AI on employment and socioeconomic equality need to be carefully addressed to ensure the responsible development and deployment of AI systems (Miller, 2022).

In conclusion, AI is a rapidly evolving field with the potential to revolutionize various aspects of our lives. It encompasses a broad range of technologies and techniques aimed at creating intelligent machines capable of performing tasks that were once exclusive to human intelligence. With continued research and development, AI holds the promise of unlocking new possibilities and transforming industries across the globe.

18.2 BENEFITS OF AI FOR SUSTAINABLE DEVELOPMENT

According to Jeffrey Sachs, a Columbia University professor of health policy and management, the world is entering a new age of sustainable development, in which nations must work together to address the most stubborn hitches such as persistent extreme poverty, social exclusion, economic injustice, poor governance, and environmental degradation. Sachs also served as a senior advisor to the UN on the SDGs and the Millennium Development Goals, and he currently serves as the director of the UN Sustainable Development Solutions Network. He suggested a paradigm for studying sustainable development through the four pillars of economic development, social development, environmental protection, and good governance at the 2002 UN World Summit on Sustainable Development in Johannesburg. Not only these four areas are interconnected, they are all necessary for global sustainable development (Goralski and Tan, 2020).

Researchers have long studied AI, from its beginnings to the present day. As innovators in numerous fields incorporate AI experimentally into theory, thought processes, and practical solutions to issues, industry trade magazines and academic journal articles have added to the amount of study. AI investment began with an abrupt increase, and then, following narrow returns, investments decreased, increased, and declined, in a reasonably regular trend throughout AI's history. AI has had a mixed record of either success or failure (Munoz and Naqvi, 2018).

18.2.1 AI FOR INCLUSIVE EDUCATION AND SUSTAINABLE DEVELOPMENT

AI's disruptive power spans all economic and social areas, including education. For instance, it can lower barriers to education, automate management processes, analyze learning patterns, and optimize learning processes to improve learning results (Pedró et al., 2019). However, the use of AI in education raises problems regarding ethics, security, and human rights. Without governmental intervention, commercial AI deployment would worsen digital gaps and deepen existing financial and learning inequalities, as marginalized and underprivileged populations are more likely to be excluded from AI-powered education. The rapid advancement of AI technologies will have a substantial impact on the skills required by various industries. Already, there is a significant skills gap in the labor market for AI-related occupations and talents.

To ensure that future graduates satisfy labor market needs and become AI-literate citizens, educational institutions and training providers will need to address these skill gaps (UNESCO, 2019). Deep learning-capable artificial intelligence in robots and robotics is already resolving cognitive issues that are typically related to human intelligence (Goralski and Tan, 2020).

Given the rapid evolution of AI and its potential to disrupt the economy—and thus people's livelihoods—it is critical that the UN position itself quickly to ensure that it can provide all of its Member States with the necessary foresight to capitalize on this technological revolution while ensuring respect for human dignity and security. AI holds immense promise for establishing inclusive knowledge societies and assisting countries in meeting their 2030 Agenda for Sustainable Development ambitions, but it also poses significant ethical problems. Underdeveloped nations are going to enjoy the advantages of the next industrial revolution if they succeed in closing the expertise gap that is stifling their national systems of innovation. To generate a technically competent, innovative workforce, they will need to introduce new, out-of-the-box teaching and training methodologies that value cross-disciplinary abilities and creative thinking. They will also need to implement policies and initiatives that foster a culture of great basic research and innovation (UNESCO, 2019).

18.2.2   AI FOR SUSTAINABLE DEVELOPMENT AND FINANCIAL INCLUSION

Researchers have also empirically investigated incorporating AI into the forthcoming industrial revolution manufacturing systems using data acquired from multiple companies to analyze the potential for sustainable Industry 4.0 (Bag et al., 2020). Finance companies must grasp the characteristics of marginalized populations to effectively help them. It is thought that using AI on historical data is crucial to assisting financial service providers in anticipating how prospective clients will respond when approached with offerings (Mogaji et al., 2021). Financial inclusion powered by AI is a crucial factor in poverty reduction, but financial service providers who do not know computer programming face obstacles in implementing AI initiatives (Jayashree et al., 2021; Mhlanga, 2021).

According to Di Vaio et al. (2020), businesses can benefit greatly from AI for achieving sustainable development. AI is capable of analyzing massive volumes of data from a variety of sources, such as gauges, social networking sites, and imagery captured by satellites to offer insightful knowledge for sustainable development decision-making. It can optimize how resources like energy, water, and materials are used. AI systems can track and regulate the utilization of resources, detect shortcomings, and minimize waste strategies thanks to their sophisticated algorithms.

18.2.3   AI FOR SUSTAINABLE URBAN DEVELOPMENT AND ENVIRONMENTAL MANAGEMENT

AI-powered systems can analyze environmental data such as sensor readings and satellite pictures to track ecosystems, biodiversity, climate change, and the depletion of natural resources (SINAY SAS, 2022). By boosting urban planning, managing energy consumption, managing trash, upgrading transportation systems, and applying precision agriculture practices, AI can aid in the creation of smart cities (Alahi et al., 2023). All stakeholders benefit from the strength of AI for achieving sustainable development objectives, for instance by improving decision-making, maximizing resource utilization, and promoting long-term sustainability objectives. To maximize the advantages of AI in sustainable development, it is essential to ensure its responsible and ethical usage, eliminate any biases, and promote collaboration among stakeholders (The Alliance Between Artificial Intelligence and Sustainable Development, 2023).

18.2.4 FOSTERING SUSTAINABLE DEVELOPMENT IN BUSINESS ECOSYSTEMS THROUGH AI

In today's interconnected world of business, sustainability and ecological conservancy are crucial components of building long-term value for people and organizations (Isabelle and Westerlund, 2022). According to research, businesses with stringent green policies foster happy work environments, draw in new clients, and see an increase in revenue. To increase environmental and sustainability awareness among businesses, as well as among governments, regulators, community groups, consumers, and other important stakeholders, it is crucial to promote eco-friendly behaviors outside of the workplace (Alomar, 2022).

Researchers have demonstrated that environmental understanding and awareness can result in environmentally conscious behavior. For example, those who received relevant training and understood more about the adverse impacts of environmental damage were more motivated to seek out eco-friendly services, reuse items, and generally act in ways that promoted environmental sustainability (Usman et al., 2022).

Sustainable development in the modern era is a multidimensional concept that seeks to balance economic growth, social well-being, and environmental protection to meet the needs of the present generation without compromising the ability of future generations to meet their own needs (D'Amore et al., 2022). In recent years, there has been increased recognition of the importance of sustainable development due to global challenges such as climate change, biodiversity loss, inequality, and resource depletion. The idea appears to have gained the widespread attention that other development theories lacked, and it is anticipated to take a long time to replace it as the dominant development paradigm (Mensah and Casadevall, 2019). Shayan et al. (2022) describe some key aspects of sustainable development in the modern era, which we discuss next.

18.2.4.1 Climate Action

Addressing climate change is a crucial component of sustainable development. This involves reducing greenhouse gas emissions, transitioning to renewable energy sources, promoting energy efficiency, and adapting to the impacts of climate change. The Paris Agreement, adopted in 2015, represents a global commitment to combat climate change and achieve sustainable development.

18.2.4.2 Circular Economy

The circular economy aims to minimize waste and maximize resource efficiency by promoting the reuse, recycling, and repurposing of materials. It focuses on designing products and systems that eliminate waste, promote durability, and ensure the responsible use of resources (Morseletto, 2020).

18.2.4.3 Sustainable Energy Transition

Shifting toward a sustainable energy system is vital for achieving sustainable development. This includes increasing the share of renewable energy sources, improving energy efficiency, and promoting decentralized energy systems (Ahmad et al., 2021). The transition to clean energy will reduce greenhouse gas emissions, improve air quality, and enhance energy access (Şerban and Lytras, 2020).

18.2.4.4 Biodiversity Conservation

Protecting biodiversity is essential for sustainable development. Efforts are needed to conserve ecosystems, preserve endangered species, and promote sustainable land use practices. Biodiversity conservation is crucial for ecosystem services, food security, and the overall health of the planet (Corlett, 2020).

18.2.4.5 Sustainable Consumption and Production

Promoting sustainable consumption and production patterns involves reducing waste generation, promoting responsible consumption, and encouraging sustainable business practices. This includes embracing sustainable lifestyles, promoting eco-friendly products, and implementing green procurement policies (Gunawan et al., 2020).

18.2.4.6 Social Equity and Inclusion

Sustainable development should prioritize social equity, ensuring that all individuals have equal access to opportunities and resources. This involves reducing poverty, promoting gender equality, providing quality education and health care, and fostering inclusive societies (Yang et al., 2020).

18.2.4.7 Technology and Innovation

Technology and innovation play vital roles in achieving sustainable development goals. Advancements in areas such as renewable energy, clean technologies, precision agriculture, and digital connectivity can drive sustainable solutions and improve the efficiency of various sectors (Khan et al., 2021).

18.2.4.8 Partnerships and Collaboration

Collaboration among governments, businesses, civil society organizations, and individuals is crucial for achieving sustainable development. Multistakeholder partnerships can facilitate knowledge sharing, resource mobilization, and collective action toward common sustainability goals (Eweje et al., 2021). Indeed, van Zanten and van Tulder (2021) emphasized the significance of integrating sustainable development principles into policy-making, business performance, and individual actions.

However, while creating such plans, we also need to consider the element of uncertainty, as in the case of COVID-19, which had an unanticipated effect on the SDGs: Some progress has been reversed, certain initiatives have been suspended, and the likelihood of meeting the 2030 objectives has decreased. Those who are already wealthy and educated may benefit most from the immense riches that AI-powered technology has the potential to produce, while others may suffer as a result of job displacement. Due to the uneven distribution of educational and computer resources worldwide, the rising economic significance of AI may lead to greater inequality on a global scale (Gulseven et al., 2020).

In addition, as a result of the biases already present in the data used to train AI algorithms, there may eventually be greater discrimination. Another related issue is the use of AI to create computational (commercial, political) propaganda based on big data (also known as "big nudging") that is disseminated through social media by independent AI agents to polarize the public's view (Ntoutsi et al. 2020). Although there is now no scientific evidence to support the technical determinism of such false information, long-term effects of AI are still conceivable (but unstudied) because there aren't enough reliable study techniques. To encourage collaboration and restrict the potential for AI to govern citizen behavior, a paradigm shift is required (Vinuesa et al., 2020).

18.2.5 Overcoming the Limitations and Possibilities at the Crossroads of AI and Sustainable Development

The goal of AI should be to help society, the economy, and the environment, to generally be socially beneficial. However, due to unequal power relations in the actual world, the greater impact of economic arguments, and the absence of emphasis on equitable political and national prioritization, sustainable development frequently prioritizes the environment and the economy over social objectives (Lal et al., 2021). The most accurate way to gauge social good is through the UN's SDGs; all 17 goals must be supported for AI to be socially beneficial. There are now more ethical questions about AI as a result of recent breakthroughs (van Zanten and van Tulder, 2021). Significant worries exist about justice, accountability, and lack of bias in AI decision-making as well as job instability.

Researchers, groups, and governments have all suggested different ethical frameworks and principles for making AI socially useful in solving these AI-related problems (de Almeida et al., 2021), and the topic of AI ethics is developing quickly to address these issues. These frameworks are largely policy statements and nonbinding for organizations that adopt or suggest them. Because of this, businesses frequently design ethical standards for "ethics washing" without having any actual authority (Nasir et al., 2023).

AI is having a rising influence on a variety of businesses. Wamba-Taguimdje et al. (2020) noted that AI technology has historically been based on the requirements and ideals of the countries where it has been created. Recent researchers have emphasized some of the possible negative effects of AI breakthroughs, such as their historical alignment with the ideals of industrialized countries and the consequent absence of ethical oversight, transparency, and democratic control in other places.

AI has the potential to be manipulative and exploit psychological flaws, which can have negative effects on human rights and societal cohesiveness, for instance some nations' AI-based citizen scores. Findings have already prompted the scientific community to emphasize the need to assess how AI systems may affect the SDGs (Jungwirth and Haluza, 2023).

The goal of sustainable development can be affected in two different ways by AI. AI may act as a catalyst, benefiting society as a whole in a variety of ways. However, if AI is overused or misused, it can pose disadvantages for some populations. Regulatory insight and supervision are required to maintain transparency, safety, and ethical norms for AI to enable sustainable development.

Practitioners, academics, and the general public have been interested in the digital transformation of social practices, business models, and goods in recent years. To improve environmental sustainability, engineers have created new digital technologies such as AI (Kraus et al., 2022). In contrast to the natural intelligence exhibited by humans and other animals, AI refers to the intelligence displayed by sophisticated machines. It encompasses skills like independent environment comprehension, experience-based learning, decision-making, decision-implementation, and sophisticated communication with people and other machines (Frank, 2021).

Vaska et al. (2021) is of the view that companies that are leading with analytics are surpassing rivals and witnessing revenue growth while their colleagues are experiencing flat or declining sales as the digital economy becomes the economy. In today's competitive business environment, utilizing data to guide choices throughout the organization is essential to success. A major advantage of AI is in data analytics, an emerging subject that combines ML and AI to produce insights, automate procedures, make predictions, and motivate actions that improve business outcomes. Data analytics is the ability of AI to analyze and learn from massive volumes of data from many sources, find patterns, and forecast future trends (Luan et al., 2020). Large amounts of data are produced by contemporary businesses, and AI data analysis provides an understanding of all that data in a sustainable way and converts it into useful knowledge (Arora, 2023).

However, as data quantities soar, the most prosperous businesses are departing from previous business analytics frameworks. Companies leading their industries are embracing AI analytics to move beyond relying on pixel-perfect dashboards meticulously curated by a team of data professionals and empower everyone with data-driven decision-making (Lu, 2019). The applications for AI in the realm of data promise to dramatically alter how company leaders perceive, assess, and act on their businesses. These applications range from natural language search and predictive capabilities to generative AI explaining discoveries as they occur (Tschang and Almirall, 2021). To fully operationalize AI in business, however, leaders must be aware of the technology's enormous potential as well as its workings, prospective advantages, and applications. Beyond achieving sustainability in businesses, AI solutions can play an ever more significant role in achieving sustainable development (Praveena et al., 2022). The foundation of applications that may significantly enhance process control behavior, efficiency, and sustainability is provided by these cutting-edge digital apps, which combine real-time data and advanced analytics for improved decision-making. (Ojokoh et al., 2020).

The use of AI in business and industry is becoming more commonplace. It has the power to completely alter how we research, learn, live, interact, and work. The potential benefits to society and the economy are enormous (Dwivedi et al., 2021). Applications for AI are quite helpful to managers and decision-makers when sorting large amounts of data to make informed business decisions. The combination of business intelligence and AI provides organization leaders with a deeper awareness of their operations, clients, rivals, and the marketplace as a whole (Sarker, 2021b). In the recent past, such tasks were very lengthy and required considerable time to think about and reach profitable business choices. Now, with the latest technological developments, organizations save considerable expenses by using AI-powered solutions for data analysis (Javaid et al., 2022).

AI allows businesses to streamline multiple data processing processes, hence minimizing the need for manual labor. Automating some processes decreases mistakes, increases accuracy, and saves human assets to concentrate on more important projects. Furthermore, leaders can use AI systems to find and eliminate ineffective practices along with lowering costs (National Artificial Intelligence . . ., 2016).

Another advantage of AI is better prediction and forecasting capabilities for achieving sustainability. AI planning and forecasting is a discipline that allows for the unsupervised production of future forecasts based on science (Nishant et al., 2020). For many industries, including sales, health care, financial services, and manufacturing, AI planning tools employ time series data to forecast future events. AI forecasting makes it simple to anticipate planning and scheduling issues (Ahmad et al., 2021).

With considerably fewer errors and consistently better results than data scientists and specialists, AI planning and forecasting employs algorithms to generate forecasts and anticipate trends without the need for human judgment (Górriz et al. 2020). In research on expert human predictions versus AI forecasts, the former are nearly always superior. Although AI and algorithms won't ever completely replace the human intellect, data scientists and forecasters will always find their capacity to analyze data to be useful tools (Vinuesa et al., 2020).

Humanity has learned a great deal about what it means to be human, how human intelligence is organized, and how people learn and acquire competence during the development of AI. By gradually replacing humans with machines that have more predictive capacity, greater efficiency, and superior outcomes, AI is quickly advancing into this field of expertise. However, as we get increasingly immersed in the systems of information and as robotics become an integral part of our human bodies and existence, homo sapiens won't be completely supplanted at once but rather gradually (Harari, 2017).

While some will fall behind, certain people will benefit intellectually and financially from the development of AI. The creation and implementation of the frameworks, tools, and processes intended to regulate artificial intelligence are already being outpaced by the technology's quick growth. With the development of technology, everyone is growing more anxious that AI may supplant humans as superior intelligence, automate combat, and displace human workers (Ashta and Herrmann, 2021). According to Stephen Hawking, because of their sluggish biological processes, people cannot compete with intelligent robots and are readily replaceable by them.

AI may cause employment losses in advanced nations, but in low-income nations, people may see technology as a means of creating new chances to end the cycle of poverty. No one will be able to avoid the terrifying development of AI because of its fast expansion. Future generations of corporate executives and national and international governments will be heavily influenced by the academic community. For scholars to understand both the present world and the future that is fast changing, AI and its benefits and drawbacks must be taught now (Goralski and Tan, 2020).

Various behavioral and societal variables influence the way people react to AI. The displacement effect of AI, for instance, will affect people's acceptance of the technology and the degree to which it is perceived as a threat to employment status and conditions because the increases in output per worker brought on by automation won't lead to a proportional expansion in demand for labor depending on where a person is situated within an organization. The overall morale and acceptance of AI may be strong if it is intended to expand or enable an employee's opportunities, as we may see in an application based on an expert system (Goralski and Tan, 2020).

However, the social-psychological effects of AI may be significant if the employee's position and potential employment are dependent on a particular task that AI replaces. Therefore, research in psychology and sociology is crucial for understanding how people react to AI. The setting of AI differs from the setting of traditional information technology functioning. Following the possibility concept, sentiments towards negative stimulus ought to provoke more intense reactions compared to favorable ones (Khogali and Mekid, 2023). Nevertheless, when AI models anticipate catastrophic global warming situations, we rarely see more intensely negative responses. It is necessary to investigate how psychology and sociology relate to AI. These two streams can be viewed via a variety of perspectives, like the theory of networks, conflict theory, cognitive models, and emotion-based theories, which can offer fresh perspectives on the function of AI (Rutenberg et al., 2021).

In addition to psychology and sociology, it is important to consider how economic factors will affect how people will react to various AI applications in terms of sustainability. Since money is essential to the evolution and operation of capitalist society, economic justification will have a significant impact on how the results of AI-based models are viewed, accepted, and employed. As a result, to comprehend the capabilities and constraints of AI in tackling environmental sustainability issues, concepts from environmental economics, conventional economics, and behavioral economics must be taken into consideration (Nishant, Kennedy, and Corbett, 2020).

Sustainable resource management practices will prevent resource depletion. Improved allocation of resources for achieving sustainable development for the benefit of both present and future generations is another advantage of AI. By dynamically varying resource mobilization, allocation parameters, available funds, and other parameters, AI can increase a system's capacity and success rate through proper resource allocation.

18.3 CHALLENGES FOR AI IN SUSTAINABLE DEVELOPMENT

While AI can help businesses and society realize their full potential for sustainable development and holds immense promise for fostering inclusive knowledge societies, it also presents serious ethical problems; it has both benefits and drawbacks. AI advantages include supporting big data analytics and creating new value for companies through authenticity, augmentation, and automation, including increased economic value and problem-solving capabilities that will help them be more resilient in the face of social and environmental concerns. AI can also be used in the workplace to improve the efficacy of corporate social responsibility programs (Truby, 2020).

However, for the use of AI to be in line with human values and beliefs, it is equally vital to look into potential risks brought on by this formidable technology and issues it raises (Zhao and Gómez, 2023). A thorough assessment of the literature reveals that the research on artificial intelligence for sustainability faces challenges like using historical data too often in ML models, unpredictable human reactions to AI-based solutions, rising threats to cybersecurity, the detrimental effects of AI applications, and measurement challenges for approaches to intervention. Even though AI solutions are intrinsically scientific, their effectiveness will depend on how successfully they negotiate and sway the emotional, social, and organizational barriers to human advancement in this field at present (Howard, 2019).

To ensure the long-term viability of AI, its application must be responsibly and morally supported by regulatory understanding. Failure to do so can result in gaps in the application of AI in comparison to responsible and ethical standards (Goralski and Tan, 2020). AI can potentially have other negative implications, such as compromising privacy and increasing algorithmic bias. Companies should analyze algorithm data, provide comparable options, and make ethical and fair recommendations. We suggest leveraging the regulatory framework to encourage more socially conscious AI by keeping an eye on and reducing the dangers involved. As a result, AI can have a wider impact across various industries (Zhao and Gómez, 2023).

Machine learning has been widely used in AI solutions for sustainability. For learning and creating predictions, ML takes advantage of the patterns and connections found in the historical data (Nishant, Kennedy, and Corbett, 2020). For instance, using AI and ML in the detection and classification of poverty may revolutionize the process of data collecting and analysis and in turn enable faster and more effective policy responses (Adadi, 2021).

In terms of AI challenges in business, there is a lack of agreement in the literature over whether and how AI will alter current corporate practices or even their very foundations. Approaches range from those that see a new paradigm of autonomous firms to those that maintain that no appreciable change will take place. Some authors assert that AI will increase the accuracy and efficacy of corporate actions while decreasing the need for human management and the costs associated with it (Tsalis et al., 2020). Some others even think that, as a result of digitalization, boards will become virtual networks of people or be entirely supplanted by AI-based solutions. Others, on the other hand, are still doubtful that technological advancements will be able to change core normative difficulties and eliminate the need for human management (Zhao and Gómez, 2023).

The difficulty facing AI is great; those making decisions still need to be sufficiently educated about the objective, worth, and potential of AI and AI-based systems, solutions, and applications. They can only address the subject and its implications in the necessary manner in this way. The academic community must accept the need for efficient, direct, simple, but not simplistic, methods of imparting AI understanding and awareness to all parts of society that is free of needless rhetoric however genuine (Hagendorff and Wezel, 2020). The purpose of AI research should be to foster dialogue on various facets of AI and its various fields of application, including in promoting inclusive and sustainable growth and development, by setting the discussion in the context of the SDGs (Castro et al., 2021).

In summary, there are many questions about how to maximize the benefits and potential of AI for our societies, and the debate over AI and its potential contribution to society has just begun. The surviving communities of scholars and practitioners must offer commensurate insights into these technological innovations as AI and its ecosystem expand (Visvizi, 2022). Amid these difficulties, AI offers an unmatched opportunity to alter how we approach environmental protection and sustainable development by processing enormous volumes of data, learning from patterns, and making decisions in real time (Rayhan, 2023).

18.4 AI-POWERED DECISION-MAKING IN SUSTAINABLE DEVELOPMENT

The UN Sustainable Development Goals have several main objective areas, and unregulated AI development runs the risk of slowing down progress in these areas. This risk is increased in developing nations. The current unchecked experimentation with unproven AI technology in markets and society casts a shadow that overpromises the practical possibilities of AI-powered progressive growth. "Big Tech" has already demonstrated that it cannot be trusted to self-regulate or adhere to voluntary standards, and in the absence of preventative control, trust will once more be broken, leading to negative outcomes and exploitation, harming progress toward the SDGs. Currently, AI is in an undeveloped and immature stage of development. Even though AI systems have shortcomings, we consistently trust them to make crucial judgments that affect our lives. Automated decision-making that is discriminatory has the potential to substantially harm economic and social progress, especially in developing nations (Vaioa, Palladinoa, Hassan, and Escobar, 2020).

Incorporating sustainability practices into an organization is advantageous for its reputation, productivity, and access to resources such as better tools, talent, and opportunities. Many businesses see sustainability efforts as opportunities to advance their long-term interests and strengthen ties with their suppliers, employees, and communities rather than as a burden that must be endured (Zhao and Gómez, 2023). Digital technologies that are powered by AI are inextricably linked to interact with, observe, and comprehend people, enterprises, economies, and lives in general. The economic and sociological effects of AI-powered digital technologies are constantly growing, and more recently, they are playing a significant part in efforts to achieve the SDGs, whose adoption is a significant decision for both wealthy and developing nations (Di Vaio et al., 2020).

AI and electronic technologies—such as massive information, blockchain technology, augmented and virtual reality, and electronic twins—create high standards of enhancing social, environmental, and economic levels of SDG-related global transformation. Researchers have identified several priorities or critical components that must be addressed for AI to best achieve the SDGs. These include alternative standards for evaluating SDG attainment, lessons learned since the panda crisis, and AI-powered digital technologies supported by widely available and trustworthy data (Pigola et al., 2021).

Joshi, Lavanchy, and Stehli (2018) have pointed out that today's world of volatility, uncertainty, complexity, and ambiguity is characterized by pervasive disruption from unlikely competitors, industry changes occurring in faster and shorter cycles, ever-changing rules such as for data protection, and a constantly dwindling time to market. In addition to acting as gatekeepers by limiting choice through precise suggestions and enabling customers to purchase goods they have never even seen, AI systems are fundamentally altering the way businesses operate, and according to research, consumers genuinely prefer these algorithmic recommendations and frequently follow them.

Because more people are using AI recommendation platforms and other systems, questions of this nature will certainly continue to be raised. The social, political, and commercial impacts of robotics, artificial intelligence, and machine learning will change many modern sectors and eliminate jobs. Past revolutions have in reality enhanced productivity and produced a net increase in jobs, but in the next few years, employees face high risk of automation. Of course, as time has gone on, the nature of employment has changed; some activities have been automated by technology, while others have become necessary. Though it is yet uncertain what occupations AI will produce, current patterns indicate that some positions will be comparatively safe, especially those needing the multiple interactions with people; interpersonal abilities; creative and unique ideas; and ease with uncertainty and inconsistency (Puntoni et al., 2021).

With the recent release of ChatGPT, a publicly available tool created by OpenAI and utilizing a huge language model, AI has made yet another advancement, this time in research. A clever technology like ChatGPT will help spread knowledge that is comparatively easy to understand and accessible, and it is a helpful tool to get us started. However, without probing, integrating, and nonlinear thinking to examine the material, the outputs may not be of much use, much as doing an immunohistochemistry experiment using an antibody of questionable specificity will produce useless results. Science has always involved people; rather than being afraid of new technology, we should embrace it. However, this shouldn't mean doing away with science's human component (Hill-Yardin et al., 2023).

Meanwhile, another aspect of incorporating ML into sustainable development is in energy management. As we described, machine learning is the process of training algorithms to automatically learn and improve from data inputs (Mehmood et al., 2020), and it can be used to analyze energy usage patterns and recommend energy-saving measures. Predictive modeling can also be used to estimate future energy demand and inform investment decisions on renewable energy sources. Despite this, since AI is still in its infancy, it is challenging to project where it will go in the future. The world must take AI requirements and expectations, such as enforcement, employment, ethics, education, entente, and evolution, into consideration to better understand and execute AI (Vaioa, Palladinoa, Hassanb, and Escobarc, 2020).

Another AI-powered decision-making application for sustainable development is natural language processing. This technology is used to understand human language and interactions, allowing for more efficient communication between stakeholders. Natural language processing can be used to analyze customer feedback, identify problem areas, and provide recommendations for improvement. Additionally, AI-powered decision-making applications can be used for forecasting and predictive modeling. This involves analyzing historical data to identify patterns and predict future outcomes (Bachmann et al., 2022). One researcher identified four main practical prospects for the application of AI for sustainable development: improving decision-making abilities, improving performance management, lessening risk factors, and more logical thinking within organizations (Colson, 2019).

18.5 CONCLUSION

In conclusion, AI-powered decision-making applications have tremendous potential for sustainable development. They are critical tools for analyzing data, identifying trends, and informing sustainable development decisions. While they are not a magic bullet, they offer significant benefits in terms of efficiency and accuracy of decision-making processes. As the capabilities of AI technologies continue to grow, we can expect them to play an increasingly important role in sustainable development (Bachmann et al., 2022).

In order to reap the benefits of artificial intelligence, AI systems must be carefully designed, developed, and implemented. They also need to be continuously monitored, evaluated, and adjusted to guarantee that they uphold moral standards and societal norms. Even though artificial intelligence-driven decision-making holds a great deal of possibilities, understanding and addressing these limitations is necessary for its responsible and effective deployment across a variety of fields.

REFERENCES

Adadi, A., (2021). A survey on data-efficient algorithms in big data era. Journal of Big Data, 8(1):24. https://doi.org/10.1186/s40537-021-00419-9

Ahmad, T., Zhang, D., Huang, C., Zhang, H., Dai, N., Song, Y., Chen, H., (2021). Artificial intelligence in sustainable energy industry: Status Quo, challenges and opportunities. Journal of Cleaner Production, 289:125834. https://doi.org/10.1016/j.jclepro.2021.125834

Alahi, M. E. E., Sukkuea, A., Tina, F. W., Nag, A., Kurdthongmee, W., Suwannarat, K., Mukhopadhyay, S. C., (2023). Integration of IoT-enabled technologies and Artificial Intelligence (AI) for smart city scenario: Recent advancements and future trends. MDPI Sensors, 23:5206. https://doi.org/10.3390/s23115206

Alomar, M. A., (2022). Performance optimization of industrial supply chain using Artificial Intelligence. Hindawi Computational Intelligence and Neuroscience, (9306265). https://doi.org/10.1155/2022/9306265

Arora, S., (2023). AI analytics explained: How it works and key industry use cases. Thought Spot. https://www.thoughtspot.com/data-trends/ai/ai-analytics. Retrieved on August 1, 2023.

Artificial Intelligence for Sustainable Development Programme, 4–8 March 2019 (UNESCO, Paris), United Nations Educational, Scientific and Cultural Organization.

Ashta, A., & Herrmann, H., (2021). Artificial intelligence and fintech: An overview of opportunities and risks for banking, investments, and microfinance. Strategic Change, 30(3):211–222. https://doi.org/10.1002/jsc.2404

Bachmann, N., Tripathi, S., Brunner, M., Jodlbauer, H., (2022). The contribution of data-driven technologies in achieving the sustainable development goals. Sustainability, 14(5):2497. https://doi.org/10.3390/su14052497

Bag, S., Yadav, G., Dhamija, P., Kataria, K. K., (2020). Key resources for industry 4.0 adoption and its effect on sustainable production and circular economy: An empirical study. Journal of Cleaner Production, 281:125233. https://doi.org/10.1016/j.jclepro.2020.125233

Castro, G. D. R., Fernandez, M. C. G., Colsa, A. U., (2021). Unleashing the convergence amid digitalization and sustainability towards pursuing the Sustainable Development Goals (SDGs): A holistic review. Journal of Cleaner Production, 280:122204. https://doi.org/10.1016/j.jclepro.2020.122204

Colson, E., (2019). What AI-powered decision-making looks like. Harvard Business Review. Retrieved on August 28, 2023 from https://hbr.org/2019/07/what-AI-powered-decision-making-looks-like

Copeland, B. J. (1993). Book, Artificial Intelligence: A Philosophical Introduction. Wiley-Blackwell.

Corlett, R. T., (2020). Safeguarding our future by protecting biodiversity. Journal of Plant Diversity, 42(4):221–228. https://doi.org/10.1016/j.pld.2020.04.002

D'Amore, G., Di Vaio, A., Balsalobre-Lorente, D., Boccia, F., (2022). Artificial intelligence in the water—energy—food model: A holistic approach towards sustainable development goals. MDPI Sustainability, 14(2):867. https://doi.org/10.3390/su14020867

Dhanabalan, T., Sathish, A. (2018). Transforming Indian Industries Through Artificial Intelligence and Robotics in Industry 4.0., International Journal of Mechanical Engineering and Technology, 9(10), 835–845. Retrieved from https://iaeme.com/Home/article_id/IJMET_09_10_087

de Almeida, P. G. R., dos Santos, C. D., Farias, J. S., (2021). Artificial intelligence regulation: A framework for governance. Ethics and Information Technology, 23(3):505–525. https://doi.org/10.1007/s10676-021-09593-z

Di Vaio, A., Palladino, R., Hassan, R., Escobar, O., (2020). Artificial intelligence and business models in the sustainable development goals perspective: A systematic literature review. Journal of Business Research, 121:283–314. https://doi.org/10.1016/j.jbusres.2020.08.019

Dwivedi, Y. K., Hughes, L., Ismagilova, E., Aarts, G., Coombs, C., Crick, T., . . . Williams, M. D., (2021). Artificial Intelligence (AI): Multidisciplinary perspectives on emerging challenges, opportunities, and agenda for research, practice and policy. International Journal of Information Management, 57:101994. https://doi.org/10.1016/j.ijinfomgt.2019.08.002

Eweje, G., Sajjad, A., Nath, S. D., Kobayashi, K., (2021). Multi-stakeholder partnerships: A catalyst to achieve sustainable development goals. Journal of Marketing Intelligence & Planning, 39(2):186–212. https://doi.org/10.1108/MIP-04-2020-0135

Frank, B., (2021). Artificial intelligence-enabled environmental sustainability of products: Marketing benefits and their variation by consumer, location, and product types. Journal of Cleaner Production, 285(125242), https://doi.org/10.1016/j.jclepro.2020.125242

Gatley, N., (2022). What is sustainable resource management and how do you achieve it? British Assessment Bureau. https://www.british-assessment.co.uk/ Retrieved on July 31, 2023.

Goodfellow, I., Bengio, Y., Courville, A., (2016). Book, Deep Learning. MIT Press.

Goralski, M. A., Tan, T. K., (2020). Artificial intelligence and sustainable development. The International Journal of Management Education, 18(1):100330. https://doi.org/10.1016/j.ijme.2019.100330

Górriz, J. M., Ramírez, J., Ortíz, A., Martinez-Murcia, F. J., Segovia, F., Suckling, J., . . . Ferrandez, J. M., (2020). Artificial intelligence within the interplay between natural and artificial computation: Advances in data science, trends and applications. Neurocomputing, 410:237–270. https://doi.org/10.1016/j.neucom.2020.05.078

Gulseven, O., Al Harmoodi, F., Al Falasi, M., ALshomali, I., (2020). How the COVID-19 pandemic will affect the UN sustainable development goals?. Available at SSRN 3592933. http://dx.doi.org/10.2139/ssrn.3592933

Gunawan, J., Permatasari, P., Tilt, C. (2020). Sustainable development goal disclosures: Do they support responsible consumption and production? Journal of Cleaner Production, 246:118989. https://doi.org/10.1016/j.jclepro.2019.118989

Hagendorff, T., Wezel, K. (2020). 15 challenges for AI: Or what AI (currently) can't do. Ai & Society, 35:355–365. https://doi.org/10.1007/s00146-019-00886-y

Harari, Y. N. (2017). Homo Deus—a brief history of tomorrow. HarperCollins Publishing.

Hill-Yardin, E. L., Hutchinson, R., Laycock, S. J., (2023). A Chat (GPT) about the future of scientific publishing. Brain, Behavior, and Immunity, 110, 152–154. https://doi.org/10.1016/j.bbi.2023.02.022

Howard, J. (2019). Artificial intelligence: Implications for the Future of Work. American Journal of Industrial Medicine, 62(11), 917–926. https://doi.org/10.1002/ajim.23037

Iqbal, M. A. B., Zafar, R., Imran, M., Khan, W. A., Batool, S., & Sadiq, W., (2023). Impact of green intellectual capital on corporate sustainability with mediating role of green innovation. Russian Law Journal, 11(3), 2707–2716.

Isabelle, D. A., Westerlund, M., (2022). A review and categorization of Artificial Intelligence-based opportunities in wildlife, ocean and land conservation. MDPI Sustainability, 14(1979). https://doi.org/10.3390/su14041979

Javaid, M., Haleem, A., Singh, R. P., Suman, R., (2022). Artificial intelligence applications for industry 4.0: A literature-based study. Journal of Industrial Integration and Management, 7(1):83–111. https://doi.org/10.1142/S2424862221300040

Jayashree, S., Reza, M. N. H., Malarvizhi, C. A. N., Mohiuddin, M., (2021). Industry 4.0 implementation and Triple Bottom Line sustainability: An empirical study on small and medium manufacturing firms. Heliyon Journal, 7(8): e07753. https://doi.org/10.1016/j.heliyon.2021.e07753

Joshi, A., Lavanchy, M., Stehli, S., (2018). Data analytics & artificial intelligence: What it means for your business and society. Insights@IMD. IMD—International Institute for Management Development 4–18. https://www.imd.org/research-knowledge/digital/articles/artificial-intelligence-real-world-impact-on-business-and-society/ Retrieved on August 28, 2023.

Jungwirth, D., Haluza, D. (2023). Artificial intelligence and the sustainable development goals: An exploratory study in the context of the society domain. Journal of Software Engineering and Applications, 16(4), 91–112. https://doi.org/10.4236/jsea.2023.164006

Khan, N., Ray, R. L., Kassem, H. S., Hussain, S., Zhang, S., Khayyam, M., Ihtisham, M., Asongu, S. A., (2021). Potential role of technology innovation in transformation of sustainable food systems: A Review. MDPI Agriculture, 11(10):984. https://doi.org/10.3390/agriculture11100984

Khan, W. A., Sarwar, K., Iqbal, S., Egos, A. Q., Phatak, S. A., Arteaga, C. S., Bashir, F., Rafique, T., Mohsin, M. (2023). Legal and ethical implications of algorithmic decision-making in human resource management in the construction industry of Pakistan. Russian Law Journal, 11(1), 122–136.

Khogali, H. O., Mekid, S., (2023). The blended future of automation and AI: Examining some long-term societal and ethical impact features. Technology in Society, 73:102232. https://doi.org/10.1016/j.techsoc.2023.102232

Kraus, S., Durst, S., Ferreira, J. J., Veiga, P., Kailer, N., Weinmann, A., (2022). Digital transformation in business and management research: An overview of the current status quo. International Journal of Information Management, 63:102466. https://doi.org/10.1016/j.ijinfomgt.2021.102466

Lal, R., Bouma, J., Brevik, E., Dawson, L., Field, D. J., Glaser, B., . . . Zhang, J. (2021). Soils and sustainable development goals of the United Nations: An international union of soil sciences perspective. Geoderma Regional, 25:e00398. https://doi.org/10.1016/j.geodrs.2021.e00398

Lu, Y. (2019). Artificial intelligence: A survey on evolution, models, applications and future trends. Journal of Management Analytics, 6(1):1–29. https://doi.org/10.1080/23270012.2019.1570365

Luan, H., Geczy, P., Lai, H., Gobert, J., Yang, S. J. H., Ogata, H., Baltes, J., Guerra, R., Li, P., Tsai, C. C. (2020). Challenges and future directions of Big Data and Artificial Intelligence in education. Frontiers in Psychology, 11:580820. https://doi.org/10.3389/fpsyg.2020.580820

McCarthy, J., Minsky, M. L., Rochester, N., & Shannon, C.E., (2006). A Proposal for the Dartmouth Summer Research Project on Artificial Intelligence, August 31, 1955. AI Magazine, 27(4), 12. https://doi.org/10.1609/aimag.v27i4.1904

Mehmood, H., Liao, D., Mahadeo, K., (2020). A review of artificial intelligence applications to achieve water-related sustainable development goals. 2020 IEEE/ITU International Conference on Artificial Intelligence for Good (AI4G):135–141. IEEE. https://doi.org/10.1109/AI4G50087.2020.9311018

Mensah, J., Casadevall, S. R., (2019). Sustainable development: Meaning, history, principles, pillars, and implications for human action: Literature review. Cogent Social Sciences, 5:1, DOI: 10.1080/23311886.2019.1653531

Mhlanga, D., (2021). Artificial Intelligence in the Industry 4.0, and its impact on poverty, innovation, infrastructure development, and the sustainable development goals: Lessons from emerging economies? MDPI Sustainability, 13, 5788. https://doi.org/10.3390/su13115788

Miller, G. J., (2022). Stakeholder roles in Artificial Intelligence projects, Project Leadership and Society, 3, https://doi.org/10.1016/j.plas.2022.100068

Minghai, Y., Wenqing, L., Khan, W. A., Kasture, A., Riasat, S., (2023). Supercharge your productivity with Artificial Intelligent. Bincang Sains dan Teknologi (BST), 2(2), 72–81. https://journal.iistr.org/index.php/BST/article/view/359

Mogaji, E., Soetan, T. O., & Kieu, T. A., (2021). The implications of artificial intelligence on the digital marketing of financial services to vulnerable customers. Australasian Marketing Journal, 29(3):235–242. https://doi.org/10.1016/j.ausmj.2020.05.003

Morseletto, P., (2020). Targets for a circular economy. Resources, Conservation and Recycling, 153:104553. https://doi.org/10.1016/j.resconrec.2019.104553

Munoz, J. M., & Naqvi, A. (Eds.), (2018). Business Strategy in the Artificial Intelligence Economy. Business Expert Press.

Nasir, O., Javed, R. T., Gupta, S., Vinuesa, R., Qadir, J., (2023). Artificial Intelligence and sustainable development Goals Nexus via four vantage points. Technology in Society, 72(102171), https://doi.org/10.1016/j.techsoc.2022.102171

National Artificial Intelligence Research and Development Strategic Plan, (2016, October). Executive office of the president of the United States. National Science and Technology Council and Networking and information technology Research and development subcommittee. https://www.nitrd.gov/PUBS/national_ai_rd_strategic_plan.pdf.

Nishant, R., Kennedy, M., & Corbett, J., (2020). Artificial intelligence for sustainability: Challenges, opportunities, and a research agenda. International Journal of Information Management, 53:102104. https://doi.org/10.1016/j.ijinfomgt.2020.102104

Ntoutsi, E., Fafalios, P., Gadiraju, U., Iosifidis, V., Nejdl, W., Vidal, M. E., . . . Staab, S., (2020). Bias in data-driven artificial intelligence systems—An introductory survey. Wiley Interdisciplinary Reviews (WIREs): Data Mining and Knowledge Discovery, 10(3): e1356. https://doi.org/10.1002/widm.1356

Ojokoh, B. A., Samuel, O. W., Omisore, O. M., Sarumi, O. A., Idowu, P. A., Chimusa, E. R., Darwish, A., Adekoya, A. F., Katsriku, F. A., (2020). Big data, analytics and artificial intelligence for sustainability. Scientific African, 9 (e00551), https://doi.org/10.1016/j.sciaf.2020.e00551

Pedro, F., Subosa, M., Rivas, A., Valverde, P., (2019). Working Papers on Education Policy; Artificial intelligence in education: challenges and opportunities for sustainable development. UNESCO Education Sector, The Global Education 2030 Agenda. https://repositorio.minedu.gob.pe/handle/20.500.12799/6533

Pigola, A., da Costa, P. R., Carvalho, L. C., Silva, L. F.d., Kniess, C. T., Maccari, E. A., (2021). Artificial Intelligence-driven digital technologies to the implementation of the sustainable development goals: A perspective from Brazil and Portugal. Sustainability, 13(13669). https://doi.org/10.3390/su132413669

Praveena, B. A., Lokesh, N., Buradi, A., Santhosh, N., Praveena, B. L., Vignesh, R., (2022). A comprehensive review of emerging additive manufacturing (3D printing technology): Methods, materials, applications, challenges, trends and future potential. Materials Today: Proceedings, 52(Part 3):1309–1313. https://doi.org/10.1016/j.matpr.2021.11.059

Puntoni, S., Reczek, R. W., Giesler, M., Botti, S., (2021). Consumers and artificial intelligence: An experiential perspective. Journal of Marketing, 85(1):131–151. https://doi.org/10.1177/0022242920953847

Rayhan, A., (2023). AI and the environment: Toward sustainable development and conservation. https://doi.org/10.13140/RG.2.2.12024.42245.

Rutenberg, I., Gwagwa, A., Omino, M., (2021). Use and impact of artificial intelligence on climate change adaptation in Africa. African Handbook of Climate Change Adaptation (pp. 1107–1126). Cham: Springer International Publishing. https://doi.org/10.1007/978-3-030-45106-6_80

Sarker, I. H., (2021a). Data science and analytics: An overview from data-driven smart computing, decision-making and applications perspective. SN Computer Science, 2(5): 377. https://doi.org/10.1007/s42979-021-00765-8

Sarker, I. H., (2021b). Machine learning: Algorithms, real-world applications and research directions. SN Computer Science, 2(160). https://doi.org/10.1007/s42979-021-00592-x

Şerban, A. C., Lytras, M. D., (2020). Artificial intelligence for smart renewable energy sector in europe—smart energy infrastructures for next generation smart cities. Journal of IEEE access, 8:77364–77377. https://doi.org/10.1109/ACCESS.2020.2990123

Shayan, N. F., Mohabbati-Kalejahi, N., Alavi, S., Zahed, M. A., (2022). Sustainable Development Goals (SDGs) as a Framework for Corporate Social Responsibility (CSR). Sustainability (MDPI), 14(3), 1222, 1–27. https://doi.org/10.3390/su14031222

Shoham, Y., Leyton-Brown, K. (2010). Book, Multiagent Systems: Algorithmic, Game-Theoretic, and Logical Foundations. Cambridge University Press.

Sustainability Article, (2022). How can Artificial Intelligence be Used to Achieve Sustainable Development Goals? SINAY SAS (2022), Project co-financed by the European Union with the European Regional Development Fund (FEDER). https://sinay.ai/en/how-can-artificial-intelligence-be-used-to-achieve-sustainable-development-goals/. Retrieved on July 11, 2023.

The 17 Goals, (2015). Department of Economic and Social Affairs, the United Nations. https://sdgs.un.org/goals. Retrieved on July 5, 2023.

Topol, E., (2019). Book, Deep Medicine: How Artificial Intelligence Can Make Healthcare Human Again. Basic Books Hachette Book Group, Inc.

Truby, J. (2020). Governing artificial intelligence to benefit the UN sustainable development goals. Sustainable Development, 28(4), 946–959. https://doi.org/10.1002/sd.2048

Tsalis, T. A., Malamateniou, K. E., Koulouriotis, D., Nikolaou, I. E., (2020). New challenges for corporate sustainability reporting: United Nations' 2030 Agenda for sustainable development and the sustainable development goals. Corporate Social Responsibility and Environmental Management, 27(4), 1617–1629. https://doi.org/10.1002/csr.1910

Tschang, F. T., Almirall, E., (2021). Artificial intelligence as augmenting automation: Implications for employment. Academy of Management Perspectives, 35(4):642–659. https://doi.org/10.5465/amp.2019.0062

Usman, M., Rofcanin, Y., Ali, M., Ogbonnaya, C., Babalola, M. T. (2022). Toward a more sustainable environment: Understanding why and when green training promotes employees' eco-friendly behaviors outside of work. Human Resource Management, 62(3), 355–371. https://doi.org/10.1002/hrm.22148

Vaioa, A. D., Palladinoa, R., Hassan, R., Escobarc, O., (2020). Artificial intelligence and business models in the sustainable development goals perspective: A systematic literature review. Journal of Business Research, 121, 283–314. https://doi.org/10.1016/j.jbusres.2020.08.019

van Zanten, J. A., van Tulder, R., (2021). Improving companies' impacts on sustainable development: A nexus approach to the SDGS. Business Strategy and the Environment, 30(8):3703–3720. https://doi.org/10.1002/bse.2835

Vaska, S., Massaro, M., Bagarotto, E. M., Mas, F. D., (2021). The digital transformation of business model innovation: A structured literature review. Frontiers in Psychology, 11:539363. https://doi.org/10.3389/fpsyg.2020.539363

Vinuesa, R., Azizpour, H., Leite, I., Balaam, M., Dignum, V., Domisch, S., Felländer, A., Langhans, S.D., Tegmark, M., Fuso Nerini, F., (2020). The role of Artificial Intelligence in achieving the sustainable development goals. Nature Communications, 11, 233, 91–112, https://doi.org/10.1038/s41467-019-14108-y

Visvizi, A., (2022). Artificial Intelligence (AI) and Sustainable Development Goals (SDGs): Exploring the impact of AI on politics and society. MDPI Sustainability, 14, 1730. https://doi.org/10.3390/su14031730

Wamba-Taguimdje, S. L., Wamba, S. F., Kamdjoug, J. R. K., Wanko, C. E. T., (2020). Influence of Artificial Intelligence (AI) on firm performance: the business value of AI-based transformation projects. Business Process Management Journal, 26(7):1893–1924. https://doi.org/10.1108/BPMJ-10-2019-0411

Yang, S., Zhao, W., Liu, Y., Cherubini, F., Fu, B., Pereira, P. (2020). Prioritizing sustainable development goals and linking them to ecosystem services: A global expert's knowledge evaluation. Geography and Sustainability, 1(4):321–330. https://doi.org/10.1016/j.geosus.2020.09.004

Zhao, J., Gómez, B. G., (2023). Artificial Intelligence and Sustainable Decisions. European Business Organization Law Review, 24, 1–39. https://doi.org/10.1007/s40804-022-00262-2

19 Artificial Intelligence Algorithms for Better Decision-Making

Ganesh Khekare, Ganesh Yenurkar,
Anil V Turukmane, Gaurav Kumar Ameta,
Pooja Sharma, and Ajay Kumar Phulre

19.1 INTRODUCTION

Researchers analyzing digital data use supervised and unsupervised learning experiments with different digital data. For instance, Ana et al. (2024) analyzed sales data to examine customer engagement based on the order volume in different cities and compared the results between supervised and unsupervised learning algorithms. Another research group trained a model to test strengths and weaknesses to identify areas of improvement. Specifically, the authors trained a model on a data set of customer age, gender, and purchasing needs using multiple unsupervised and supervised learning algorithms (Boutaba et al., 2018).

The results from this chapter will be useful for determining the best-suited algorithms for machine learning (ML; Dipo et al., 2024) based on multiple parameters such as deviation from real value, mean scale prediction, the accuracy of the model, and computation performance of algorithms. The rest of this chapter proceeds as follows: Section 19.2 is the literature survey, and in Section 19.3, we describe our methodology. In Section 19.4, we describe the project implementation, in Section 19.5, we give the results with a discussion, and in Section 19.6, we give our conclusions; we also discuss future scopes.

19.2 LITERATURE REVIEW

The importance of ML is that it has different types of models for deciding and factoring in the results with set rules and algorithms (Khekare et al., 2023). Thus, it provides results with a clear scope and is based on the functions provided (Gustavo et al., 2022). ML models have many applications in various fields (Ilya et al., 2024), efficiently trained according to how they're needed (Trillo et al., 2023). Embedded models use classifiers and other methods to provide results based on multiple factors such as accuracy and efficiency (Zhan et al., 2022).

DOI: 10.1201/9781032635170-19

19.2.1 Software Reliability

System software reliability is the probability that the system and servers will not fail over time and will continue to run without issues (Khekare et al., 2022). Failure occurs within a predefined time frame and under explicit conditions. Information on programming dependability is exceptionally crucial in basic frameworks since it shows plan flawlessness (Kim et al., 2022). In this, the objective is to improve the dependability of the product as well as the system that is being used to host the ML model (Nakagome et al., 2020). The model is given specific details and is required to provide results, and if it provides wrong information, it is refined (Khekare et al., 2022).

19.2.2 Supervised Learning

Supervised learning is a well-known type of AI plot utilized in planning many input factors with many output factors to deduce capacity from the information (Jahan et al., 2023). The preparation data set contains the necessary occasion values for features (Perdikis et al., 2021). The model's prediction accuracy is tested and refined. The aim of the inferred function could also be to unravel a regression or classification problem (Khekare et al., 2023). There are a few measurements utilized in the estimation of the learning task like exactness, affectability, particularity, kappa, and the region under the curve (Shashank et al., 2023). In this work, the point is to give reports and examine the available business information (Tessa et al., 2024). The aim of the inferred function could also be to unravel a regression or classification problem (Vaccaro et al., 2021).

19.2.3 Unsupervised Learning

Unsupervised learning is the model that helps in training machines (Ozdenizci et al., 2020) using data that are neither classified nor attributed and facilitates the system to work using the model (Khekare et al., 2021). The job of the model and the machine is to collect random data based on its shapes, resemblance, specimen, and variance with no earlier training on predefined algorithms or functions (Wang et al., 2023). Thus, the machine has no idea about sales, gender, value, etc. (Nguyen et al., 2024). Unsupervised learning algorithms are divided into two categories: Clustering groups data based on multiple factors such as customer purchasing behaviors or where they are from (Sarkar et al., 2021), whereas association in an unsupervised learning model describes patterns such as Person X tends to buy Item Y.

19.3 METHODOLOGY

19.3.1 Classification

In information mining and AI, classification refers to allocating a pre-indicated class to obscure information that is given to the model. A learning model is built depending on the relationship correlation, the indicator characteristics, and the estimation of the objective. The challenge is to perfectly predict the class based on its findings on the previous data. This is supervised learning. The accuracy of the classification system depends highly on the quality of the information provided to the model for learning and the ML model used.

19.3.2 Optimization

All ML calculations have a fixed strategy of learning that relies on estimating their boundaries. When boundaries are dissimilar, the grouping exactness fluctuates for each situation. The test in AI is to locate the most appropriate boundary estimations of the calculations for meeting the objective. In this way, one needs to tweak the calculation boundaries that best suit the issue. There are a few enhancement procedures like genetic algorithms, particle swarm optimization, and search methods.

19.3.3 Development Cycle

Our detailed development cycle for the platform we propose is shown in Figure 19.1. It consists of four main parts: data collection and model design, testing, and comparison.

19.4 IMPLEMENTATION

The interface we developed is a web-based application. Our work featured supervised and unsupervised ML algorithms, and in this section, we describe each algorithm along with their uses. Figure 19.2 gives the flow chart of our model development process.

First, we briefly describe predictive modeling. These models use past data and other attributes to help predict the future using supervised or unsupervised learning. In a supervised learning model, the algorithm already has a target variable, and it's been trained on the training data set; in an unsupervised learning model, the algorithm does not have a target.

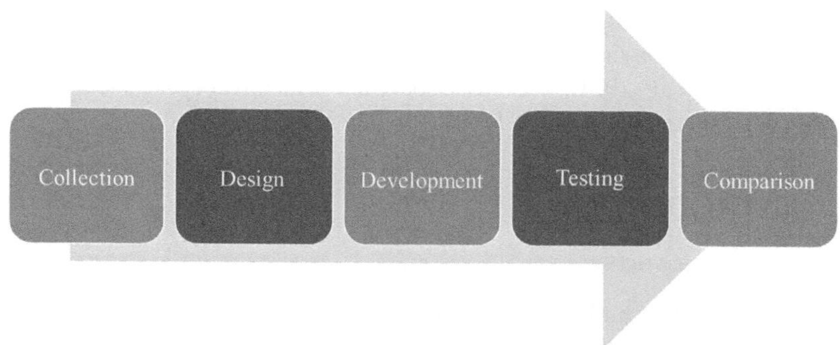

FIGURE 19.1 Development cycle.

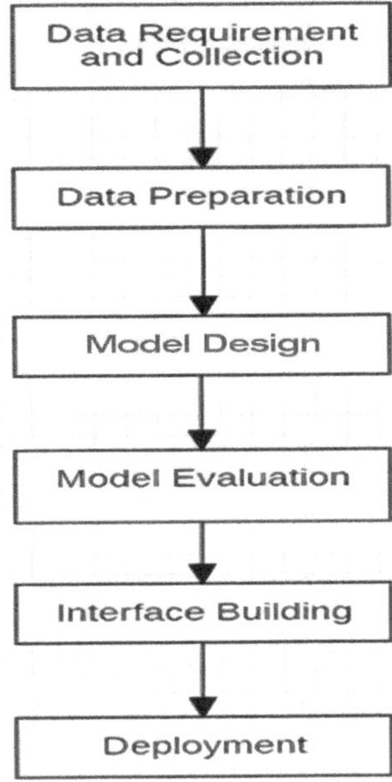

FIGURE 19.2 Flow chart of the model development process.

19.4.1 DATA REQUIREMENTS

Common supervised learning algorithms include logistic regression, which uses the logistic function to model a binary variable, and naïve Bayes classifier; naïve Bayes is rooted in Bayes' theorem, which provides full freedom to agents. A common unsupervised learning technique is hierarchical clustering, which creates a hierarchical structure of clusters with the help of any kind of tree. Another is k-Means clustering, which divides whole data into k different clusters concerning the centroid.

19.4.2 DATA PREPARATION

Preparing the data entailed identifying errors in the collected data sets, removing missing, invalid, and duplicate values and ensuring that it is properly formatted for modeling.

19.4.3 MODEL DEVELOPMENT

We used two distinct data sets for experimentation: one for binary classification (city vs. sell) and another for multiclass classification of cell types. First we developed a custom three-layer convolutional neural network that demonstrated impressive accuracy in binary classification. Then, we pretrained models DenseNet, ResNet, VGG16, and VGG19 for the more intricate task of tumor type classification. Then, we implemented an algorithm.

In this phase, we were determining how the data could be visualized to obtain the required result; we designed and tested all the algorithms in this stage. We used a training set for the prediction systems because this allowed for determining if the system needs to be computed. However, our developed module would help to process various data sets by using algorithms available in the interface.

19.4.4 MODEL EVALUATION

During this phase we tested the module algorithms to determine the correct analytical approach or method to solve the problem and determine if the model met the initial objective of building the research groundwork. We determined that the model generated the required results.

19.4.5 BUILDING THE INTERFACE

The model has a front-end that helps in determining the result. We will establish a connection between the interface and the systems built in historical steps.

19.4.6 MODEL DEPLOYMENT

Once the back end was complete and the complete model was tested, the model provided at least 90% accuracy and provided different results for deep analytics. We used the following technology stack to design the interface: Node.js, React, MongoDB, Python, Canvas.js. We provided the application with data on 25,000 sales transactions after processing the data using ML models with supervised and unsupervised learning. The front end of the application showed the factors that were being measured with the application: item sales data, sales locations for multiple stores, comparison of algorithms, the accuracy of the model, and server performance while processing data.

19.5 RESULTS AND DISCUSSION

The user selects one supervised learning algorithm and one from unsupervised learning as shown in Figure 19.3. After the user has selected both algorithms, the user can move forward to compare the models.

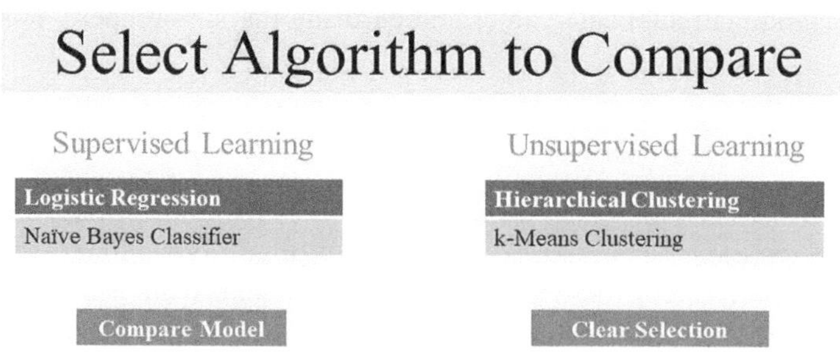

FIGURE 19.3 Selection of algorithms for comparison.

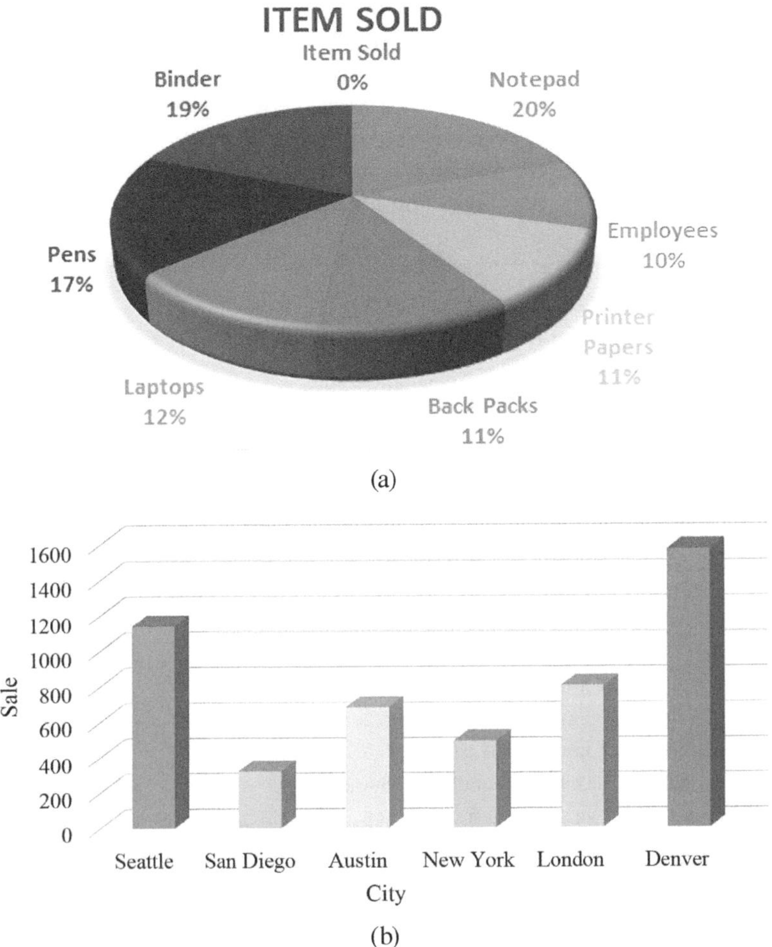

FIGURE 19.4   Visualization of the data from the sales data set: a) items sold; b) sales locations.

Figure 19.4a graphically displays the sales item data set, and Figure 19.4b displays the sales locations per the attributes. Visualization helps the user gain a different understanding of the data.

Attributes and other details from the data are shown in Table 19.1. We used 5,000 records as the data set with 15 attributes per record, and we found 95% accuracy with a 95% confidence interval. Figure 19.5 shows the model accuracy before (a) and after (b) data is provided to the model; accuracy increases as more data is provided.

We also compared the per-user deviation within the models, graphed in Figure 19.6.

We also examined server performance for each model. Because not users performed the experiments using different hardware, the graph in Figure 19.6 shows estimated server response times, which can be used to determine the hardware required for the specific model. In Figure 19.7, we graphed overall server performance compared with response time. The novel interface we developed empirically analyzes different algorithms based on such combinations of parameters, which automatically provides the user with the best algorithm for their data set.

TABLE 19.1

Attributes and Other Features of the Sample Data

Sample Data

#	Store Location	Sale Date	Payment	Customer Email	Gender	Age	Satis-faction	Item Bought	Coupon Used
1	London	4/24/2015	Online	er@ki.nz	M	20	5	1	No
2	Denver	11/6/2013	Instore	hol@12ugi12ami12.li	M	47	5	1	No
3	Seattle	5/30/2013	Instore	amis@kof.vn	M	41	4	8	No
4	Denver	4/16/2016	Instore	kiEa@zutbe.ae	F	46	3	10	No
5	San Francisco	8/4/2013	Instore	fofadoca@aluvoog.sv	M	20	5	5	No
6	London	10/25/2013	Instore	beme@ganoe.kw	F	49	2	9	No
7	Seattle	12/8/2016	Instore	zoz@awjawro.2t	M	57	4	9	No
8	Seattle	11/26/2017	Instore	fi12@luw.jo	F	27	4	9	No
9	Denver	1/112015	Online	kofa112ub@duwarsa.mo	F	31	4	2	No
10	Austin	3/31/2013	Instore	ga@siw.sr	M	65	3	5	No

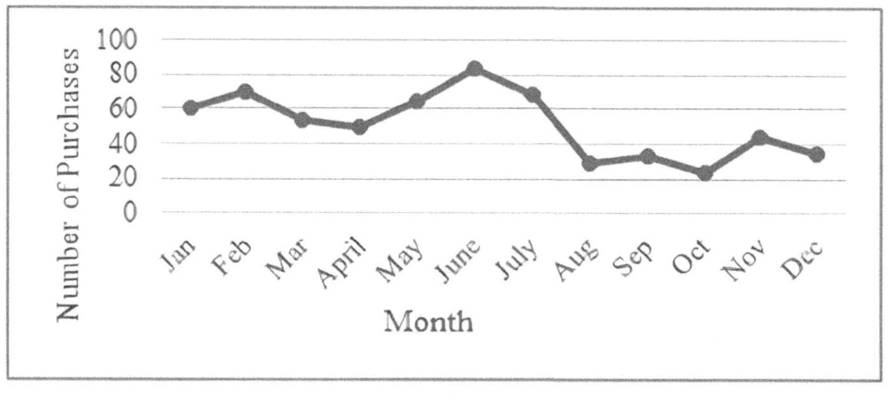

(a)

FIGURE 19.5   Mean sales predictions from the data.

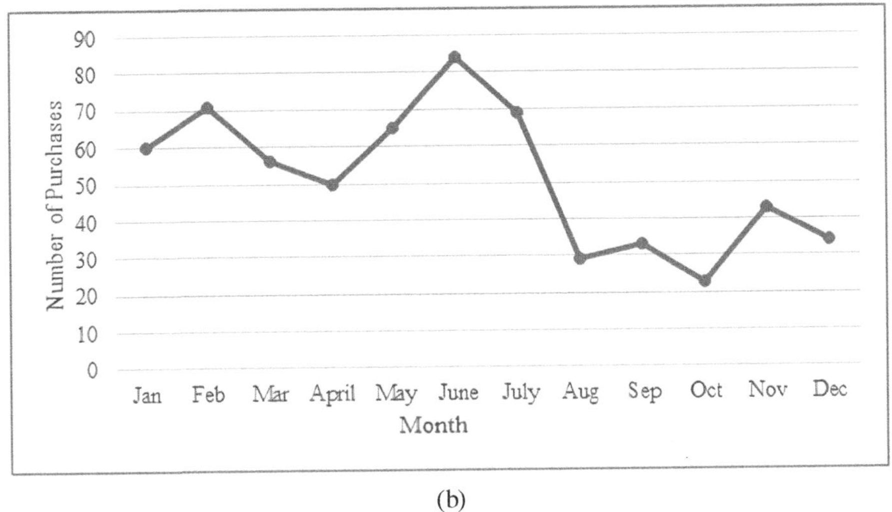

(b)

FIGURE 19.5   (Continued)

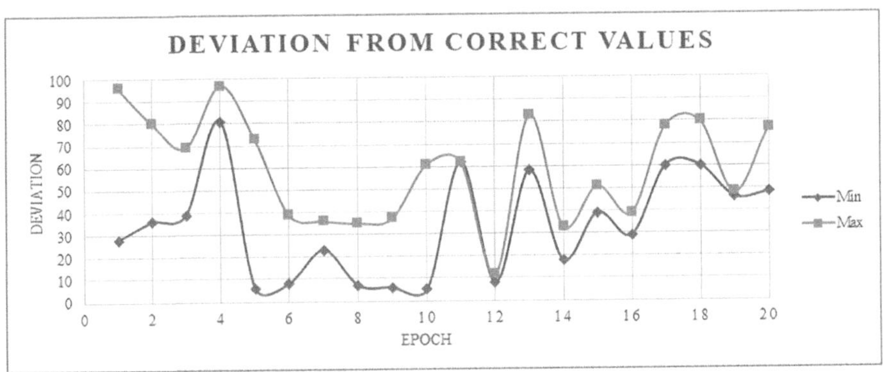

FIGURE 19.6   Deviation in the values.

19.6 CONCLUSION

The algorithms we compared in the research for this chapter provide enough information to determine what could be expected from the model. We compared one supervised and one unsupervised learning algorithm on multiple factors based on the attributes provided to the algorithm via the data set. The interface we developed provided empirical analysis on the following factors: attributes, data visualization, mean sales prediction, model accuracy and deviation from real values, and server performance. The proposed interface provides better accuracy for decision-making.

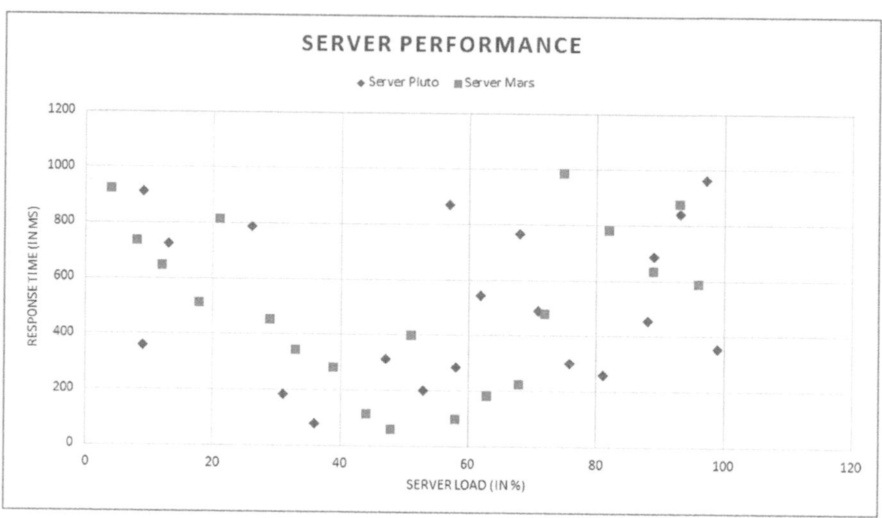

FIGURE 19.7 The performance of the tested servers.

Based on our experiment and analyses, we concluded that the proposed system allows for the desired analysis of data sets in machine learning. This research will save time for researchers seeking the optimal algorithms for analyzing their data sets. New features can be added per end-user requirements and remain as a future research scope.

REFERENCES

Boutaba, R., Salahuddin, M. A., Limam, N. et al. (2018). A Comprehensive Survey on Machine Learning for Networking: Evolution, Applications and Research Opportunities. J Internet Serv Appl, vol. 9, no. 16. https://doi.org/10.1186/s13174-018-0087-2.

Dunsin, D., Ghanem, M. C., Ouazzane, K., & Vassilev, V. (2024). A Comprehensive Analysis of the Role of Artificial Intelligence and Machine Learning in Modern Digital Forensics and Incident Response. Forensic Science International: Digital Investigation, vol. 48, p. 301675, ISSN 2666-2817. https://doi.org/10.1016/j.fsidi.2023.301675.

Ganesh Khekare, K. P. K., Prasanthi, K. N., Godla, S. R., Rachapudi, V., Al Ansari, M. S., & El-Ebiary, Y. A. B. (2023). Optimizing Network Security and Performance Through the Integration of Hybrid GAN-RNN Models in SDN-based Access Control and Traffic Engineering. International Journal of Advanced Computer Science and Applications (IJACSA), vol. 14, no. 12. http://dx.doi.org/10.14569/IJACSA.2023.0141262.

Gonçalves, A. R., Pinto, D. C., Shuqair, S., Dalmoro, M., Mattila, A. S. (2024). Artificial Intelligence vs. Autonomous Decision-Making in Streaming Platforms: A Mixed-Method Approach. International Journal of Information Management, p. 102748, ISSN 0268-4012, https://doi.org/10.1016/j.ijinfomgt.2023.102748.

Gustavo, N., Mbunge, E., Belo, M., Fashoto, S. G., Pronto, J. M., Metfula, A. S., Carvalho, L. C., Akinnuwesi, B. A., & Chiremba, T. R. (2022). Emphasizing the Digital Shift of Hospitality Towards Hyper-Personalization: Application of Machine Learning

Clustering Algorithms to Analyze Travelers. In N. Gustavo, J. Pronto, L. Carvalho, & M. Belo (Eds.), Optimizing Digital Solutions for Hyper-Personalization in Tourism and Hospitality (pp. 1–19). IGI Global. https://doi.org/10.4018/978-1-7998-8306-7.ch001.

Haesevoets, T., Verschuere, B., Van Severen, R., & Roets, A. (2024). How do Citizens Perceive the Use of Artificial Intelligence in Public Sector Decisions? Government Information Quarterly, vol. 41, no. 1, p. 101906, ISSN 0740-624X. https://doi.org/10.1016/j.giq.2023.101906.

Jackson, I., Ivanov, D., Dolgui, A., & Namdar, J. (2024). Generative Artificial Intelligence in Supply Chain and Operations Management: A Capability-based Framework for Analysis and Implementation. International Journal of Production Research. https://doi.org/10.1080/00207543.2024.2309309.

Jahan S. et al. (2023). AI-Based Epileptic Seizure Detection and Prediction in Internet of Healthcare Things: A Systematic Review. IEEE Access, vol. 11, pp. 30690–30725. https://doi.org/10.1109/ACCESS.2023.3251105.

Khekare, G., & Midhunchakkravarthy. (2023). Smart Image Recognition System for The Visually Impaired People. 2023 International Conference on Energy, Materials and Communication Engineering (ICEMCE), Madurai, India, pp. 1–6. https://doi.org/10.1109/ICEMCE57940.2023.10434130.

Khekare, G., & Verma, P. (2021). Prophetic Probe of Accidents in Indian Smart Cities Using Machine Learning. In V. Bhateja, S. C. Satapathy, C. M. Travieso-González, & V. N. M. Aradhya (Eds), Data Engineering and Intelligent Computing. Advances in Intelligent Systems and Computing (vol 1407). Springer. https://doi.org/10.1007/978-981-16-0171-2_18

Khekare, G., Verma, P., & Raut, S. (2022). The Smart Accident Predictor System using Internet of Things. Cloud IoT. Chapman and Hall/CRC, pp. 163–175.

Kim, H., Cheon, S., Jeong, I. et al. (2022). Enhanced Model Reduction Method Via Combined Supervised and Unsupervised Learning for Real-Time Solution of Nonlinear Structural Dynamics. Nonlinear Dyn, vol. 110, pp. 2165–2195. https://doi.org/10.1007/s11071-022-07733-8.

Nakagome, S., Luu, T. P., He, Y. et al. (2020). An Empirical Comparison of Neural Networks and Machine Learning Algorithms for EEG Gait Decoding. Sci Rep, vol. 10, p. 4372. https://doi.org/10.1038/s41598-020-60932-4.

Nguyen, T. V. T., & Vo, N. T. M. (Eds.). (2024). Using Traditional Design Methods to Enhance AI-Driven Decision Making. Hershey, PA: IGI Global. https://doi.org/10.4018/979-8-3693-0639-0.

Ozdenizci, O., & Erdoğmuş, D. (2020 January). Information Theoretic Feature Transformation Learning for Brain Interfaces. IEEE Transactions on Biomedical Engineering, vol. 67, no. 1, pp. 69–78. https://doi.org/10.1109/TBME.2019.2908099.

Perdikis, S., Leeb, R., Chavarriaga, R., & Millán, J. D. R. (2021 August). Context-Aware Learning for Generative Models. IEEE Transactions on Neural Networks and Learning Systems, vol. 32, no. 8, pp. 3471–3483. https://doi.org/10.1109/TNNLS.2020.3011671.

Sarker, I. H. (2021). Machine Learning: Algorithms, Real-World Applications and Research Directions. SN Comput Sci, vol. 2, p. 160. https://doi.org/10.1007/s42979-021-00592-x.

Trillo, J. R., Herrera-Viedma, E., Morente-Molinera, J. A., & Cabrerizo, F. J. (2023 September). A Group Decision-Making Method Based on the Experts' Behavior During the Debate. IEEE Transactions on Systems, Man, and Cybernetics: Systems, vol. 53, no. 9, pp. 5796–5808. https://doi.org/10.1109/TSMC.2023.3275056.

Vaccaro, L., Sansonetti, G., & Micarelli, A. (2021). An Empirical Review of Automated Machine Learning. Computers, vol. 10, no. 1, p. 11. https://doi.org/10.3390/computers10010011.

Vaid, S., Puntoni, S., & Khodr, A. (2023). Artificial Intelligence and Empirical Consumer Research: A Topic Modeling Analysis. Journal of Business Research, vol. 166, p. 114110, ISSN 0148-2963. https://doi.org/10.1016/j.jbusres.2023.114110.

Wang, Y. et al. (2023 October). Decision-Making Driven by Driver Intelligence and Environment Reasoning for High-Level Autonomous Vehicles: A Survey. IEEE Transactions on Intelligent Transportation Systems, vol. 24, no. 10, pp. 10362–10381. https://doi.org/10.1109/TITS.2023.3275792.

Zhan, J., Fang, W., Love, P. E. D., & Luo, H. Explainable Artificial Intelligence: Counterfactual Explanations for Risk-Based Decision-Making in Construction. IEEE Transactions on Engineering Management. https://doi.org/10.1109/TEM.2023.3325951.

20 Multicriterion Analysis of Fusion Sort

A Hybrid Approach to Sorting Algorithms

Sathvik V Koushik, Shamanth Showri N R,
Shreya C R, Urjitha P, and Divya C D

20.1 INTRODUCTION

Efficient sorting algorithms play a pivotal role in diverse data processing tasks, directly influencing overall computational efficiency (Satish et al., 2009). This research addresses the challenge of sorting data sets with intermediate sizes by designing a novel algorithm.

Previous researchers have extensively explored various sorting algorithms and their performance under different scenarios. Hoare (1962) conducted an empirical study comparing Quick Sort, Merge Sort, and Insertion Sort on large data sets, shedding light on each algorithm's relative strengths and weaknesses. Bentley (1980) introduced a divide and conquer approach optimized for large data sets, but its efficiency diminished with smaller arrays.

Estivill-Castro et al. (1992) explored adaptive sorting algorithms, discussing theoretical contributions in complexity and disorder measures, practical insights into adaptivity, and faster alternatives for nearly sorted sequences. Al-Kharabsheh et al. (2013) investigated sorting algorithms, comparing Grouping Comparison Sort with conventional methods for improved performance. While established sorting techniques like Quick Sort (Hoare, 1962; Steier et al., 1989) and Merge Sort (Cole, 1988) are widely used and respected for their efficacy, they may not always be the most efficient choices for data sets falling within this specific size range.

Sedgewick (1978) presented the practical implementation of Quick Sort and its variants on real computers to optimize performance while minimizing extra storage usage and summarized the analytic performance results. Rösler (1991) examined the number of comparisons performed by Quick Sort and established the convergence of the random variable $(Xn—E(Xn))/n$ to Y, and Y exhibited exponential tails, indicating efficient sorting performance. TimSort's adoption as the standard sorting algorithm in Java and Python challenges the perception of merge algorithms' inefficiency. Other researchers developed a new framework and a simpler competitive algorithm and confirmed O(n log n) running time for TimSort (Auger et al., 2015). One group

DOI: 10.1201/9781032635170-20

compared five merge sorting algorithms for large data sets on FPGA, determining the best algorithm based on resource utilization, delay, and area comparisons (Lobo et al., 2020). Another found that a hybrid sort algorithm outperformed bitonic merge sort and bitonic odd–even sort with increased input bit width, achieving three times faster processing time through pipelining and parallel processing (Harshini et al., 2019).

De Micco et al. (2019) introduced a hybrid data ordering algorithm, combining serial and parallel instructions for sorting, performing complexity analysis, and comparing it with other sorting algorithms are presented. Lakshmivarahan et al. (1984) studied parallel sorting algorithms, vital for commercial applications, in relation to parallel computer architecture, covering fixed interconnection networks and global memory models. Special-purpose network-sorting and SIMD machine algorithms are included.

Blelloch et al. (1991) compared three parallel sorting algorithms on the CM-2 Supercomputer and found that sample sort outperformed the others on large data sets, while radix sort was stable, and bitonic sort was space efficient. Estivill-Castro et al. (1992) explored adaptive sorting algorithms' contributions in theory and practice, discussing complexity descriptions and introducing faster algorithms for nearly sorted sequences, and a different group found that Selection Sort outperformed Quick Sort for integer and string arrays, with an upper bound running time of $O(n)$ (Aliyu et al., 2013). Khandelwal (2021) discussed the significance of efficient sorting algorithms in various fields, presenting an analysis of parallel and sequential algorithms on different GPU and CPU architectures, exploiting task parallelism.

In this work, we present Fusion Sort, a novel hybrid sorting algorithm tailored to effectively manage data sets of intermediate sizes. By harnessing the advantages of both Insertion Sort and divide and conquer, Fusion Sort dynamically adapts its sorting approach based on a carefully selected threshold. For smaller subarrays, it efficiently employs Insertion Sort, capitalizing on its superior performance for such data; for larger subarrays, it seamlessly switches to divide and conquer, leveraging its efficiency in handling bigger data sets.

Extensive experimental evaluations underscore the superiority of Fusion Sort over Quick Sort and Merge Sort for intermediate-sized data sets. Its adaptive nature empowers Fusion Sort to outperform traditional sorting algorithms in this specific context, making it an appealing choice for real-world applications that frequently encounter data sets of moderate sizes. This study provides an in-depth analysis of Fusion Sort's algorithmic steps, complexity, and experimental results, substantiating its standing as a practical and robust sorting solution for data sets within the intermediate size range.

20.2　PROBLEM STATEMENT

Efficient sorting algorithms serve as the backbone of various data processing tasks, significantly impacting the overall computational efficiency (Satish et al., 2009). Sorting algorithms, such as Quick Sort and Merge Sort, have been extensively studied and are well established for large data sets. However, there exists a unique challenge in the realm of data sorting: data sets of intermediate sizes. While large data sets have their dedicated sorting strategies, and small data sets can be efficiently managed with algorithms like Insertion Sort, the sorting of intermediate-sized data sets has received relatively less attention. We sought to fill this gap in the research by presenting Fusion Sort and conducting a comprehensive comparative analysis to demonstrate its efficacy and superiority in handling data sets within the intermediate size range.

20.3 THE PROPOSED WORK

20.3.1 METHODOLOGY

20.3.1.1 Experimental Setup

To evaluate the performance of Fusion Sort, we set up a series of experiments using data sets of different sizes and characteristics. We describe the sets we used, including synthetic data sets with random elements and real-world data sets representing diverse data types.

20.3.1.2 Threshold Selection

The threshold plays a crucial role in the performance of Fusion Sort. We experimented with different thresholds to identify the best one for intermediate input sizes. After thorough analysis, we found that a threshold of 34 yielded the optimal results for medium-sized data sets.

20.3.1.3 Comparative Algorithms

To assess Fusion Sort's efficiency, we select two well-established sorting algorithms, Quick Sort and Merge Sort, as benchmarks for comparison. We briefly outline the working principles of each algorithm and their respective complexities.

20.3.1.4 Performance Metrics

We define the performance metrics used to evaluate the sorting algorithms' efficiency. These metrics include average-case time complexity and the number of comparisons performed during the sorting process.

20.3.1.5 Experimental Procedure

We conduct a series of experiments with varying input sizes to measure the performance of Fusion Sort, Quick Sort, and Merge Sort. Each experiment is repeated multiple times to ensure statistical significance. We describe the steps taken to eliminate biases and ensure fair comparisons between the algorithms.

20.3.1.6 Data Analysis

After conducting the experiments, we compare the accuracy of Fusion Sort with that of Quick Sort and Merge Sort. We present the experimental data in graphical form, highlighting the performance differences among the algorithms for different data set sizes.

20.3.2 ALGORITHMS

Fusion Sort is a hybrid sorting algorithm that combines the strengths of Insertion Sort and divide and conquer to achieve accurate and effective sorting performance.

This algorithm optimizes the sorting process by employing Insertion Sort on smaller subarrays and divide and conquer on larger subarrays. The algorithm uses a user-specified threshold value to determine when to switch between the two strategies.

20.3.2.1 Algorithm: Fusion Sort

Input: An unsorted array of comparable elements.

Threshold: An integer representing the threshold value that determines when to use Insertion Sort on smaller subarrays.

Output: A sorted array in ascending order.

```
function fusion_sort(arr, threshold)
  if the length of arr ≤ threshold then
    return insertion_sort(arr[:]) # Create a new list
      to sort and return
    end if
  pivot = choose_pivot(arr)
  smaller_subarray, larger_subarray = partition(arr,
    pivot)
  smaller_subarray = fusion_sort(smaller_subar-
    ray, threshold)
  larger_subarray = fusion_sort(larger_subarray,
    threshold)
  merged_arr = merge(smaller_subarray, [pivot],
    larger_subarray)
  return merged_arr
end function
```

20.3.2.2 Algorithm: Choose Pivot

Input: An array 'arr' for which the pivot needs to be chosen.

Output: The pivot element (e.g., the median of the first, middle, and last elements in the array).

```
function choose_pivot(arr)
  # Choose the pivot element (e.g., median of the
    first, middle, and last elements)
  return sorted([arr[0], arr[length of arr // 2], arr[-
    1]])[1]
end function
```

20.3.2.3 Algorithm: Partition

Input: An array 'arr' and a pivot element 'pivot' for which the array needs to be partitioned.

Output: Two subarrays: 'smaller_subarray' containing elements less than the pivot, and 'larger_subarray' containing elements greater than the pivot.

```
function partition(arr, pivot)
  smaller_subarray = [ ]
  larger_subarray = []
```

```
    for x in arr do
      if x < pivot then
        append x to smaller _ subarray
      else if x > pivot then
        append x to larger _ subarray
      end if
    end for
   return smaller _ subarray, larger _ subarray
  end function
```

20.3.2.4 Algorithm: Insertion Sort

Input: An unsorted array 'arr'.
Output: The sorted array 'arr' in ascending order.

```
    function insertion _ sort(arr)
      for i from 1 to length of arr - 1 do
        key = arr[i]
        j = i - 1
        while j ≥ 0 and arr[j] > key do
          arr[j + 1] = arr[j]
          j = j - 1
        end while
        arr[j + 1] = key
      end for
      return arr
    end function
```

20.3.2.5 Algorithm: Merge Arrays

Input: Three sorted arrays—arr1, arr2, and arr3
Output: Merged sorted array containing all elements from arr1, arr2, and arr3

```
    function merge(arr1, arr2, arr3)
      merged _ arr = []
      i = j = k = 0
      while i < length of arr1 and j < length of arr2 and
        k < length of arr3 do
        if arr1[i] ≤ arr2[j] and arr1[i] ≤ arr3[k] then
          append arr1[i] to merged _ arr
          i = i + 1
        else if arr2[j] ≤ arr1[i] and arr2[j] ≤ arr3[k] then
          append arr2[j] to merged _ arr
          j = j + 1
        else
          append arr3[k] to merged _ arr
          k = k + 1
        end if
      end while
      while i < length of arr1 do
```

```
      append arr1[i] to merged_arr
      i = i + 1
    end while
    while j < length of arr2 do
      append arr2[j] to merged_arr
      j = j + 1
    end while
    while k < length of arr3 do
      append arr3[k] to merged_arr
      k = k + 1
    end while
    return merged_arr
  end function
```

20.4 RESULTS AND DISCUSSION

We present the experimental results obtained from evaluating the Fusion Sort algorithm and compare it with Quick Sort and Merge Sort. We conducted a comprehensive set of experiments using data sets of different sizes and characteristics and repeated the experiments multiple times to ensure statistical significance, and the average results are reported here.

20.5 PERFORMANCE ANALYSIS

20.5.1 Execution Time

The average time taken by each algorithm to sort the small, medium, and large data sets (Figures 20.1 to 20.3).

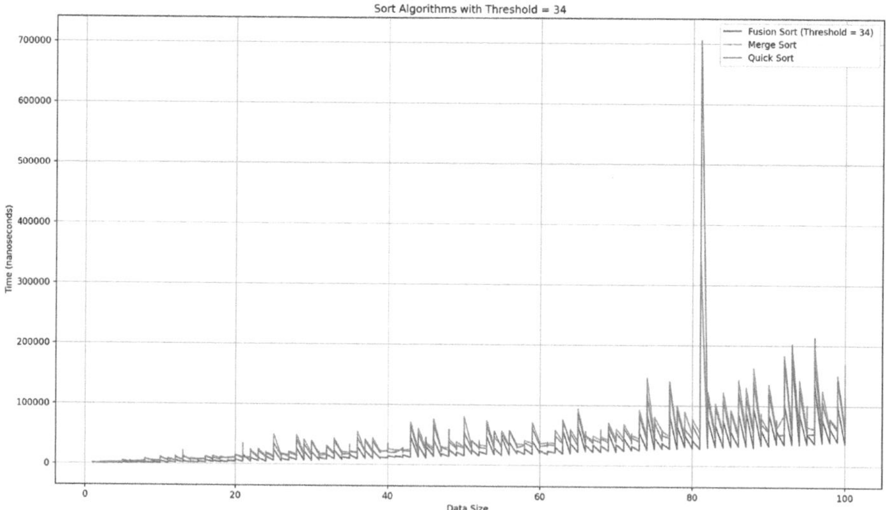

FIGURE 20.1 Execution time for small input sizes.

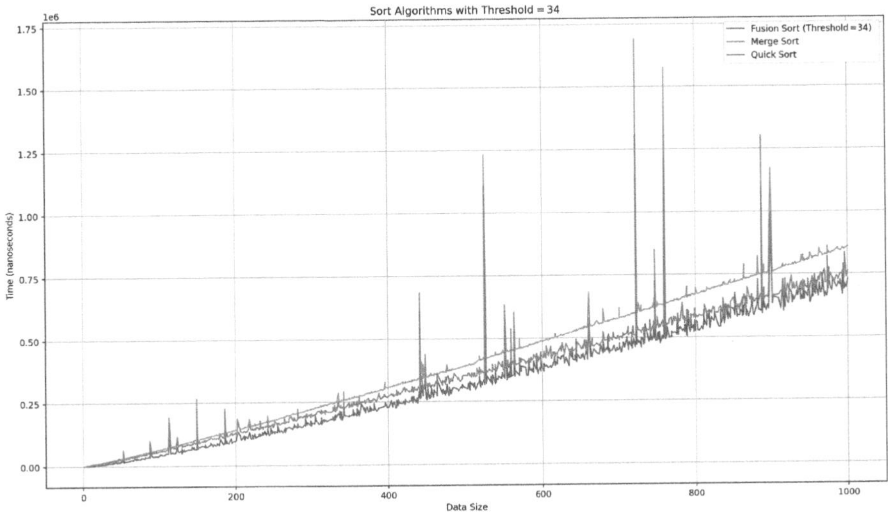

FIGURE 20.2   Execution time for medium input sizes.

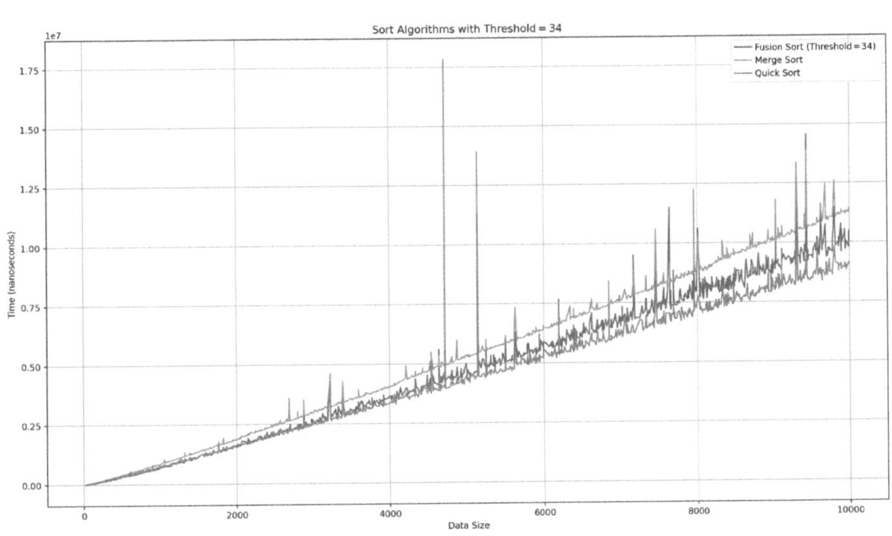

FIGURE 20.3   Execution time for large input sizes.

20.5.2   NUMBER OF COMPARISONS

The total number of element comparisons made during the sorting process for the small, medium, and large data sets (Figures 20.4 to 20.6); fewer comparisons indicate higher accuracy.

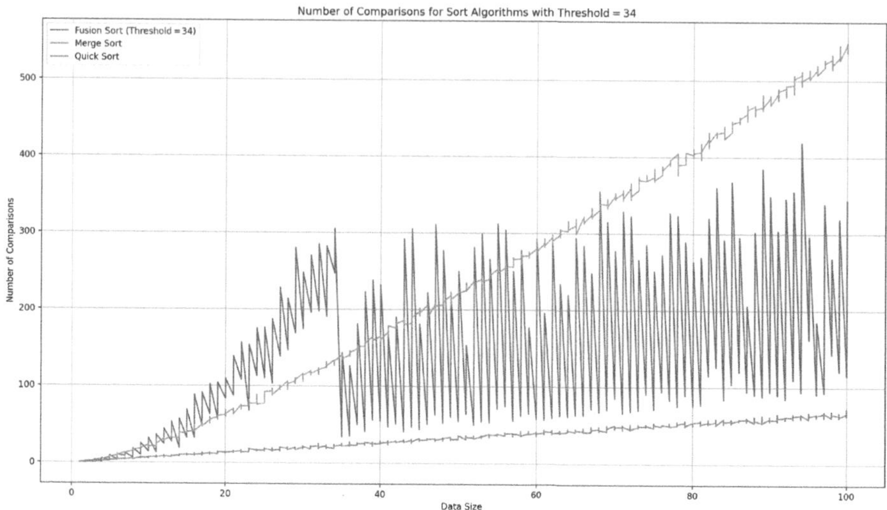

FIGURE 20.4   Number of comparisons for small input sizes.

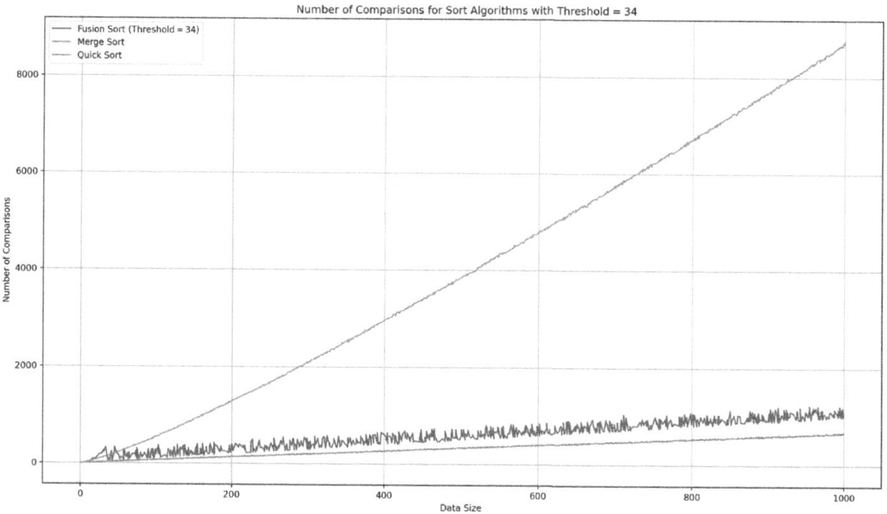

FIGURE 20.5   Number of comparisons for medium input sizes.

20.5.3 TIME COMPLEXITY

To derive the time complexity of Fusion Sort, we need to analyze the number of operations performed at each step of the algorithm. Let's break down the time complexity analysis step by step:

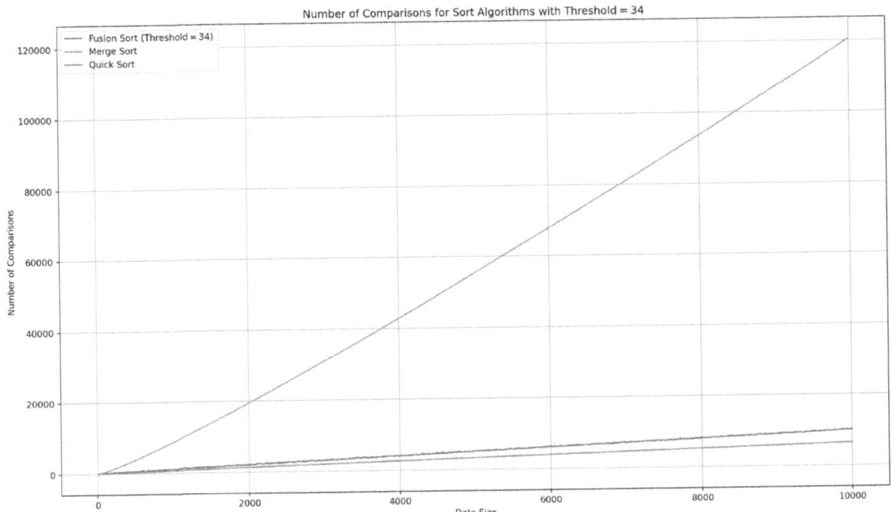

FIGURE 20.6 Number of comparisons for large input sizes.

Choosing the Pivot:
- The 'choose_pivot' function selects the pivot element in O(1) time as it only involves basic array indexing and comparisons.

Partitioning the Array:
- The 'partition' function divides the input array into two subarrays based on the pivot element in linear time, O(n), where n is the size of the input array. It iterates through the array once and assigns each element to one of the two subarrays.

Recursive Calls:
- Fusion Sort recursively calls itself on 'smaller_subarray' and 'larger_subarray'. The size of these subarrays is proportional to the size of the original array.

Merging the Subarrays:
- 'Merge' combines the sorted subarrays and the pivot element into a single sorted array in linear time, O(n), where n is the total number of elements in the input array.

Let T(n) represent the time complexity of Fusion Sort for an input array of size n. The time complexity can be expressed as a recurrence relation:

$$T(n) = 2 \times T(n/2) + O(n) \tag{20.1}$$

The recurrence relation is derived from the two recursive calls in Step 1, and the time complexity for merging the subarrays in Step 5.

Using the Master Theorem, we can solve the recurrence relation.

For the given recurrence relationship:

a = 2 (number of recursive calls),

b = 2 (size of the subproblems),

f(n) = O(n) (time complexity for merging the subarrays).

Since f(n) = O(n) falls under Case 1 of the Master Theorem (where f(n) = O(n^c) and c < $\log_b(a)$), the overall time complexity of Fusion Sort can be determined as follows:

$$T(n) = \Theta(n\log n) \tag{20.2}$$

Thus, the time complexity of Fusion Sort is $\Theta(n\log n)$, which makes it more efficient than traditional sorting algorithms like Quick Sort and Merge Sort for data sets of intermediate sizes. For small subarrays (size ≤ threshold), the time complexity is $O(k^2)$, where k is the size of the subarray, due to the use of Insertion Sort. This adaptive nature of Fusion Sort contributes to its superior performance for data sets of varying sizes and characteristics.

20.6 CONCLUSION

Fusion Sort emerges as a highly efficient sorting algorithm tailored for data sets with intermediate sizes. By judiciously combining Insertion Sort with divide and conquer, Fusion Sort performed well in scenarios where sorting intermediate-sized data sets is a common requirement. Its dynamic, seamless handling of varying input sizes makes Fusion Sort a valuable and versatile addition to the domain of sorting algorithms, offering a better solution for the real-world sorting of data sets of moderate dimensions. The algorithm's ability to adapt to different data set characteristics and deliver superior performance sets it apart as a valuable tool for enhancing data processing accuracy and efficiency across diverse computational tasks.

REFERENCES

Aliyu, Ahmed M., and P. B. Zirra. "A comparative analysis of sorting algorithms on integer and character arrays." *The International Journal of Engineering and Science* (2013): 25–30. ISSN(e): 2319–1813, https://citeseerx.ist.psu.edu/document? repid=rep1&type=pdf&doi=b8b872f7cb7b08cd61088cbf55aa2e42221b23a7

Al-Kharabsheh, Khalid Suleiman, et al. "Review on sorting algorithms a comparative study." *International Journal of Computer Science and Security (IJCSS)* 7.3 (2013): 120–126, https://irjet.net/archives/V3/i12/IRJET-V3I12115.pdf

Auger, Nicolas, Cyril Nicaud, and Carine Pivoteau. "Merge strategies: From merge sort to Timsort." (2015), https://hal.science/hal-01212839

Bentley, Jon Louis. "Multidimensional divide-and-conquer." *Communications of the ACM* 23.4 (1980): 214–229, https://doi.org/10.1145/358841.358850

Blelloch, Guy E., et al. "A comparison of sorting algorithms for the connection machine CM-2." *Proceedings of the Third Annual ACM Symposium on Parallel algorithms and Architectures.* (1991), https://dl.acm.org/doi/pdf/10.1145/113379.113380

Cole, Richard. "Parallel merge sort." *SIAM Journal on Computing* 17.4 (1988): 770–785, https://doi.org/10.1137/0217049

De Micco, Luciana, Mariano L. Acosta, and Maximiliano Antonelli. "Hybrid sorting algorithm implemented by high level synthesis." *IEEE Latin America Transactions* 18.02 (2019): 430–437, https://doi.org/10.1109/TLA.2020.9085300

Estivill-Castro, Vladmir, and Derick Wood. "A survey of adaptive sorting algorithms." *ACM Computing Surveys (CSUR)* 24.4 (1992): 441–476, https://doi.org/10.1145/146370.146381

Harshini, V. S., and KK Senthil Kumar. "Design of hybrid sorting unit." *2019 International Conference on Smart Structures and Systems (ICSSS)*. IEEE (2019), https://doi.org/10.1109/ICSSS.2019.8882866

Hoare, Charles A. R. "Quicksort." *The Computer Journal* 5.1 (1962): 10–16, https://doi.org/10.1093/comjnl/5.1.10

Khandelwal, Vikram. "Analysis and review of sorting algorithms." *International Journal of Algorithms Design and Analysis* 7.2 (2021): 1–5, https://computers.journalspub.info/index.php?journal=JADA&page=article&op=view&path[]=740

Lakshmivarahan, S., Sudarshan K. Dhall, and Leslie L. Miller. "Parallel sorting algorithms." *Advances in Computers* 23. Elsevier (1984): 295–354, https://doi.org/10.1016/S0065-2458(08)60467-2

Lobo, Joella, and Sonia Kuwelkar. "Performance analysis of merge sort algorithms." *2020 International Conference on Electronics and Sustainable Communication Systems (ICESC)*. IEEE (2020), https://doi.org/10.1109/ICESC48915.2020.9155623

Rösler, Uwe. "A limit theorem for 'Quicksort'." *RAIRO-Theoretical Informatics and Applications* 25.1 (1991): 85–100, http://www.numdam.org/item?id=ITA_1991__25_1_85_0

Satish, Nadathur, Mark Harris, and Michael Garland. "Designing efficient sorting algorithms for manycore GPUs." *2009 IEEE International Symposium on Parallel & Distributed Processing*. IEEE (2009), https://doi.org/10.1109/IPDPS.2009.5161005

Sedgewick, Robert. "Implementing quicksort programs." *Communications of the ACM* 21.10 (1978): 847–857, https://doi.org/10.1145/359619.359631

Steier, D. M., et al. "Quicksort." *Algorithm Synthesis: A Comparative Study* (1989): 24–38, https://doi.org/10.1007/978-1-4613-8877-7_3

21 Cruising through the Choices

Unraveling Destination Decision-Making Dilemmas with Social Networks via MCDM

Hai-Ninh Do

21.1 INTRODUCTION

Over the past two decades, the Internet has transformed information seekers into cocreators, enabling global sharing and peer-to-peer applications (Frías et al., 2008; Volo, 2010; Akehurst, 2009). However, web 2.0 applications have, among other benefits, led to the rapid growth of online social networking sites with tourism (Werthner & Ricci, 2004). Gretzel and Yoo (2008) noted that travel information searches are becoming more common online as web technology evolves. According to Zheng et al. (2008), the Internet has transformed tourism information distribution and communication, allowing e-travelers and potential tourists to learn about destinations from other tourists (Frías et al., 2008).

Tourists today plan their trips using web searches in combination with traditional sources. Frías et al. (2008) and O'Connor and Murphy (2004) found that YouTube, blogs, Facebook, and Flickr, among others, have altered the role of travel agents in terms of providing information. Gretzel and Yoo (2008) agreed with this notion, highlighting the significance of social media content in searches for tourist destination information. Social media platforms allow tourists to engage with both tourism operators and experienced tourists (Jenkin, 2010), making it easier to share travel-related comments, thoughts, personal experiences, and videos and images.

Tourist decision-making has always been an important in tourism research (Wong & Yeh, 2009), and information searching is crucial to decision-making. Tourists invariably need knowledge about potential destinations to make informed choices (Sheldon, 1993). Bieger and Laesser (2004) noted that most visitors' information-searching and decision-making models include pre-purchase information searches. Therefore, the tourism sector must make precise and deliberate marketing strategy choices.

DOI: 10.1201/9781032635170-21

The Decision-Making Trial and Evaluation Laboratory (DEMATEL) method (Fontela & Gabus, 1974) uses structural modeling to solve complicated problems with consumers' destination choices by revealing cause-and-effect relationships or criteria for measuring a problem (Chen-Yi et al., 2007). For this study, I applied multicriteria decision-making (MCDM)-tree model to examine decision-making processes related to tourist destinations, with a specific emphasis on the role of information searching through social network sites (SNSs). I began with the proposal that motivation, information searching, and the decision-making process are perceptual variables influencing tourist destination choice, and I give an interdisciplinary perspective on a tourist's destination decision-making as a social network user. Expanding on the research backdrop, I test the destination decision-making tree using DEMATEL MCDM methodology. We empirically tested the research model using interviews and surveys.

Three goals drove this study:

1) Investigate why tourists choose a destination on social media and what elements significantly influence their decision-making. We examined tourists' searches and motivation-based decision-making.
2) Examine how social media affects travel motivation and perceived limits for tourists using social media as a tool for destination selection. We examined the information tourists used on social media.
3) Use the MCDM model to weigh criteria and suggest ways for industry stakeholders to improve tourist destination decision-making.

The following reasons make this work empirically significant. First, most studies on social media and social marketing strategies use SNSs, and these need to be examined as a category of research instrument. With this study, I aimed to assist tourism businesses in understanding the role of tourist-social network users in destination decision-making, marketing strategy development, and location promotion.

Second, I investigated whether internal or external factors affect tourism destination decision-making. This study could help with recent motivational theory attempts to understand SNSs as external or push factors of information-seeking demand. Third, SNSs and tourism have been studied extensively, but tourist–SNS user relationships have not. This study could help national tourism organizations and the tourism sector more broadly construct a more compelling online social network to attract more tourists.

21.2 LITERATURE REVIEW

21.2.1 THEORETICAL FOUNDATION

21.2.1.1 Motivation: Push–Pull Theories

Maslow's hierarchy of needs, a well-known theory in sociology and social psychology, encompasses various human needs and offers a comprehensive framework for understanding human motivation and behavior (Chon, 1989). Other researchers have explored sociopsychological and humanizing motives through different conceptual models (Crompton, 1979), such as the push–pull model (Dann, 1981) and four primary motivation concepts (Murphy, 1985). Others have studied tourist motivation, including Mill and Morrison (1985), Mansfeld (1992), McIntosh et al. (1995), Gnoth (1997), Goossens (2000), and Nicoletta and Servidio (2012). Moscardo et al. (1996) described tourist motivation as a combination of internal psychological factors (needs, wants, and goals), while Middleton (1994) asserted that motivation must align with long-term, multimotive, and nondeterministic needs and personal goals for travel groups.

Kim et al. (1996) indicated that tourism motivation is a key determinant of visitor behavior; to promote a destination, marketers must understand what motivates tourists' actions and behavior (Gee et al., 1989). Tourist motivation is a complex topic that describes people's decisions and behaviors before and during vacations. Fulfilling tourist desires requires understanding their primary motivations (Huang & Hsu, 2009). Some academics employed Maslow's hierarchy of needs theory to analyze tourist motivation and found travel to be a need or wish satisfaction (Mill & Morrison, 1985). Another Maslow-inspired researcher, Dann (1977), examined tourist motivators and identified push–pull elements when choosing a destination. As Pearce and Caltabiano (1983) concluded, tourists choose destinations to meet their needs and wants.

Tourism motivation literature widely accepts the distinction between push and pull influences. Dann (1977) and Crompton (1979) pioneered tourist motivation research by categorizing it into push and pull factors and categorized the latter as the underlying notion of travel (Crompton, 1979). The push elements are tied to tourists' desire to leave an area and seek an intrinsic benefit, whereas the pull aspects are related to destination qualities. According to Wolfe and Hsu (2004), individuals' choices are controlled by push and pull influences, and I employed the push–pull construct as the backdrop that guides tourist decision-making from motivation to final decision.

21.2.1.2 Decision-Making Theories

Travel behavior experts have adapted the broad theories of consumer behavior to understand travel behavior. According to Woodside and Jacobs (1985), travelers' decision-making is complex and impacted by their behavior. Moutinho (1987) also proposed extensive conceptual and empirical efforts to define the model of tourist decision-making processes. Fuller et al. (2007) used multiple decision theories to outline the influences of internal and external factors on tourists' trip-buying decisions; they outlined personality, motivation, and lifestyle as internal elements in tourist behavior and situational, cultural, and family influences as external. Middleton (1994) also presented a trip decision model that identified stimulus input, communication channels, buyer characteristics and decision process, and purchase output as components of the travel decision process; the author found that incentives drive travel behavior, bridging the gap between need and action or purchase. According to Mill and Morrison (2002), travelers' decisions are influenced by four factors: tourism forces, motivation, destination selection, and travel purchase.

21.2.2 Tourist Motivation with Destination Decision-Making

Tourist motivation is a dynamic process of the interactions of push and pull motivations. Kim and Baum (2007) and Devesa et al. (2010) identified that tourist characteristics are the push factors that explain the desire to go on vacation, while the pull factors are the destination characteristics that attract tourists.

21.2.2.1 Tourist Characteristics

Individuals' decisions to travel are guided in part by their sociodemographic and personality traits (Wolfe & Hsu, 2004), and researchers have extensively examined sociodemographic factors as explanations for evoked alternative destinations and information-processing antecedents (Woodside & Lysonski, 1989). Gender, age, occupation, education, and socioeconomic class have been identified as push factors that influence destination perceptions, with age and education being particularly influential in final destination decisions (Um & Crompton, 1990).

Bagolu and Mangaloglu (2001) showed small differences in US destination image based on age, marital status, and occupation, and McIntosh and Goeldner (1984) suggested that travel incentives were influenced by sex, age, and education. Figure 21.1 illustrates the relationship between tourist characteristics, travel motivations, and destination types. For instance, higher education levels influence travel motivation and the use of the Internet for travel-related transactions and seeking information (Bonn et al., 1999), and age, education, and marital status are associated with travel decision-makers' resources and constraints (Gretzel et al., 2012). Sociodemographic factors also affect tourist selection (Grønflaten, 2009; Snepenger et al., 1990).

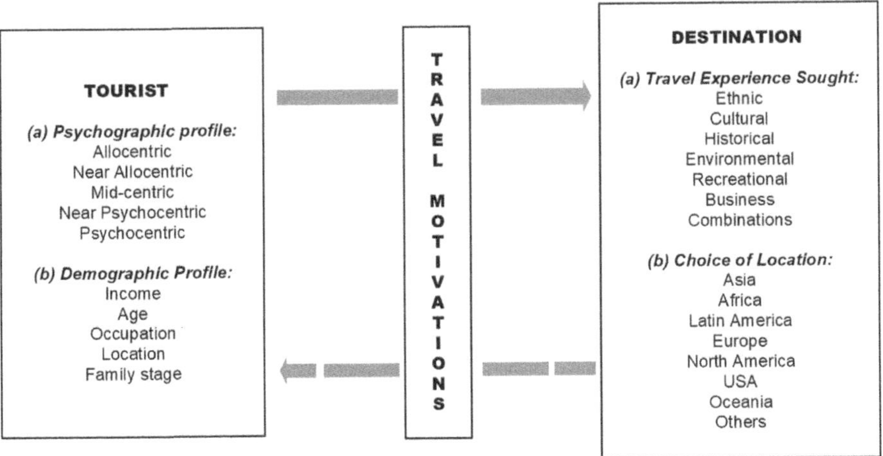

FIGURE 21.1 Relationships between types of tourists, travel motivations, and types of destination.

21.2.2.2   Destination Characteristics

21.2.2.2.1   *Destination Environment*

Fridgen (1991) defined a destination environment as the social and physical surroundings at a given destination, and Murphy et al. (2000) identified destination environment characteristics as including physical variables like amenities, weather, climate, scenic and natural resources and social factors like culture, language, social structure, and residents' friendliness. Potentially negative social factors including culture, safety, security, and socioeconomics can pose significant barriers to tourism (Chon et al., 1999). Mihalič (2000) found that destination environment influences travel decisions.

21.2.2.2.2   *Destination Image*

A destination image represents a complete set of beliefs about a place, an overall view of a destination, and a fundamental part of destination marketing (Chen & Tsai, 2007). Tourists tend to have preconceived opinions and expectations of a destination before visiting, and destination image influences tourist behavior and attracts them to a destination (Beerli & Martin, 2004; Woodside & Lysonski, 1989; Hunt, 1975). The most crucial stage in tourists' destination choosing is image formation before the trip. Chon (1991) demonstrated that good images significantly influence tourist destination decisions, while negative images do not.

21.2.2.2.3   *Destination Brand*

Destination brands are "perceptions about a place as reflected by the associations held in tourist memory" (Cai, 2002). Destination branding can be expanding into tourism despite little tourism literature coverage. Marzano and Scott (2009) studied destination brands for their external influence on tourist behavior and decisions, and Murphy et al. (2007) confirmed that destination brand and image are linked to customer decisions. Lim et al. (2012) argue that the Internet and social media, such as Facebook and Twitter, significantly influence decision-making, so marketers tend to combine branding techniques with social media. According to Cai (2002), vacation spot decision-making processes are complex, and destination branding must aim to build good reputations, boost competitiveness, and attract tourists. Gartner and Ruzzier (2011), Chon (1991), Echtner and Ritchie (1993), and Prebensen (2007) have all highlighted the differences between destination image and destination brand. Prebensen (2007) particularly emphasized the sustainability of destination brand over destination image, although while acknowledging similarities between the two.

21.2.3   Information Searching in Tourists' Destination Decision-Making

Information is crucial to destination selection and behavior at a destination, according to tourist decision models. Consumer behavior researchers aim to understand how customers process information, which can impact their buying decisions (Frías et al., 2008). Fodness and Murray (1997) determined that information search affects and is affected by tourist characteristics, with some motivation and search components directly affecting destination intention.

Consumers commonly employ search methods to aid in their purchase decisions while tourists rely on diverse sources of information that differ based on the type of holiday and type of traveler. Common sources of information for travel planning include recommendations from friends and family, destination-specific literature such as guidebooks and travel brochures, professional advisors such as travel agencies, and, increasingly, the Internet (Nolan, 1976). Tourism information search uses internal and external sources (Ho et al., 2012). Sharon & Smith (1987) claimed that people first examine their memory for information, then go out and find it. This section discusses the second stage of tourist decision-making, information search, which involves motivated seeking of information from memory or the surroundings.

21.2.3.1 Internal Search

Whenever travelers realize that they need to decide, an information search is likely to occur. Initially, the decision-making takes place internally based on previous experience and knowledge (Fodness & Murray, 1997).

21.2.3.1.1 Memory

Experience with different places and activities is key to leisure tourism decision-making (Snepenger et al., 1990). Murray (1991) asserted that consumers review past purchase memories and product knowledge via environmental learning before making a new purchase. Internal search refers to the process of retrieving decision-relevant information from long-term memory. Interior sources also include personal encounters with a specific location or a similar one (Vogt & Fesenmaier, 1998). Additional research indicates that experience affects trip planning knowledge acquisition.

21.2.3.1.2 Knowledge

Prior knowledge is fundamental for information searches because it strongly affects the decision-making process (Kerstetter & Cho, 2004). Gursoy and Terry Umbreit (2004) found that decision-makers' knowledge, attitude, and motivation determine strategy selection; to make decisions efficiently, individuals must come across, remember, arrange, and analyze information because memory-based knowledge is essential for information processing. Consumer behavior researchers have extensively investigated variables such as familiarity and product experience; as individuals learn more about a product, their decisions are frequently influenced by memory, prior knowledge, and familiarity rather than by new external information (Bettman, 1979).

21.2.3.2 External Search

While investigating the effect of prior knowledge on information search behavior, DiPietro et al. (2007) found a high probability of using external information sources to make decisions by travelers during the first stage of product knowledge. Furthermore, Hyde (2008) found that tourists are likely to seek information from external sources when it is not already present in their knowledge and memory. That is, external search is necessary for individuals who are unfamiliar with the product, including travelers deciding on destinations. Travelers utilize the following four broad external information sources when planning their trips: word-of-mouth (WOM) advice from family and friends, destination-specific literature, media and travel consultants (third-party), and more recently, the Internet (Hyde, 2008).

21.2.3.2.1 WOM from Friends and Family

According to research, word of mouth from friends and family is the most influential pre-purchase information source; in WOM, consumers share their experiences with and opinions about a company or product.

Consumers use WOM to reduce risk and uncertainty (Zhang et al., 2010). The fundamental challenge in tourism is the absence of direct experience with a destination, making it difficult to assess its quality and benefits before consumption (Schmallegger & Carson, 2008), and advice from consumers who have used a tourist product such as a destination is influential in travel decision-making. WOM has long been a critical external source for vacation planning and goods purchases (Shankar et al., 2003).

21.2.3.2.2 Tourism Intermediaries

Tourism products are delivered through numerous different intermediaries such as travel agencies (Akehurst, 2009). Although the leading sources of information for making destination decisions are typically informal recommendations from friends and family, it can be helpful to seek professional information from tour operators or travel companies before a trip. People choose third-party trip planners, especially for large groups, to reduce the risks involved when searching for information and making travel arrangements. Goossens (2000) stated that features provided by travel agencies and the information provided in tour brochures improve tourists' external searches, although travelers rated SNS information searches as more influential than the information from agencies.

21.2.3.2.3 Social Networks

Researchers define online social networking as "a platform that enables users to share personal information and connect with others who have similar interests and activities" or Internet-based applications that facilitate the sharing of consumer-generated content (CGC) (Boyd & Ellison, 2007). One "mega-trend" on the internet is that nearly half of trip buyers visit a forum or online community before making a decision (Zhang et al., 2010). Social networking sites hosting CGC such as blogs, Facebook, and LinkedIn and media platforms like YouTube and Flickr are particularly popular among online travelers.

Social media allows consumers to share their experiences in different ways, such as by posting their stories, comments, photos, and videos, and potential travelers can learn from tourists and make informed decisions about operators based on feedback from close friends, family, coworkers, and even strangers worldwide (Xiang & Gretzel, 2010). Gretzel and Yoo (2008) discovered that readers trust traveler reviews more than travel service provider information. Researchers have extensively studied online traveler reviews on social media because of their importance to both visitors and tourism firms (Ye et al., 2009).

21.2.4 Travel Planning and Destination Decision-Making

Travel decision-making can be explained partly by a hierarchical structure with decisions and subdecisions spread across layers. These include trip characteristics (travel purpose, distance, and length of the trip) and destination choices (evaluate choice, purchase, etc.) (Moutinho, 1987).

21.2.4.1 Travel Characteristics

The purpose of the trip is considered the most significant influence on the traveler's decision-making. Moscardo et al. (1996) divided the primary purpose of travel into four main categories: business, visiting friends or relatives, personal issues, and pleasure, although Dwyer et al. (2000) found that product type and purpose and information sources influence consumers' purchase decisions. Travel distance is also an essential dimension of any travel planning; overcoming distance imposes a major constraint on travel behavior (Chon et al., 1999). Becken and Schiff (2011) identified that the influence of travel distance on tourists' decision to travel to New Zealand decreased slightly with longer stays in the country and with greater total distances during their stay. However, Cook and McCleary (1983) suggest that most consumer travel decision models contain a distance variable, often representing distance estimates from traveler origin to potential destination. Meanwhile, Decrop and Snelders (2005) and Money and Crotts (2003) found that length of stay at a destination is one of the key elements in tourism decision-making and purchase behavior.

21.2.4.2 Travel Choices

Personal preferences and objectives play a significant role in determining the choice of holiday destination (Decrop, 2010; Moutinho, 1987). Careful travelers gather relevant facts and information from diverse sources of information and evaluate their options to make well-informed decisions, beginning with destination. Personality and sociodemographic traits, such as age, education, occupation, and the approach to gathering information, also all contribute to the destination decision process. Tourists face challenges when evaluating travel and tourism services, leading to an increased perception of decision-making risk as consumers value efficiency in time and expense for making rapid and cost-effective travel plans (Arsal et al., 2010). Finally, the consumer weighs all the inputs and makes the final travel choices and purchases.

21.3 RESEARCH DESIGN AND METHODOLOGY

21.3.1 The MCDM Framework and Validation

MCDM was introduced in the mid-1960s for enabling groups or individuals to rank, select, and/or compare different alternatives by allowing several criteria to be considered simultaneously (Tzeng et al., 2010). Tourist decision-making involves multiple criteria, which makes MCDM techniques effective decision-making aids. For this study, the criteria I examined were tourist characteristics (sociodemographic features and personality traits), destination characteristics (environment, image, brand), internal search (memory, knowledge), external search (WOM, Internet, tourism intermediaries), planning characteristics (travel purpose, distance, length of stay), and choice set (evaluation and purchase); see Figure 21.2.

MCDM addresses decision problems by analyzing and comparing alternatives based on competing criteria to find the optimal solution (Alptekin & Büyüközkan, 2011). For this study, the framework focused on factor qualities and priority weights. We derived the attribute importance weights from both tourism professional and traveler feedback in accordance with theories on tourism destination decision-making.

21.3.2 DEMATEL

Decisions are made on a daily basis, yet most entail complicated, competing, and interactive criteria (Tzeng et al., 2010). Human decision-making is often uncertain and inadequate for estimating numeric values, and DEMATEL with fuzzy logic is used to resolve human perceptions and uncertainties. The Science and Human

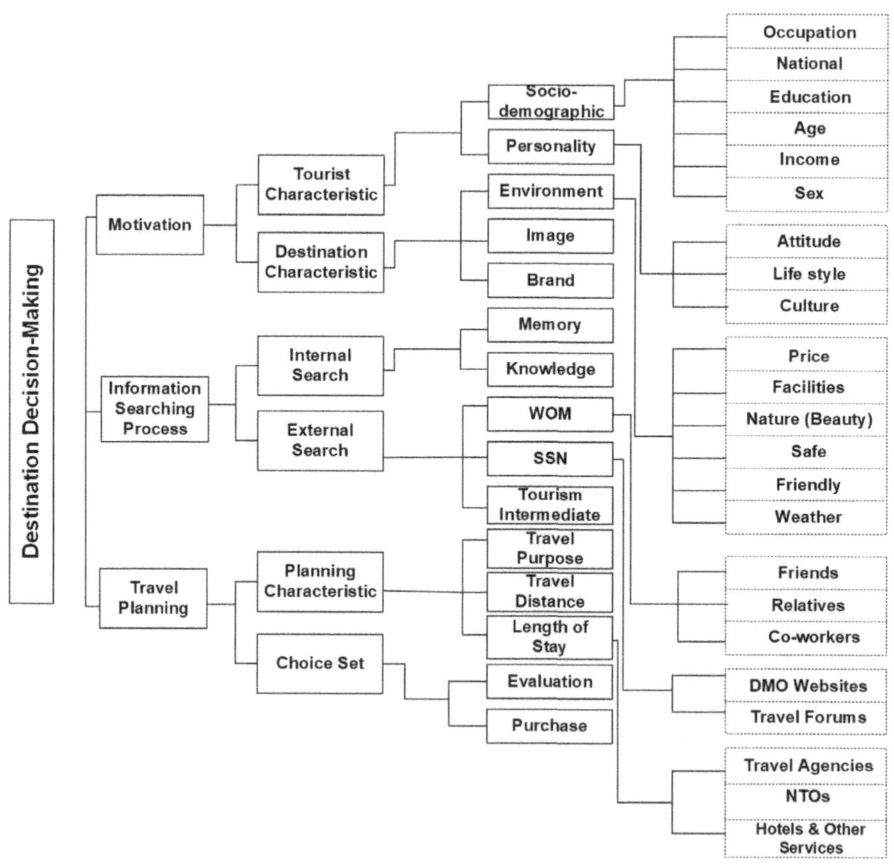

FIGURE 21.2 Framework for multicriteria tourism destination decision-making.

Affairs Program of the Battelle Association of Geneva Research Center developed DEMATEL between 1973 and 1976 to analyze and solve complex problems (Fontela & Gabus, 1974, 1976).

By creating structural models, DEMATEL helps groups make decisions and solve significant challenges (Wu & Lee, 2007; Tseng, 2009). Since each criterion in a decision might affect any other, the DEMATEL technique is used to structure MCDM problems via pairwise comparison. DEMATEL was developed using matrices and related math theories and divides components into cause-and-effect groups. The technique was designed with the assumption that scientific research methods might help with understanding a specific problem, a cluster of interconnected difficulties, and a hierarchical structure of practical solutions (Jeng & Tzeng, 2012). Significant issues are conceptually difficult to solve; DEMATEL breaks down large problems into smaller ones and analyzes their relationships using collective data.

In the context of this study, the interaction between tourist destination decision-making and social media users is complex, and improvements should be directed towards the most important aspects. DEMATEL matrices or digraphs show the contextual relationships between system elements, with numerals representing influence or relational network strength (Chen-Yi et al., 2007). For this study, I used DEMATEL to investigate direct and indirect causal links between criteria following the steps I discuss next.

21.3.2.1 Step 1: Calculate the Original Average Matrix

For the study, I asked a set of experts to estimate the individual impacts of each of our criteria on the others with respect to making travel choices, where the criteria were tourist characteristics, destination characteristics, internal search, external search, planning characteristics, and choice set. On scales ranging from 0 to 4, where 0 = no influence and 4 = strong influence, I asked each respondent to indicate the degree to which they believed that a given factor i affected factor j, notated as x_{ij}; when i = j, $x_{ij} = 0$. The experts' scores gave us an x n non-negative answer matrix, with $1 \leq k \leq Z$; x, x2, x3, . . ., xn were the answer matrices for each of the Z experts, and each element xk was an integer. The arithmetic average method incorporated all expert opinions; the initial average matrix X in Equation (21.1) assumed that n variable impacts the decision-making model:

$$X = \begin{bmatrix} 0 & \cdots x_{1j} \cdots & x_{1n} \\ \vdots & \ddots & \vdots \\ x_{n1} & \cdots x_{1i} \cdots & 0 \end{bmatrix} \tag{21.1}$$

21.3.2.2 Step 2: Establish the Structure Model and Calculate the Direct-Influence Matrix

We obtained the next matrix, N, by normalizing the average matrix X. We separately calculated the sums of the rows and the columns and obtained the maximums using Equation (21.2). Then I divided X by N to get the normalized initial direct-relation matrix A in Equation (21.3). Each value in matrix A is between 0 and 1:

$$N = \max\left\{\max \sum\nolimits_{j=0}^{n}(x_{ij}), \max \sum\nolimits_{i=0}^{n} x_{ij}\right\}, i,j = (1,2,\ldots\ldots,n) \tag{21.2}$$

$$A = \frac{X}{N} \tag{21.3}$$

21.3.2.3   Step 3: Calculate the Full Relationship Matrix

Once I arrived at matrix A, I calculated the full relationship matrix T with Equation (21.4), and Equation (21.5) with I is the n × n identity matrix:

$$T = \sum\nolimits_{1}^{l} A^{n} \tag{21.4}$$

$$T = A + A^{2} + A^{3} + \ldots + A^{l} \tag{21.5}$$

The requirements is that when q → ∞,

$$0 \le x_{ij} < 1, 0 < \left(\sum_{j=1}^{n} x_{ij}, \sum_{j=1}^{n} x_{ij}\right) \le 1$$

21.3.2.4   Step 4: Calculate and Analyze the Results

Let t_{ij} (i, j = 1, 2, . . ., n) be the elements of matrix T and the sum of the rows and the sum of the columns be respectively denoted as vector d_i and r_j using Equations (21.6) and (21.7):

$$T = \left[t_{ij} \right]_{n \times n}, \text{ ij } (\in: 1, 2, \ldots, n)$$

$$d = \left[\sum_{j=1}^{n} t_{ij} \right]_{n \times 1} \tag{21.6}$$

$$r = \left[\sum_{i=1}^{n} t_{ij} \right]_{1 \times n} \tag{21.7}$$

where d_i represents the sum when I influences other elements and r_j represents the sum when element j is influenced by other elements. The horizontal axis vector (d + r) is made by adding vector d to vector r, called prominence, which indicates the element's degree of influence. Similarly, the vertical axis vector (d − r) is made by deducting vector d from vector r, called a relationship, which separates criteria into cause or effect. Conceptually, when (d − r) is positive, the criterion is a cause, and when (d − r) is negative, it's an effect. Diagrams can visualize the complicated causal relationships between criteria as visible structural models, providing valuable problem-solving insight based on recognizing the differences between causes and effects.

21.3.3   The Research Procedure

The research procedure is shown in Figure 21.3. We devised a questionnaire for this study measured on a DEMATEL scale of absolute numbers to answer the

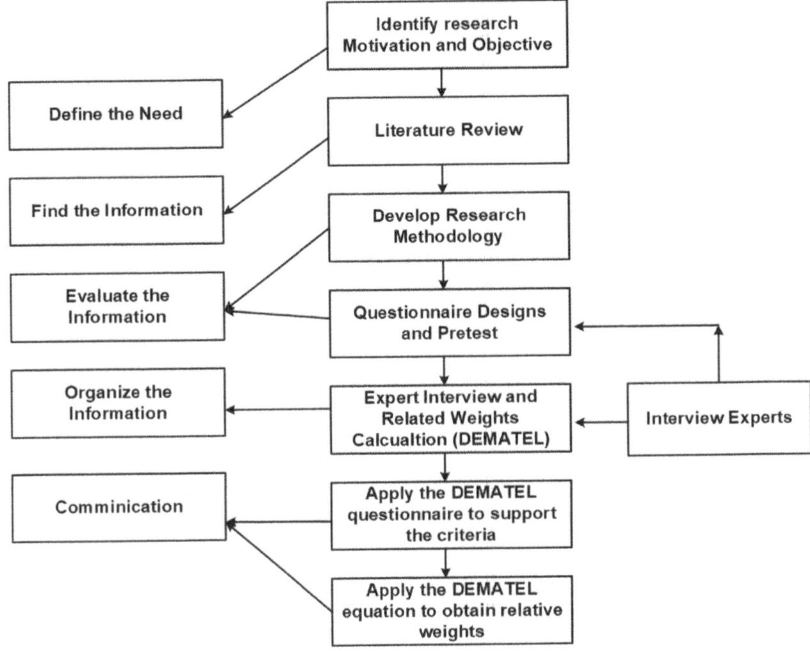

FIGURE 21.3   The research procedure for this study.

fundamental question in all pairwise comparisons. The decision-makers could verbally express their preferences between each pair of elements as equally important, weak or slight, moderate importance, moderate plus, and strong importance.

21.3.4   QUESTIONNAIRE DESIGN AND SAMPLING

For this research, I examined how social network users decide on tourist destinations using DEMATEL to decrease the complexity of addressing the differences between social network users and nonusers.

Based on our literature review findings, the 15 criteria I asked all subjects rate were consumer sociodemographic features and personality traits; destination environment, image, and brand; memory, knowledge, WOM, SNS usage, tourism intermediary, travel purpose, travel distance, length of stay, evaluation, and purchase. Because I conducted the study in Vietnam, the questionnaire was in English and Vietnamese. I asked SNS users with three travel dates per year to rate the elements of destination decision-making that influence each other on 5-point scales from 0 = no influence to 4 = very strong influence for each research criterion.

For the first part of the study, I administered the DEMATEL survey to 200 tourists to have them rate the relevant criteria for them for choosing a vacation destination. To find these tourists, I contacted travel agents, hotels, and famous travel destinations about interviewing international tourists. Then, in the second phase, I interviewed 20 tourism agency managers (experts) to evaluate alternative strategies based on their positions within their respective fields and their experience in international travel. I asked the experts to perform pairwise comparisons of the influences of criteria i on criteria j and compared the experts' opinion data with the survey results.

21.4   RESULTS AND DISCUSSION

The respondent's demographic information is introduced in Tables 21.1 and 21.2. The survey questionnaires are divided into two parts to compare the weight difference between the results from the tourism manager expert and the tourist. Experts were interviewed after questionnaires were dispatched to tourists by tourism organizations and tour guides.

21.4.1   EXPERTS' DEMATEL RESULTS

Although I invited 20 experts to evaluate the influence of criteria in the DEMATEL survey, I could only meet an interview 16 because of time constraints. However, the quality of the respondents and their feedback exceeded expectations. I averaged the responses into the direct influence matrix X, adding the total impacts and dividing that total by the number of respondents.

TABLE 21.1

Features of the Respondents

Tourist Variable to Weight the Criteria	Travel
Number of trips per year	3 times
Time	2 days
SNS user	200

TABLE 21.2

Features of the Experts

Number of Experts	Title	Expertise	Years(At Least)
2	Operation Manager	Tourism agency	5
3	Director	Tourism agency	5
1	Director	Vietnam Tourism Association (Website)	7
1	General Manager	Hanoi Tourism Association	15
1	Chief of Travel department	Vietnam Tourism (Government)	12
2	General Manager	Tourism Magazine	17
3	Backpacker Tourist	Also operation manager of largest website about backpacker tourism in Vietnam	8
3	Professor	Tourism department, tourism college	14

I added the initial average influence matrix to the normalized direct–influence matrix N from Equations (21.2) and (21.3) and then inverted this result to produce the total influence matrix T using Equations (21.4) and (21.5). The results for T showed that all elements I identified in this study had at least some relationship with each other. However, to reduce the complexity of the elements of the matrix, I carefully calculated threshold α as follows (Sumrit & Anuntavoranich, 2012; Tzeng et al., 2007):

$$\alpha = \frac{\sum_{i=1}^{n}\sum_{j=1}^{n}\left[t_{ij}\right]}{N}$$

Based on a calculated $\alpha = 0.86625$, the influence matrix after threshold T' is given in Table 21.3. Table 21.4 gives the influence relationships: d represents influence on other factors, while r represents received influence; (d + r) reflects the factor relationship, while (d − r) represents the effect. A high (d + r) shows prominence, while a high (d − r) indicates a strong link with other criteria.

Table 21.4 and Figure 21.4 show that the experts rated SNSs as having the highest (d + r) at 27.5145, followed by WOM (26.9316), travel purpose (26.7513), and purchase (24.8877). Seven aspects were positive and therefore net influencers (personality traits, memory, knowledge, WOM, tourist intermediary, travel purpose, and length of stay), and eight were negative and therefore net receivers (sociodemographic features, environment, image, brand, SNS use, evaluation, and purchase).

21.4.2 Tourists' DEMATEL Results

While serving as a tour guide, I distributed the questionnaires through the tourism company at the same time as I conducted the expert interviews. I distributed 300 questionnaires, but only 185 qualified for this research, and I used these 185 for the DEMATEL survey. Following the same calculation step as for the expert survey but with the threshold of $\alpha = 0.651$, I obtained the matrix in Table 21.5. Table 21.6 gives the influence relationships as rated by the experts.

Table 21.4 and Table 21.6 show that the tourists' and experts' DEMATEL survey results were the same, with SNSs playing the most crucial role (20.8556 for the tourists) and second and third being WOM (19.7972) and travel purpose (19.5969). The lowest value for tourists was sociodemographic features at 18.6440. Destination characteristics, external search, and choice set are the strong receivers because each aspect has two negative values for (d − r). On the other hand, tourist and planning characteristics are strong causes because (d − r) is positive for both.

At the criterion level, Figure 21.5 shows that the criteria with positive (d − r) were sociodemographic features (C1), personality traits (C2), environment (C3), memory (C6), tourism intermediary (C10), travel purpose (C11), and length of stay (C13); these are net causes. However, (d − r) was near 0 for environment (0.1254), memory (0.1477), tourism intermediary (0.1823), travel purpose (0.0570), and length of stay (0.0306), indicating their weak influences on the other criteria. On the other side, image (C4), brand (C5), knowledge (C7), WOM (C8), SNS use (C9), travel distance (C12), evaluation (C14), and purchase (C15) all had negative (d − r) and were net receivers. Particularly, WOM (−0.3609), SNS use (−0.5177), and purchase (−0.3724) were strongly affected by other criteria.

TABLE 21.3

Experts Total Influence Matrix T with Threshold α = 0.8663

T'	C1	C2	C3	C4	C5	C6	C7	C8	C9	C10	C11	C12	C13	C14	C15	d
C1	0	0	0	0	0	0	0	0.878	0.915	0	0	0	0	0	0	1.79296
C2	0	0	0.867	0.874	0.873	0	0.872	0.893	0.945	0.871	0.871	0	0.87	0	0	7.93536
C3	0	0	0	0.885	0	0	0.868	0.87	0.927	0.88	0.873	0	0	0	0	5.30302
C4	0	0	0	0	0	0	0	0.869	0.913	0	0	0	0	0	0	1.78157
C5	0	0	0	0	0	0	0	0	0	0	0	0	0	0	0	0
C6	0.872	0.876	0.874	0.895	0.881	0	0.894	0.907	0.943	0.894	0.897	0.869	0.884	0.883	0.876	12.4463
C7	0.887	0.896	0.909	0.917	0.914	0	0	0.926	0.976	0.91	0.916	0.891	0.907	0.9	0.888	11.8362
C8	0.892	0.899	0.92	0.917	0.911	0.881	0.911	0.868	0.974	0.925	0.916	0.9	0.909	0.903	0.898	13.6224
C9	0.891	0.9	0.904	0.919	0.916	0	0.903	0.932	0.906	0.915	0.928	0.902	0.905	0.911	0.891	12.725
C10	0.887	0.889	0.894	0.914	0.897	0.868	0.898	0.915	0.967	0	0.902	0.886	0.876	0.896	0.885	12.5743
C11	0.897	0.905	0.918	0.923	0.908	0.886	0.92	0.934	0.982	0.921	0	0.909	0.909	0.912	0.898	12.8242
C12	0	0	0	0.872	0	0	0	0.878	0.912	0	0	0	0	0	0	2.66184
C13	0.89	0.89	0.906	0.915	0.905	0.877	0.907	0.937	0.978	0.918	0.91	0.892	0	0.897	0.877	12.7006
C14	0	0	0	0	0	0	0	0	0	0	0	0	0	0	0	0
C15	0	0	0	0	0	0	0	0	0.876	0	0	0	0	0	0	0.87592
r	6.217	6.254	7.192	9.032	7.205	3.512	7.173	10.81	12.21	7.234	7.215	6.25	6.261	6.303	6.213	109.08

TABLE 21.4

Experts' Criterion Influences—DEMATEL Survey

Criterion/Influence	d	r	d + r	d − r
Tourist Characteristics				
Sociodemographic features	12.71131	12.72849	25.4398	−0.01718
Personality traits	13.00863	12.85735	25.86598	0.15127
Destination Characteristics				
Environment	12.95815	12.97272	25.93087	−0.01457
Image	12.69172	13.10347	25.79519	−0.41175
Brand	11.89689	13.01106	24.90795	−1.11417
Internal Search				
Memory	13.24119	12.5384	25.77958	0.702791
Knowledge	13.54295	12.97649	26.51945	0.566457
External Search				
WOM	13.62242	13.30925	26.93167	0.313171
SNS use	13.58336	13.93114	27.5145	−0.34778
Tourism intermediary	13.41628	13.11207	26.52834	0.304213
Planning Characteristics				
Travel purpose	13.68034	13.07102	26.75136	0.609325
Travel distance	12.79508	12.79805	25.59314	−0.00297
Length of stay	13.53817	12.92438	26.46255	0.613782
Choice Set				
Evaluation	12.02329	12.87894	24.90224	−0.85565
Purchase	12.19542	12.69235	24.88776	−0.49693

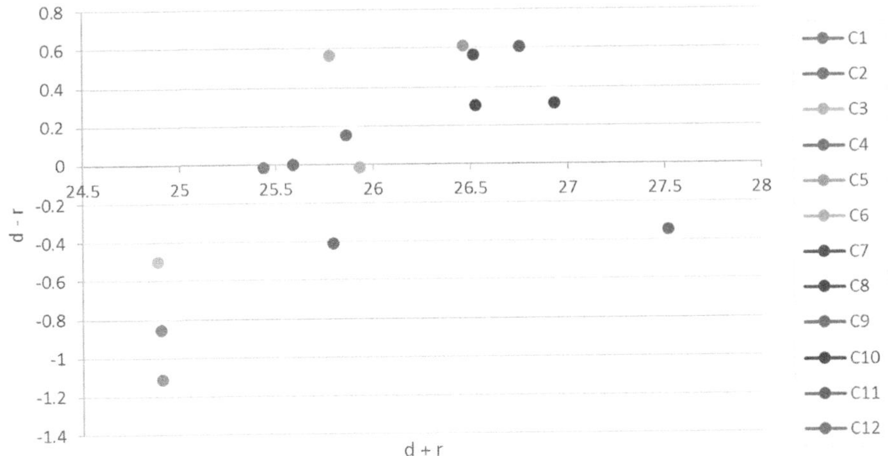

FIGURE 21.4 Causal map of the dependent relationships among elements.

TABLE 21.5

Tourists Total Influence Matrix T' with Threshold $\alpha = 0.651$

T'	C1	C2	C3	C4	C5	C6	C7	C8	C9	C10	C11	C12	C13	C14	C15	d
C1	0	0	0.651	0.661	0.662	0.652	0.658	0.6807	0.726	0.654	0.661	0.663	0.659	0.667	0.671	8.6663
C2	0	0	0.653	0.667	0.659	0.652	0.657	0.6798	0.723	0	0.657	0.664	0.657	0.663	0.666	7.996
C3	0	0	0	0.662	0.656	0.654	0.657	0.674	0.717	0.651	0.656	0.658	0.653	0.661	0.668	7.96626
C4	0	0	0	0	0.658	0	0	0.6717	0.715	0	0.651	0.654	0	0.655	0.657	4.66111
C5	0	0	0	0.655	0	0	0	0.6735	0.711	0.651	0	0.655	0.652	0.656	0.661	5.31455
C6	0	0	0	0.659	0.656	0	0.661	0.6796	0.719	0.654	0.658	0.663	0.655	0.666	0.672	7.34123
C7	0	0	0	0.653	0	0	0	0.6766	0.717	0	0.652	0.661	0.654	0.664	0.664	5.34134
C8	0	0	0	0.656	0	0	0.653	0	0.712	0	0.656	0.657	0.653	0.66	0.665	5.31165
C9	0	0.655	0.673	0.684	0.683	0.673	0.679	0.7052	0.678	0.679	0.686	0.69	0.683	0.689	0.696	9.55201
C10	0	0	0.655	0.668	0.662	0.652	0.662	0.6871	0.729	0	0.665	0.668	0.661	0.673	0.674	8.05738
C11	0	0	0.654	0.661	0.659	0.65	0.657	0.6765	0.722	0.656	0	0.669	0.657	0.668	0.669	7.99895
C12	0	0	0	0	0	0	0	0.6604	0.697	0	0	0	0	0	0	1.3576
C13	0	0	0	0.658	0.656	0	0.655	0.6769	0.718	0.655	0.655	0.665	0	0.665	0.667	6.6722
C14	0	0	0	0	0	0	0	0.6656	0.702	0	0	0	0	0	0.652	2.01931
C15	0	0	0	0	0	0	0	0.6608	0.703	0	0	0	0	0	0	1.36384
r	0	0.655	3.286	7.284	5.951	3.932	5.938	9.4684	10.69	4.6	6.599	7.967	6.585	7.987	8.682	

TABLE 21.6

Experts' Criterion Influences—DEMATEL Survey

Criterion/Influence	d	r	d + r	d – r
Tourist Characteristics				
Sociodemographic features	9.841271	8.802781	18.64405	1.03849
Personality traits	9.809124	9.349225	19.15835	0.459898
Destination Characteristics				
Environment	9.770502	9.645042	19.41554	0.125459
Image	9.700295	9.804407	19.5047	–0.10411
Brand	9.704722	9.75594	19.46066	–0.05122
Internal Search				
Memory	9.800498	9.652764	19.45326	0.147734
Knowledge	9.730648	9.748096	19.47874	–0.01745
External Search				
WOM	9.718181	10.07911	19.79729	–0.36093
SNS use	10.16897	10.68671	20.85568	–0.51774
Tourism intermediary	9.886044	9.703649	19.58969	0.182395
Planning Characteristics				
Travel purpose	9.827012	9.769923	19.59693	0.057088
Travel distance	9.515729	9.84398	19.35971	–0.32825
Length of stay	9.787709	9.757088	19.5448	0.030622
Choice Set				
Evaluation	9.577875	9.867416	19.44529	–0.28954
Purchase	9.55034	9.922784	19.47312	–0.37244

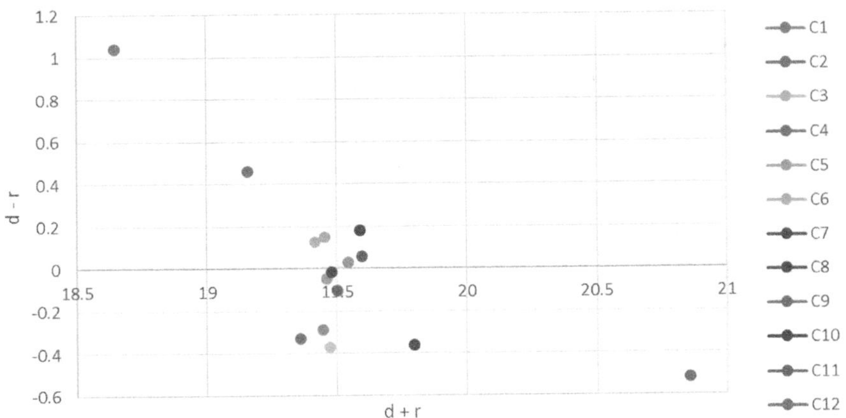

FIGURE 21.5 Causal map of dependent relationship among elements

21.5 CONCLUSIONS AND SUGGESTIONS

21.5.1 CONCLUSIONS

The main purpose of this study was to provide an interdisciplinary contribution to the research on tourists' decision-making. We applied a systematic MCDM approach in order to (1) investigate what factors significantly impact tourists' decision-making; (2) identify and investigate how social networks influence travel motivation and perceived constraints by tourist SNS users as information search tools before travel; and (3) determine the weight for each criterion and recommend strategic alternatives for parties involved in this industry.

Based on the results of this study, we drew several conclusions. First, regarding the first purpose, identify the factors that most influence tourists' decision-making when they choose a destination, we identified from both the tourist and the expert surveys that external search played the most important role. This supports the findings of Kattiyapornpong and Miller (2007), who developed and tested a model of pre-vacation decision-making and discovered that tourists are most likely to research, plan, and pre-book their touring itinerary in advance using external search information. Second, these findings support Zheng Xiang (2010) finding that SNSs are one of the most powerful search engines and play an important role in trip planning and Kerstetter and Cho's (2004) finding that information search is a main influence on destination decision-making. We did find differences between the experts and the tourists based on their DEMATEL survey findings. For instance, the experts rated destination characteristics as the second most important factor, whereas tourists rated choice set second and experts rated choice set fifth.

21.5.2 PRACTICAL IMPLICATIONS AND SUGGESTIONS

Tourism industry practitioners, such as marketers, managers, and experts, can benefit from the framework presented in this study. Marketing managers have used various analyses to identify tourists' decision-making for segmentation and product positioning and to develop targeted promotion strategies for different destination markets (Lilien & Kotler, 1983). For this study, we adopted an MCDM approach to explore concepts and enhance understanding of the tourist decision-making process. Such a goal is critical for the marketer, who seeks to better understand the different needs that drive tourist decision-making processes and associated information searches to communicate product benefits more effectively and efficiently.

This framework provides a theoretical and empirical basis for developing promotional strategies related to the tourist decision-making process. Therefore, a better understanding of tourist decision-making may contribute to developing more effective customer-driven marketing concepts and campaigns, such as maintaining a competitive position for their destination. Understanding this study's findings and taking actions that resonate with the tourism market will help travel agencies and related companies and better design marketing strategies and effectively target tourists by meeting their information needs and desires.

Since it has been argued that the future of Internet-based tourism will be focused on consumer-centric technologies that will support tourism organizations in interacting with the tourist (Buhalis, 2000), this study provides valuable insight into the tourism marketing strategies that marketers need to pursue to achieve the goal. The findings indicate a great need for a tourist destination decision-making process that prioritizes external search to ensure that social network websites are at the top of the ranking list. Therefore, tourism marketers need to understand the key factors to effectively reach out and promote their businesses and destinations through online social networks and websites.

This study suggests future research in various directions. First, I conducted this study in Vietnam, so future researchers could apply the framework in this study to examine tourists in other national contexts. Second, whether within Vietnam or in other countries, researchers could collect responses from domestic tourists and measure the differences between foreign and domestic tourists. Third, I only studied tourists' decision-making before the trip; future researchers could examine tourists' decision-making processes during trips and repurchase intentions following trips. Lastly, future researchers could investigate the most influential factors within tourism information search characteristics. Tourism operators, agents, managers, marketers, and other stakeholders can use these combined findings to better target their marketing efforts depending on their desired destination consumers.

REFERENCES

Alptekin, G. I., & Büyüközkan, G. (2011). An integrated case-based reasoning and MCDM system for web based tourism destination planning. *Expert Systems with Applications*, *38*(3), 2125–2132. https://doi.org/10.1016/j.eswa.2010.07.153

Akehurst, G. (2009). User generated content: The use of blogs for tourism organisations and tourism consumers. *Service Business*, *3*(1), 51–61. https://doi.org/10.1007/s11628-008-0054-2

Arsal, I., Woosnam, K. M., Baldwin, E. D., & Backman, S. J. (2010). Residents as travel destination information providers: An online community perspective. *Journal of Travel Research*, *49*(4), 400–413. https://doi.org/10.1177/0047287509346856

Baloglu, S., & Mangaloglu, M. (2001). Tourism destination images of Turkey, Egypt, Greece, and Italy as perceived by US-based tour operators and travel agents. *Tourism Management*, *22*(1), 1–9. https://doi.org/10.1016/S0261-5177(00)00030-3

Becken, S., & Schiff, A. (2011). Distance models for New Zealand international tourists and the role of transport prices. *Journal of Travel Research*, *50*(3), 303–320. https://doi.org/10.1177/0047287510362919

Beerli, A., & Martín, J. D. (2004). Factors influencing destination image. *Annals of Tourism Research*, *31*(3), 657–681. https://doi.org/10.1016/j.annals.2004.01.010

Bettman, J. R. (1979). An information processing theory of consumer choice. ISBN: 978-0201008340, Addison-Wesley.

Bieger, T., & Laesser, C. (2004). Information sources for travel decisions: Toward a source process model. *Journal of Travel Research*, *42*(4), 357–371. https://doi.org/10.1177/0047287504263030

Bonn, M. A., Furr, H. L., & Susskind, A. M. (1998). Using the internet as a pleasure travel planning tool: An examination of the sociodemographic and behavioral characteristics among internet users and nonusers. *Journal of Hospitality & Tourism Research*, *22*(3), 303–317. https://doi.org/10.1177/109634809802200307

Bonn, M. A., Furr, H. L., & Susskind, A. M. (1999). Predicting a behavioral profile for pleasure Travelers on the basis of internet use segmentation. *Journal of Travel Research, 37*(4), 333–340. https://doi.org/10.1177/004728759903700403

Boyd, D. M., & Ellison, N. B. (2007). Social network sites: Definition, history, and scholarship. *Journal of Computer-Mediated Communication, 13*(1), 210–230. https://doi.org/10.1111/j.1083-6101.2007.00393.x

Buhalis, D. (2000). Marketing the competitive destination of the future. *Tourism Management, 21*(1), 97–116. https://doi.org/10.1016/S0261-5177(99)00095-3

Cai, L. A. (2002). Cooperative branding for rural destinations. *Annals of Tourism Research, 29*(3), 720–742. https://doi.org/10.1016/S0160-7383(01)00080-9

Chen, C.-F., & Tsai, D. (2007). How destination image and evaluative factors affect behavioral intentions? *Tourism Management, 28*(4), 1115–1122. https://doi.org/10.1016/j.tourman.2006.07.007

Chen-Yi, H., Ke-Ting, C., & Gwo-Hshiung, T. (2007). FMCDM with fuzzy DEMATEL approach for customers' choice behavior model. *International Journal of Fuzzy Systems, 9*(4), 236. https://doi.org/10.30000/IJFS.200712.0007

Chon, K.-S. (1989). Understanding recreational traveler's motivation, attitude and satisfaction. *Tourism Review, 44*(1), 3–7. https://doi.org/10.1108/eb058009

Chon, K.-S. (1991). Tourism destination image modification process: Marketing impliations. *Tourism Management, 12*(1), 68–72. https://doi.org/10.1016/0261-5177(91)90030-W

Chon, K. S., Pizam, A., & Mansfeld, Y. (1999). *Consumer Behavior in Travel and Tourism.* New York: Haworth Hospitality Press.

Cook, R. L., & Mccleary, K. W. (1983). Redefining vacation distances in consumer minds. *Journal of Travel Research, 22*(2), 31–34. https://doi.org/10.1177/004728758302200207

Crompton, J. L. (1979). Motivations for pleasure vacation. *Annals of Tourism Research, 6*(4), 408–424. https://doi.org/10.1016/0160-7383(79)90004-5

Crompton, J. L. (1992). Structure of vacation destination choice sets. *Annals of Tourism Research, 19*(3), 420–434. https://doi.org/10.1016/0160-7383(92)90128-C

Dann, G. M. S. (1977). Anomie, ego-enhancement and tourism. *Annals of Tourism Research, 4*(4), 184–194. https://doi.org/10.1016/0160-7383(77)90037-8

Dann, G. M. S. (1981). Tourist motivation an appraisal. *Annals of Tourism Research, 8*(2), 187–219. https://doi.org/10.1016/0160-7383(81)90082-7

Decrop, A. (2010). Destination choice sets: An inductive longitudinal approach. *Annals of Tourism Research, 37*(1), 93–115.

Decrop, A., & Snelders, D. (2005). A grounded typology of vacation decision-making. *Tourism Management, 26*(2), 121–132. https://doi.org/10.1016/j.tourman.2003.11.011

Devesa, M., Laguna, M., & Palacios, A. (2010). The role of motivation in visitor satisfaction: Empirical evidence in rural tourism. *Tourism Management, 31*(4), 547–552. https://doi.org/10.1016/j.tourman.2009.06.006

DiPietro, R. B., Wang, Y., Rompf, P., & Severt, D. (2007). At-destination visitor information search and venue decision strategies. *International Journal of Tourism Research, 9*(3), 175–188. https://doi.org/10.1002/jtr.600

Dwyer, L., Forsyth, P., & Rao, P. (2000). The price competitiveness of travel and tourism: A comparison of 19 destinations. *Tourism Management, 21*(1), 9–22. https://doi.org/10.1016/S0261-5177(99)00081-3

Echtner, C. M., & Ritchie, J. R. B. (1993). The measurement of destination image: An empirical assessment. *Journal of Travel Research, 31*(4), 3–13. https://doi.org/10.1177/004728759303100402

Fontela, E., & Gabus, A. (1974). DEMATEL: Progress achieved. *Futures, 6*(4), 361–363.

Fontela, E., & Gabus, A. (1976). *The DEMATEL observer.* Battelle Geneva Research Center, Geneva.

Fodness, D., & Murray, B. (1997). Tourist information search. *Annals of Tourism Research, 24*(3), 503–523. https://doi.org/10.1016/S0160-7383(97)00009-1

Frías, D. M., Rodríguez, M. A., & Castañeda, J. A. (2008). Internet vs. travel agencies on pre-visit destination image formation: An information processing view. *Tourism Management*, 29(1), 163–179. https://doi.org/10.1016/j.tourman.2007.02.020

Fridgen, J. (1991). *Dimensions of Tourism*. Educational Institute, American Hotel & Motel Association, Library of Congress.

Fuller, D., Wilder, S., Hanlan, J., & Mason, S. (2007). Destination decision making in tourism regions on Australia's coast. *Flinders Business School Research PaperSeries*, 07(07).

Gartner, W. C., & Ruzzier, M. K. (2011). Tourism destination brand equity dimensions: Renewal versus repeat market. *Journal of Travel Research*, 50(5), 471–481. https://doi.org/10.1177/0047287510379157

Gee, C. Y., Makens, J. C., & Choy, D. J. L. (1989). *The Travel Industry* (3rd ed.). New York: Van Nostrand Reinhold.

Gnoth, J. (1997). Tourism motivation and expectation formation. *Annals of Tourism Research*, 24(2), 283–304. https://doi.org/10.1016/S0160-7383(97)80002-3

Goossens, C. (2000). Tourism information and pleasure motivation. *Annals of Tourism Research*, 27(2), 301–321. https://doi.org/10.1016/S0160-7383(99)00067-5

Gretzel, U., Hwang, Y.-H., & Fesenmaier, D. R. (2012). Informing destination recommender systems design and evaluation through quantitative research. *International Journal of Culture, Tourism and Hospitality Research*, 6(4), 297–315. https://doi.org/10.1108/17506181211265040

Gretzel, U., & Yoo, K. (2008). Use and impact of online travel reviews. In P. O'Connor, W. Höpken & U. Gretzel (Eds.), *Information and Communication Technologies in Tourism 2008* (pp. 35–46). Vienna: Springer.

Grønflaten, Ø. (2009). Predicting travelers' choice of information sources and information channels. *Journal of Travel Research*, 48(2), 230–244. https://doi.org/10.1177/0047287509332333

Gursoy, D., & Terry Umbreit, W. (2004). Tourist information search behavior: cross-cultural comparison of European union member states. *International Journal of Hospitality Management*, 23(1), 55–70. https://doi.org/10.1016/j.ijhm.2003.07.004

Ho, C.-I., Lin, M.-H., & Chen, H.-M. (2012). Web users' behavioural patterns of tourism information search: From online to offline. *Tourism Management*, 33(6), 1468–1482. https://doi.org/10.1016/j.tourman.2012.01.016

Huang, S. S., & Hsu, C. H. C. (2009). Travel motivation: Linking theory to practice. *International Journal of Culture*, 3(4), 287–295. https://doi.org/10.1108/17506180910994505

Hunt, J. D. (1975). Image as a factor in tourism development. *Journal of Travel Research*, 13(3), 1–7. https://doi.org/10.1177/004728757501300301

Hyde, K. F. (2008). Information processing and touring planning theory. *Annals of Tourism Research*, 35(3), 712–731. https://doi.org/10.1016/j.annals.2008.05.001

Jeng, D. J.-F., & Tzeng, G.-H. (2012). Social influence on the use of clinical decision support systems: Revisiting the unified theory of acceptance and use of technology by the fuzzy DEMATEL technique. *Computers & Industrial Engineering*, 62(3), 819–828. https://doi.org/10.1016/j.cie.2011.12.016

Jenkin, H. (2010). *The Impact of Social Networking Mediums on the Decision Making Process of Tourists: A Case Study of Stray LTD and Spaceships New Zealand LTD*. New Zealand: The University of Waikato.

Kattiyapornpong, U., & Miller, K. E. (2007). Differences within and between travel preference, planned travel and choice behavior of Australians travelling to Asian and overseas destinations. CAUTHE 2007: Tourism: past achievements, future challenges (pp. 1–12). Sydney: University of Technology Sydney.

Kerstetter, D., & Cho, M.-H. (2004). Prior knowledge, credibility and information search. *Annals of Tourism Research*, 31(4), 961–985. https://doi.org/10.1016/j.annals.2004.04.002

Kim, Y., Weaver, P., & McCleary, K. (1996). A structural equation model: The relationship between travel motivation and information sources in the senior travel market. *Journal of Vacation Marketing*, 3(1), 55–66. https://doi.org/10.1177/135676679600300105

Kim Lian Chan, J., & Baum, T. (2007). Motivation factors of ecotourists in ecolodge accommodation: The push and pull factors. *Asia Pacific Journal of Tourism Research, 12*(4), 349–364. https://doi.org/10.1080/10941660701761027

Lilien, G. L., & Kotler, P. (1983). *Marketing Decision Making: A Model-Building Approach.* New York [etc.]: Harper and Row.

Lim, Y., Chung, Y., & Weaver, P. (2012). The impact of social media on destination branding: consumer-generated videos versus destination marketer-generated videos. *Journal of Vacation Marketing, 18*(3), 197–206. https://doi.org/10.1177/1356766712449366

Mansfeld, Y. (1992). From motivation to actual travel. *Annals of Tourism Research, 19*(3), 399–419. https://doi.org/10.1016/0160-7383(92)90127-B

Marzano, G., & Scott, N. (2009). Power in destination branding. *Annals of Tourism Research, 36*(2), 247–267. https://doi.org/10.1016/j.annals.2009.01.004

McIntosh, R. W., & Goeldner, C. R. (1984). *Tourism Principles, Practices, and Philosophies* (4th ed.). Columbus, OH: Grid, inc.

McIntosh, R. W., Goeldner, C. R., & Ritchie, J. R. B. (1995). *Tourism: Principles, Practices, Philosophies.* Wiley & Sons.

Middleton, V. T. C. (1994). *Marketing in Travel and Tourism* (2nd ed.). Oxford: Heinemann.

Mihalič, T. (2000). Environmental management of a tourist destination: A factor of tourism competitiveness. *Tourism Management, 21*(1), 65–78. https://doi.org/10.1016/S0261-5177(99)00096-5

Mill, R. C., & Morrison, A. M. (1985). *The Tourism System: an Introductory Text.* Englewood Cliffs, NJ: New Jersey Prentice-Hall.

Mill, R. C., & Morrison, A. M. (2002). *The Tourism System.* (4th ed.). Dubuque, IA: Kendall/Hunt Publishing Company.

Money, R. B., & Crotts, J. C. (2003). The effect of uncertainty avoidance on information search, planning, and purchases of international travel vacations. *Tourism Management, 24*(2), 191–202. https://doi.org/10.1016/S0261-5177(02)00057-2

Moscardo, G., Morrison, A. M., Pearce, P. L., Lang, C.-T., & O'Leary, J. T. (1996). Understanding vacation destination choice through travel motivation and activities. *Journal of Vacation Marketing, 2*(2), 109–122. https://doi.org/10.1177/135676679600200202

Moutinho, L. (1987). Consumer behaviour in tourism. *European Journal of Marketing, 21*(10), 5–44. https://doi.org/10.1108/EUM0000000004718

Murphy, L., Moscardo, G., & Benckendorff, P. (2007). Using brand personality to differentiate regional tourism destinations. *Journal of Travel Research, 46*(1), 5–14. https://doi.org/10.1177/0047287507302371

Murphy, P. E. (1985). *Tourism: A Community Approach/Peter E. Murphy.* New York: Methuen.

Murphy, P. E., Pritchard, M. P., & Smith, B. (2000). The destination product and its impact on traveller perceptions. *Tourism Management, 21*(1), 43–52. https://doi.org/10.1016/S0261-5177(99)00080-1

Murray, K. B. (1991). A test of services marketing theory: Consumer information acquisition activities. *Journal of Marketing, 55*(1), 10–25. https://doi.org/10.2307/1252200

Nicoletta, R., & Servidio, R. (2012). Tourists' opinions and their selection of tourism destination images: An affective and motivational evaluation. *Tourism Management Perspectives, 4*(0), 19–27. https://doi.org/10.1016/j.tmp.2012.04.004

Nolan, S. D. (1976). Tourists' use and evaluation of travel information sources: Summary and conclusions. *Journal of Travel Research, 14*(3), 6–8. https://doi.org/10.1177/004728757601400302

O'Connor, P., & Murphy, J. (2004). Research on information technology in the hospitality industry. *International Journal of Hospitality Management, 23*(5), 473–484. https://doi.org/10.1016/j.ijhm.2004.10.002

Pearce, P. L., & Caltabiano, M. L. (1983). Inferring travel motivation from travelers' experiences. *Journal of Travel Research, 22*(2), 16–20. https://doi.org/10.1177/004728758302200203

Prebensen, N. K. (2007). Exploring tourists' images of a distant destination. *Tourism Management, 28*(3), 747–756. https://doi.org/10.1177/004728758302200203

Schmallegger, D., & Carson, D. (2008). Blogs in tourism: Changing approaches to information exchange. *Journal of Vacation Marketing, 14*(2), 99–110. https://doi.org/10.1177/1356766707087519

Shankar, V., Smith, A. K., & Rangaswamy, A. (2003). Customer satisfaction and loyalty in online and offline environments. *International Journal of Research in Marketing, 20*(2), 153–175. https://doi.org/10.1016/S0167-8116(03)00016-8

Sharon, E. B., & Smith, S. M. (1987). External search effort: An investigation across several product categories. *Journal of Consumer Research, 14*(1), 83–95.

Sheldon, P. J. (1993). Destination information systems. *Annals of Tourism Research, 20*(4), 633–649. https://doi.org/10.1016/0160-7383(93)90088-K

Snepenger, D., Meged, K., Snelling, M., & Worrall, K. (1990). Information search strategies B y destination-naive tourists. *Journal of Travel Research, 29*(1), 13–16. https://doi.org/10.1177/004728759002900104

Sumrit, D., & Anuntavoranich, P. (2012). Using DEMATEL method to analyze the causal relations on technological innovation capability evaluation factors in thai technology-based firms. *International Transaction Journal of Engineering, Management, & Applied Sciences & Technologies, 4.*

Tseng, M.-L. (2009). A causal and effect decision making model of service quality expectation using grey-fuzzy DEMATEL approach. *Expert Systems with Applications, 36*(4), 7738–7748. https://doi.org/10.1016/j.eswa.2008.09.011

Tzeng, G.-H., Chen, W.-H., Yu, R., & Shih, M.-L. (2010). Fuzzy decision maps: A generalization of the DEMATEL methods. *Soft Computing, 14*(11), 1141–1150. https://doi.org/10.1007/s00500-009-0507-0

Tzeng, G.-H., Chiang, C.-H., & Li, C.-W. (2007). Evaluating intertwined effects in e-learning programs: A novel hybrid MCDM model based on factor analysis and DEMATEL. *Expert Systems with Applications, 32*(4), 1028–1044. https://doi.org/10.1016/j.eswa.2006.02.004

Um, S., & Crompton, J. L. (1990). Attitude determinants in tourism destination choice. *Annals of Tourism Research, 17*(3), 432–448. https://doi.org/10.1016/0160-7383(90)90008-F

Vogt, C. A., & Fesenmaier, D. R. (1998). Expanding the functional information search model. *Annals of Tourism Research, 25*(3), 551–578. https://doi.org/10.1016/S0160-7383(98)00010-3

Volo, S. (2010). Bloggers' reported tourist experiences: Their utility as a tourism data source and their effect on prospective tourists. *Journal of Vacation Marketing, 16*(4), 297–311. https://doi.org/10.1177/1356766710380884

Werthner, H., & Ricci, F. (2004). E-commerce and tourism. *Commun. ACM, 47*(12), 101–105. https://doi.org/10.1145/1035134.1035141

Wolfe, K., & Hsu, C. H. C. (2004). An application of the social psychological model of tourism motivation. *International Journal of Hospitality & Tourism Administration, 5*(1), 29–47. https://doi.org/10.1300/J149v05n01_02

Wong, J.-Y., & Yeh, C. (2009). Tourist hesitation in destination decision making. *Annals of Tourism Research, 36*(1), 6–23. https://doi.org/10.1016/j.annals.2008.09.005

Woodside, A. G., & Jacobs, L. W. (1985). Ste p two in benefit segmentation: Learning the benefits realized by major travel markets. *Journal of Travel Research, 24*(1), 7–13. https://doi.org/10.1177/004728758502400102

Woodside, A. G., & Lysonski, S. (1989). A general model of traveler destination choice. *Journal of Travel Research, 27*(4), 8–14. https://doi.org/10.1177/004728758902700402

Wu, W.-W., & Lee, Y.-T. (2007). Developing global managers' competencies using the fuzzy DEMATEL method. *Expert Systems with Applications, 32*(2), 499–507. https://doi.org/10.1016/j.eswa.2005.12.005

Xiang, Z. (2010). Modeling the persuasive effects of search engine results. *Information Technology & Tourism, 12*(3), 233–248.

Xiang, Z., & Gretzel, U. (2010). Role of social media in online travel information search. *Tourism Management, 31*(2), 179–188. https://doi.org/10.1016/j.tourman.2009.02.016

Ye, Q., Law, R., & Gu, B. (2009). The impact of online user reviews on hotel room sales. *International Journal of Hospitality Management, 28*(1), 180–182. https://doi.org/10.1016/j.ijhm.2008.06.011

Zhang, Z., Ye, Q., Law, R., & Li, Y. (2010). The impact of e-word-of-mouth on the online popularity of restaurants: A comparison of consumer reviews and editor reviews. *International Journal of Hospitality Management, 29*(4), 694–700. https://doi.org/10.1016/j.ijhm.2010.02.002

Zheng, X., Wöber, K., & Fesenmaier, D. R. (2008). Representation of the online tourism domain in search engines. *Journal of Travel Research, 47*(2), 137–150. https://doi.org/10.1177/0047287508321193

22 Analyzing Outcome-Based Education Using Multicriteria Decision-Making

*Sathish Kumar Kumaravel, Regan Murugesan,
Nagadevi Bala Nagaram, Kala Raja Mohan,
and R. Narmada Devi*

22.1 INTRODUCTION

In the realm of contemporary education, outcome-based education (OBE) prioritizes the results of learning experiences. Rooted in the belief that education should be purposeful and geared toward specific outcomes, OBE has gained prominence in shaping curricula and instructional methodologies. In this chapter, we examine the intersection of OBE with invaluable input from students, particularly focusing on the positive feedback loop that emerges from this dynamic.

At its core, OBE shifts the traditional focus from what is taught to what is learned. The emphasis is on clearly defined learning outcomes that students are expected to achieve. This shift aligns with a broader educational philosophy that aims to equip learners with practical skills and knowledge relevant to their future endeavors. As we embark on this exploration, it is crucial to recognize the symbiotic relationship between OBE and student feedback, a relationship that holds the potential to enhance the educational landscape significantly.

Student feedback in its various forms serves as a compass guiding educators and institutions toward effective teaching practices. It encapsulates the diverse experiences, perspectives, and challenges encountered by students throughout their educational journeys. When coupled with the principles of OBE, this feedback becomes a catalyst for continuous improvement. Positive feedback in particular unveils the strengths and successes of educational approaches, affirming that the intended outcomes are not only met but often exceeded. One of the key advantages of OBE is its inherent adaptability based on student feedback. By actively seeking and incorporating input from learners, educators can fine-tune their strategies to better resonate with the needs of student populations. Positive feedback becomes a beacon that signals areas of success within the curriculum, teaching methods, and overall learning environment. This iterative process fosters a sense of collaboration between educators and students, creating a shared responsibility for the learning journey.

In essence, positive feedback in the context of OBE serves as a reinforcement mechanism, validating that the educational system is not just delivering content but facilitating genuine understanding and skill acquisition. Students, feeling heard and acknowledged, are more likely to actively engage in their learning processes, creating a positive cycle of motivation and achievement. Students' perceptions of alignment between their educational experiences and real-world applications cultivates a sense of purpose and relevance.

Moreover, the positive feedback loop in OBE extends beyond individual achievements to impact institutional effectiveness. Institutions that embrace OBE and leverage student feedback as a tool for improvement cultivate environments of continuous enhancement. This proactive approach positions educational institutions as responsive entities that evolve in tandem with the changing needs of their learners and the broader society.

The method of Multi-Criteria Decision Making method has attracted most researchers since it serves as an efficient method in identifying the selection of finalists. This decision-identifying method comes with different forms of analysis. A few of them are AHP, ANP, TOPSIS, CBA, ranking and correlation.

Reza Khalesi et al. (2010) utilized fuzzy logic in decision-making analysis. Ermatita et al. (2011) conducted a study on the detection of gene mutations by applying the ELECTRE method, focusing on bioinformatics systems and group decision-making. Mark Velasquez et al. (2013) presented a study exploring various decision-making methods. Martin Aruldoss et al. (2013) provided an analysis of decision-making methods in a survey report that has been discussed. Jafar Rezaei (2015) conducted a review on reverse logistics within the context of MCDM. V. Alpagut Yavuz (2016) carried out an analysis on making informed decisions regarding job changes using the MCDM method. M. Gökhan Yücel (2016) presented the application of the ELECTRE method in the decision-making involved in the acquisition of a company. Wenshuai Wu (2017) conducted an analysis of credit risk in group decision-making using the grey relational analysis method. Nitesh Kumar et al. (2018) employed the outranking method in the selection of suppliers. Ansar Daghouri et al. (2019) introduced the MCDM method to study firm performance analysis in the investment of the IT sector. Edmundas Kazimieras Zavadskas et al. (2019) applied the MCDM method to develop an information management system for businesses. Vakkas Ulucay et al. (2019) introduced neutrosophic multi-sets in decision-making using the outranking method has been discussed. Ahmed Abdel-Monem et al. (2023) developed a framework for evaluating risk in the excavation system. Aysha Meshaal Alshamsi et al. (2023) conducted a selection of the best microemulsion oil, considering performance and evaluation for analysis. Ertugrul Ayyildiz et al. (2023) focused on students' performance in finding the best university for their studies. Hamed Taherdoost et al. (2023a) utilized the VIKOR method with rank analysis and provided a comprehensive overview of decision-making methods. Kala Raja Mohan et al. (2023) analyzed the gold rate. Reda M Hussien et al. (2023) employed MCDM for the selection of wheat suppliers.

Numerous researchers have employed TOPSIS in various fields in decision making analysis. The increased flexibility in the teaching and learning implementations now TOPSIS allows for better adaptation to make optimal choices in diverse situations.

22.2　STANDARD DEFINITIONS

Various definitions involved in the decision-making algorithm are listed below for reference.

22.2.1　ATTRIBUTES

In decision-making, attributes pertain to the objectives that guide the selection process. Criteria are defined according to the specific requirements of the desired solution, with crucial objectives deemed beneficial attributes. Conversely, attributes requiring minimal consideration are categorized as non-beneficial attributes.

22.2.2　CRITERIA

Elements associated with the decision problem are designated as criteria. Criteria encompass various rules, values, and conditions that enhance the impartiality and fairness of the decision-making process

22.2.3　DECISION MATRIX

A decision matrix is a matrix with dimensions m × n, where the values are determined by the relationships between attributes and criteria. In this context, m denotes the number of attributes, and n represents the number of criteria. The decision matrix serves as a structured representation of how each attribute relates to each criterion in the decision-making process.

22.2.4　NORMALIZED DECISION MATRIX

The matrix obtained by dividing the elements of the decision matrix by its Euclidean norm is referred to as a normalized matrix. The Euclidean norm is calculated by taking the square root of the sum of squares of the elements within the decision matrix. This normalization process helps standardize the values in the matrix, making it easier to compare and analyze different elements.

22.2.5　WEIGHTED NORMALIZED MATRIX

Weights represent the preference assigned to each attribute in a decision-making process. The matrix obtained by multiplying the normalized decision matrix by the weighted matrix is the weighted normalized matrix. This matrix combines the weights assigned to each attribute with the normalized values derived from the decision matrix, providing a comprehensive representation that considers both the significance of each attribute and the normalized relationships between attributes and criteria.

22.2.6　TOPSIS SCORE

For each alternative, calculate the geometric mean of its normalized distance to the ideal and anti-ideal solutions.

22.2.7 RANKING

Rank the alternatives based on their TOPSIS scores. Higher scores indicate better performance.

22.3 THE OUTRANKING ALGORITHM

The outranking algorithm involves a number of steps: building a decision matrix and a normalized decision matrix; calculating weights; building a weighted normalized matrix; calculating the positive and negative ideal solutions; and building the ranking table.

22.4 THE TOPSIS ALGORITHM

TOPSIS involves scaling quantitative criteria using their respective real numbers. To account for the imprecision of spatial data and human cognition regarding the criteria, the algorithm employs linguistic variables; a linguistic variable is characterized by values expressed in words or sentences in either a natural or artificial language. This approach is particularly valuable when traditional quantification methods struggle to articulate situations that are inherently complex or challenging to describe. Linguistic variables are essential and beneficial for rating qualitative criteria and comparing evaluation criteria. The range of alternative ratings for a qualitative criterion in the classical TOPSIS method involves assessing and assigning values to different alternatives based on their performance against the given criterion. In Table 22.1, Group 1, Group 2, Group 3, and Group 4 represent the first-, second-, third-, and fourth-year students, respectively.

Table 22.2 gives the decision matrix for the criteria. Learning outcomes referred to how well students understood the intended learning outcomes. Assessment methods referred to whether fairness and equity were maintained when providing feedback. The focus with teaching methods was on engagement strategies and clarity of instruction, and the focus with curriculum design was on flexibility to customize and relevance to industry needs. With student engagement, we assessed interest, motivation, and active participation. The values obtained from the student's feedback are reported in the matrix as averages.

TABLE 22.1

Alternatives Concerning Criteria

Criterion	Group 1	Group 2	Group 3	Group 4
Learning outcomes	6	5	8	7
Assessment methods	7	6	5	6
Teaching methods	9	6	7	6
Curriculum design	7	8	6	5
Student engagement	8	7	5	9

TABLE 22.2

Decision Matrix

Criterion	Theory Course	Integrated Course	LaboratoryCourse	Weight
Learning Outcomes	70.75	67.25	68	118.9627
Assessment Methods	60	69.5	68	114.2552
Teaching Methods	68	64.5	61.25	111.9634
Curriculum Design	76.75	71.25	71.25	126.6637
Student Engagement	63.5	62.5	60.5	107.6975

After we generated the decision matrix, we calculated the standardized decision matrix using Equation (22.1):

$$a_{ij} = \frac{y_{ij}}{\sqrt{\sum_{c=1}^{n} x_{c_j}^2}} \quad \text{For every } i = 1,...m \text{ and } j = 1,... \tag{22.1}$$

In MCDM, attributes' weights reflect their relative importance in the decision-making process; it is inappropriate to assume equal importance for each criterion or attribute. Therefore, we used two weighting methods, subjective and objective. For the subjective weighting, we determined the weights based solely on the decision-makers' discretion; for the objective weighting, we applied mathematical techniques to compute the overall rating for each decision-maker. In this step, if any member of the decision matrix is zero, then the value in the standardized decision matrix from Equation (22.1) is also zero; see Table 22.3 for the standard normalization matrix.

This procedure involves multiplying preestablished criteria weights (where the sum of the weights for the evaluation criteria equals 1 with the elements of the R matrix, as depicted in Table 22.4. Through this method, w_{ij}, a weighted normalized decision matrix, is derived. Then each column of the normalized decision matrix is multiplied by its respective associated criterion values.

The elements of the matrix are used to calculate the positive ideal solution Si* and the negative ideal solution Si*. An optimal solution Si* signifies the hypothetical alternative that possesses the most favorable values across all criteria. In reality, such an ideal solution is typically unattainable, and decision-makers seek alternatives that closely approximate this ideal. Conversely, the negative-ideal solution Si' comprises the lowest ratings for each criterion, representing the least desirable qualities across all criteria (Tables 22.5 and 22.6).

Calculate the distances of the existing alternatives from the positive ideal and negative ideal solutions by determining two Euclidean distances for each alternative:

$$S_i^* = \sqrt{\sum_{j=1}^{m} \left(v_{ij} - v_j^+\right)^2} \quad \text{and } S_i' = \sqrt{\sum_{j=1}^{m} \left(v_{ij} - v_j^-\right)^2} \tag{22.2}$$

where $i = 1, 2,....m.$

TABLE 22.3
Standardized Decision Matrix

Criterion	Theory Course	Integrated Course	Laboratory Course
Learning Outcomes	0.594724	0.565303	0.571608
Assessment Methods	0.50436	0.584217	0.571608
Teaching Methods	0.571608	0.542187	0.514867
Curriculum Design	0.64516	0.598927	0.598927
Student Engagement	0.533781	0.525375	0.508563

TABLE 22.4
Standardized Weighted Decision Matrix

Criterion	Theory Course	Integrated Course	Laboratory Course	Maximum	Minimum
Learning Outcomes	3.86571	3.67447	3.71545	3.865707349	3.674470943
Assessment Methods	3.02616	3.5053	3.42965	3.505300276	3.026158512
Teaching Methods	4.00125	3.79531	3.60407	4.001254032	3.604070727
Curriculum Design	4.19354	3.89303	3.89303	4.193541187	3.893026835
Student Engagement	3.86991	3.80897	3.68708	3.869910347	3.687079937

TABLE 22.5
Positive Ideal Solutions

Criterion	Theory Course	Integrated Course	Laboratory Course
Learning Outcomes	0	0.036571	0.022577
Assessment Methods	0.2295768	0	0.005724
Teaching Methods	0	0.042414	0.157755
Curriculum Design	0	0.090309	0.090309
Student Engagement	0	0.003714	0.033427
Si*	**0.4791418**	**0.415943**	**0.556589**

TABLE 22.6
Negative Ideal Solutions

Criterion	Theory Course	Integrated Course	Laboratory Course
Learning Outcomes	0.036571	0	0.001679
Assessment Methods	0	0.229577	0.162802
Teaching Methods	0.157755	0.036571	0
Curriculum Design	0.090309	0	0
Student Engagement	0.033427	0.014856	0
Si˘	**0.56397**	**0.530099**	**0.405563**

TABLE 22.7
Ranking the Alternatives

Criteria	Theory Course	Integrated Course	Laboratory Course
Si*	0.479142	0.415943	0.556589
Si'	0.56397	0.530099	0.405563
Si*+Si'	1.043111	0.946042	0.962152
R = Si'/(Si* + Si')	**0.5407**	**0.5603**	**0.4215**

Determine the relative closeness to the ideal alternative by utilizing the calculated Euclidean distances from the positive ideal and negative ideal solutions from Equation (22.2). Rank the alternatives as in Table 22.7. The table shows that most students selected the integrated course.

22.5 CONCLUSION

Multicriteria decision-making methods find applications in various everyday situations and offer effective solutions. The selection process is instrumental in arriving at a distinguished solution. In this chapter, we used diverse MCDM methods including outranking to help a hypothetical group of students select their best-suited career course paths. We identified multiple criteria and attributes and implemented a scoring system, and for most students, the integrated course was the most suitable for outcome-based education. Our work showcased the versatility and real-time applicability of this method in providing precise solutions across different contexts. The integrated method proved invaluable in pinpointing the most suitable course for outcome-based education in student feedback.

REFERENCES

Abdel-Monem, Ahmed, Mohammed K. Hassan, Ahmed Abdelhafeez and Shimaa S. Mohamed, Neutrosophic Set Hybrid MCDM Methodology for Choosing Best Surfactant-Free Microemulsion Oils within Performance and Emission Criteria Over a Wide Range of Engine Loads, Neutrosophic Sets and Systems, Vol. 56, PP. 190–199, 2023.

A multi-criteria decision-making (MCDM) approach for data-driven distance learning recommendations | Education and Information Technologies (springer.com)

Alshamsi, Aysha Meshaal, Hadeel ElKassabi, Mohamed Adel Serhani and Chafk Bouhaddioui, A Multicriteria Decisionmaking (MCDM) Approach for Datadriven Distance Learning Recommendations, Education and Information Technologies, Vol. 28, PP. 10421–10458, 2023.

Daghouri, Ansar, Khalifa Mansouri, Mohammed Qbadou, The impact of IT investment on firm performance based on MCDM techniques, International Journal of Electrical and Computer Engineering, Vol. 9, No. 5, PP. 4344–4354, 2019. DOI: 10.11591/ijece.v9i5.pp4344–4354

Gökhan Yücel, M. and Ali Görener, Decision Making for Company Acquisition by ELECTRE Method, International Journal of Supply Chain. Management, Vol. 5(1), PP. 75–83, 2016.

Kala, Raja Mohan, R. Narmada Devi, Nagadevi Bala Nagaram, T. Bharathi and Suresh Rasappan, Neutrosophic Statistical Analysis on Gold Rate, Neutrosophic Sets and Systems, Vol. 60, PP. 113–123, 2023. DOI: 10.5281/zenodo.10224140

Kumar, N., Tarun Soota, Neetesh Gupta and Sunil Kumar Rajput, Multi attribute outranking approach for supplier selection, IOP Conf. Series: Materials Science and Engineering, Vol. 404, 012008, PP. 1–11, 2018. DOI: 10.1088/1757-899X/404/1/012008

Martin, Aruldoss, T. Miranda Lakshmi and V. Prasanna Venkatesan, A Survey on Multi Criteria Decision Making Methods and Its Applications, American Journal of Information Systems, Vol. 1(1), PP. 31–43, 2013. A Survey on Multi Criteria Decision Making Methods and Its Applications (sciepub.com)

Velasquez, Mark and Patrick T. Hester, An Analysis of Multi-Criteria Decision Making Methods, International Journal of Operations Research, Vol. 10(2), PP. 56–66, 2013.

23 Selecting a Best Professor Awardee Using Multicriteria Decision-Making

Kala Raja Mohan, R. Narmada Devi,
Nagadevi Bala Nagaram, Regan Murugesan,
and Sathish Kumar Kumaravel

23.1 INTRODUCTION

Multicriteria decision-making (MCDM) refers to making decisions when there are multiple factors to consider that conflict with each other, and MCDM techniques have to date been applied extensively across many fields, beginning with testing the applicability of different decision-finding methods themselves (Martin Aruldoss et al., 2013; Jafar Rezaei, 2015).

For instance, researchers have applied fuzzy MCDM in civil engineering (Zhi Wen et al., 2021) and used an MCDM algorithm to evaluate the performance of Indonesia's defense ministry (A. P. Sumarno et al., 2021). Researchers have used VIKOR for ranking (Hamed Taherdoost et al., 2023a) and used TODIM to analyze weights (A. Sahaya Sudha et al., 2020). Other researchers used outranking for decision-making in neutrosophic multi-sets (Vakkas Ulucay et al., 2019) and described using interval-valued neutrosophic soft sets in certain operations (Rana Muhammad Zulqarnain et al., 2022). One group compared fuzzy cognitive maps and neutrosophic cognitive maps with respect to COVID-19 variables (Regan Murugesan et al., 2023). Another applied neutrosophic cognitive maps with SWOT analysis (Jagan Obbineni et al., 2023).

In business, researchers have used MCDM methods to evaluate commercial bank performance (Mohamed Abdel-Basset et al., 2021), evaluate the risks with an excavation system (Ahmed Abdel-Monem et al., 2023), select suppliers using outranking (Nitesh Kumar et al., 2018), manage information systems (Edmundas Kazimieras Zavadskas et al., 2019), evaluate the financial performance of food and drink indices using fuzzy MCDM (Eyad Aldalou and Selcuk Percin, 2020), analyze blockchain rates (Ahmed Abdel-Monem et al., 2020) and the rate of gold (Kala Raja Mohan et al., 2023), select wheat suppliers (Reda M Hussien et al., 2023) and

DOI: 10.1201/9781032635170-23

microemulsion oils (Ahmed Abdel-Monem et al., 2023), evaluate employee performance (in Taiwan; Zon-Yau Lee et al., 2020), rank energy resources (Mahammad Nuriyev, 2020), and analyze firm performance in the IT sector (Ansar Daghouri et al., 2019). On the human resources side, researchers used the MULTIMOORA MCDM method to analyze employee performance for appraisals (Abteen Ijadi Maghsoodi et al., 2018) and investigated using MCDM to decide on changing jobs (V. Alpagut Yavuz, 2016).

With respect to the context of this study, using MCDM techniques to identify a recipient for a best instructor award, MCDM has seen wide application in education as well (William Ho et al., 2006; Durrani Aimi Abdul Malik et al., 2021). Researchers have used MCDM algorithms to measure lecturers' productivity (Nguyen Anh Tuan et al., 2020) and student performance (Pavani Sirigiri et al., 2015), and Aysha Meshaal Alshamsi et al. (2023) presented findings on distance learning. On the student side, researchers have investigated students' use of MCDM to find the best university (Ertugrul Ayyildiz et al., 2023) and use of a rough set to choose between two universities (Somen Debnath, 2023).

Teaching is a selfless and noble profession. In nearly every field, people think of their individual development, but teachers aim to develop their students rather than themselves; their real happiness lies in elevating their students. Teachers mold their students and encourage them to be the best they can be and prove themselves in society. However, teachers themselves benefit from acknowledgement of their efforts, and one way to recognize high performers is with awards. Awards are common forms of recognition across almost all professions and can stimulate professional development in pursuit of further recognition.

In the context of this study, assessing higher education instructor performance in recognition for an award will be guided by many separate criteria including experience, results, publications, projects, and funding. For this study, the organization that will grant the award will encounter applicants with many different profiles who will need to be scored according to the multiple criteria. Accordingly, this is a highly appropriate scenario for examining techniques of multicriteria decision-making (MCDM) in the process of selecting finalists. For this chapter, we adopt the MCDM procedure called outranking to select the best instructor.

23.2 STANDARD DEFINITIONS

23.2.1 ATTRIBUTES

Attributes refer to the objective responsible for the decision; they are fixed based on the requirements for the solution. Attributes that are required for the decision are termed beneficial, and those that require little consideration are called nonbeneficial.

23.2.2 CRITERIA

Criteria are the various rules and values that determine the conditions of the decision problem.

23.2.3 Decision Matrix

A decision matrix is a matrix of dimensions $m \times n$ in which the values are fixed based on the relationship between the attributes and the criteria. Here m represents the number of attributes, and n is the number of criteria.

23.2.4 Normalized Decision Matrix

A normalized decision matrix is derived by dividing the elements by their Euclidian norm, the square root of the sum of the squares of the elements.

23.2.5 Variance

Variance is a measure of the amount of spread in data. It is obtained by dividing the sum of squares of deviation of data from the mean by the number of data.

23.2.6 Weighted Normalized Matrix

Weight is a measure of the preference for given attributes or criteria, and the matrix obtained from the product of the normalized decision matrix and the weighted matrix is the weighted normalized matrix.

23.2.7 Concordance Matrix

In a concordance matrix, the diagonal elements are left free. Each of the remaining elements a_{ij} is obtained by adding the weights obtained for the corresponding column $a_i > a_j$. The row elements consisting of a_i and a_j are considered for the calculation. When there is equality between a_i and a_j, half the weightage is considered.

23.2.8 Disconcordance Matrix

The diagonal elements in a disconcordance matrix d_{ij} are also left free. The remaining elements are calculated using Equation (23.1):

$$d_{ij} = \begin{cases} 0 & \text{if } g_k(a_i) \geq g_k(a_j) \text{ for all } k \\ \dfrac{\max\left[g_k(j) - g_k(i)\right]}{\max\left|g_k(j) - g_k(i)\right|} & \text{if } g_k(a_i) \geq g_k(a_j) \text{ for all } k \end{cases} \tag{23.1}$$

23.2.9 Pure Concordance Index

The pure concordance index is calculated using Equation (23.2):

$$C_j = \sum_{k=1}^{n} c(j,k) - \sum_{j=1}^{n} c(k,j) \text{ where } j \neq k \tag{23.2}$$

23.2.10 Pure Disconcordance Index

The pure disconcordance index is calculated using Equation (23.3):

$$D_j = \sum_{k=1}^{n} d(j,k) - \sum_{j=1}^{n} d(k,j)\, where\ j \neq k \qquad (23.3)$$

23.2.11 Ranking Table

Ranking involves sorting elements according to preference in ascending or descending order, and in the final outranking step, a table is produced.

23.3 METHODOLOGY

The education field involves teachers with many aspects of development. The first and foremost eligibility for a teacher is their educational qualification: For this study, the three qualification criteria for best university instructor were M.Phil., Ph.D., or Postdoctoral Fellow (PDF). The next attribute was years of experience, which the organization ranked as less than 10, 10–15, 15–20, or >20. The organization also tallied each applicant's number of publications in reputed journals, categorized as fewer than 20, 10–20, or more than 20 and number of scholars they supervised or produced, scored as 0, fewer than 5, 5–10, or more than 10. The final scores were based on the importance of each criteria.

23.4 MCDM RESULTS AND DISCUSSION

For this scenario, a university department of mathematics was recognizing its best-performing professor with an award. The decision-making committee considered four attributes and four criteria for four applicants D1, D2, D3, and D4.

Although as we noted earlier, teachers are the building pillars of students, they continue developing themselves throughout their careers to keep their skills current. For this study, the math department considered four attributes of professor development in considering a best math instructor: A1, applicant's education qualification; A2, applicant's years of experience; A3, applicant's number of articles published; and A4, applicant's number of PhD graduates supervised.

The award committee also considered four criteria for selecting a best professor: C1, applicant's education qualification; C2, applicant's years of experience; C3, applicant's number of articles published; and C4, applicant's number of PhD graduates supervised. Table 23.1 presents the scores for attributes versus criteria for making the decision matrix; in the scoring, A1, A2, and A3 are beneficial attributes and A4 is nonbeneficial.

Following earlier rounds, the committee shortlisted four contenders for best math professor, and their scores are shown in Table 23.2.

With the finite set of finalists, we produced the following decision matrix to help the committee choose:

TABLE 23.1

Scoring for Attributes with Respect to Criteria

Scoring Table

A1		Education		
	C1	M.Phil.	PhD	PDF
		0.3	0.5	0.9

A2		Experience			
C2		<10	10–15	15–20	>20
		0.4	0.5	0.7	0.9

A3		Publication		
	C3	<20	20–25	>25
		0.4	0.5	0.9

A4		PhD Graduates Supervised		
C4	**0**	1–4	5–10	>10
	0.1	04	0.5	0.9

TABLE 23.2

The Four Applicants' Attribute Scores

Applicant/Attribute	A1	A2	A3	A4
D1	PhD	12	21	6
D2	PDF	18	26	8
D3	M.Phil.	9	15	0
D4	PDF	12	15	0

$$\begin{bmatrix} 0.5 & 0.5 & 0.5 & 0.5 \\ 0.9 & 0.7 & 0.9 & 0.5 \\ 0.3 & 0.4 & 0.4 & 0.1 \\ 0.9 & 0.5 & 0.4 & 0.1 \end{bmatrix}$$

Next we produced the normalized decision matrix that follows based on beneficial versus nonbeneficial attributes; for beneficial attributes, the values were taken as such for calculation, and for nonbeneficial attributes, the reciprocals of the values were taken:

$$\begin{bmatrix} 0.3571 & 0.4662 & 0.4256 & 0.0347 \\ 0.6429 & 0.6527 & 0.7662 & 0.0347 \\ 0.2143 & 0.3730 & 0.3405 & 0.0069 \\ 0.6429 & 0.4662 & 0.3405 & 0.0069 \end{bmatrix}$$

The variances obtained for the specified attributes are given in Table 23.3, and the attribute weights are shown in Table 23.4.

The weighted normalized matrix was then

$$\begin{bmatrix} 0.16255 & 0.06340 & 0.17290 & 0.00009 \\ 0.29260 & 0.08870 & 0.31130 & 0.00009 \\ 0.09750 & 0.05070 & 0.13830 & 0.00001 \\ 0.29260 & 0.06340 & 0.13830 & 0.00001 \end{bmatrix}$$

The concordance set was then as given as follows:

$$C_{12} = \{4\}; C_{13} = \{1,2,4\}; C_{14} = \{2,3\};$$

$$C_{21} = \{1,3,4\}; C_{23} = \{1,2,4\}; C_{24} = \{1,2,3\}$$

$$C_{31} = \{\ \}; C_{32} = \{\ \}; C_{34} = \{3\}$$

$$C_{41} = \{1,2\}; C_{42} = \{1\}; C_{43} = \{1,2,3\}$$

The disconcordance set obtained was given as follows:

$$D_{12} = \{1,2,3\}; D_{13} = \{3\}; D_{14} = \{1,4\};$$

$$D_{21} = \{2\}; D_{23} = \{3\}; D_{24} = \{4\}$$

$$D_{31} = \{1,2,3,4\}; D_{32} = \{1,2,3,4\}; D_{34} = \{1,2,4\}$$

$$D_{41} = \{3,4\}; D_{42} = \{2,3,4\}; D_{43} = \{4\}$$

TABLE 23.3

Variances of the Attributes

Attributes	A1	A2	A3	A4
Variance	0.0345	0.0103	0.0308	0.0002

TABLE 23.4

Weights of the Attributes

Attributes	A1	A2	A3	A4
Weights	0.4552	0.1359	0.4063	0.0026

The following is the concordance matrix:

$$\begin{bmatrix} - & 0.0013 & 0.5937 & 0.4743 \\ 0.8623 & - & 0.8641 & 0.7698 \\ 0 & 0 & - & 0.2082 \\ 0.5232 & 0.2276 & 0.7943 & - \end{bmatrix}$$

Following is the disconcordance matrix derived from the relationships calculated in Equation (23.1):

$$\begin{bmatrix} - & 1 & 0.00123 & 1 \\ 0 & - & 0.00041 & 0 \\ 1 & 1 & - & 1 \\ 0.26605 & 1 & 0 & - \end{bmatrix}$$

As a follow-up, we calculated the concordance and disconcordance indices by applying the outputs of Equations (23.2) and (23.3) and ranked the indices. Then we averaged these ranks to produce the final rank list. Table 23.5 displays the indices and the ranks for applicants D1 to D4. In the table, PCI = pure concordance index, PR = PCI rank, PDCI = pure disconcordance index, DR = PDCI rank, AR = average of PR and DR, and FR = final rank for decision.

23.5 RESULTS AND DISCUSSION

The ranks in Table 23.5 show that applicant D2 earned the highest PCI and thereby the highest rank; the final rank order was D2 > D4 > D1 > D3, with D1 and D3 having a tie score. Based on the ranking, applicant D2 should receive the best math professor award.

23.6 CONCLUSION

MCDM is widely applied in various daily life situations, with outranking providing an eminent solution to decision problems. For this chapter, we used MCDM

TABLE 23.5

The Final Ranking of Applicants

D	PCI	PR	PDCI	DR	AR	FR
D1	−0.3162	3	0.73395	4	3.5	3
D2	2.2673	1	−2.9996	1	1	1
D3	−0.7943	4	2.99836	3	3.5	4
D4	−0.2083	2	−0.73395	2	2	2

outranking to help a fictional university math department select an instructor to receive a best professor award. Using the ratings from the award committee, we built various decision matrices and followed outranking procedures to arrive at a final professor who should receive the award. Future researchers can use the procedures in this study to perform ranking in any applicable multicriterion decisions.

REFERENCES

Abteen, Ijadi Maghsoodi, Gelayol Abouhamzeh, Mohammad Khalilzadeh and Edmundas Kazimieras Zavadskas, Ranking and selecting the best performance appraisal method using the MULTIMOORA approach integrated Shannon's entropy, Frontiers of Business Research in China, Vol. 12, No. 2, PP. 1–21, 2018. https://doi.org/10.1186/s11782-017-0022-6

Ahmed, Abdel-Monem, Amal Abdel Gawad and Heba Rashad, Blockchain risk evaluation on enterprise systems using an intelligent MCDM based model, Neutrosophic Sets and Systems, Vol. 83, PP. 368–382, 2020. https://doi.org/10.5281/zenodo.4306907

Ahmed, Abdel-Monem, Mohammed K. Hassan, Ahmed Abdelhafeez, and Shimaa S. Mohamed, Neutrosophic set hybrid MCDM methodology for choosing best surfactant-free microemulsion oils within performance and emission criteria over a wide range of engine loads, Neutrosophic Sets and Systems, Vol. 56, PP. 190–199, 2023. https://doi.org/10.5281/zenodo.8194761

Ansar, Daghouri, Khalifa Mansouri and Mohammed Qbadou, The impact of IT investment on firm performance based on MCDM techniques, International Journal of Electrical and Computer Engineering, Vol. 9, No. 5, PP. 4344–4354, 2019. https://doi.org/10.11591/ijece.v9i5

Aysha, Meshaal Alshamsi, Hadeel ElKassabi, Mohamed Adel Serhani, and Chafk Bouhaddioui, A Multicriteria Decisionmaking (MCDM) approach for datadriven distance learning recommendations, Education and Information Technologies, Vol. 28, PP. 10421–10458, 2023. https://doi.org/10.1007/s10639-023-11589-9

Durrani, Aimi Abdul Malik, Yuhani Yusof and Ku Muhammad Na'im Ku Khalif, A view of MCDM application in education, Journal of Physics: Conference Series, Vol. 0120603, PP. 1–19, 2021. https://doi.org/10.1088/1742-6596/1988/1/012063

Edmundas, Kazimieras Zavadskas, Jurgita Antucheviciene and Prasenjit Chatterjee, Multiple-Criteria Decision-Making (MCDM) Techniques for Business Processes Information Management, Information, Vol. 10, No. 4, PP. 1–7, 2019. https://doi.org/10.3390/info10010004

Ertugrul, Ayyildiz, Mirac Murat, Gul Imamoglu and Yildiz Kose, A novel hybrid MCDM approach to evaluate universities based on student perspective, Scientometrics, Vol. 128, PP. 55–86, 2023. https://doi.org/10.1007/s11192-022-04534-z

Eyad, Aldalou and Selcuk Percin, Financial performance evaluation of food and drink index using fuzzy MCDM approach, International Journal of Economics and Innovation, Vol. 6, No. 1, PP. 1–19, 2020. https://doi.org/10.20979/ueyd.650422

Gökhan Yücel, M. and Ali Görener, Decision making for company acquisition by ELECTRE method, International Journal of Supply Chain. Management, Vol. 5, No. 1, PP. 75–83, 2016. https://doi.org/10.59160/ijscm.v5i1.1166

Hamed, Taherdoost and Mitra Madanchian, VIKOR method—An effective compromising ranking technique for decision making, Macro Management & Public Policies, Vol. 5, No. 2, PP. 27–33, 2023a. https://doi.org/10.30564/mmpp.v5i2.5578

Hussien, Reda M, Amr A. Abohany, Karam M. Sallam and Ahmed Salem, Smart assessment of wheat suppliers via MARCOS-based MCDM modelling under a neutrosophic scenario, Neutrosophic Sets and Systems, Vol. 57, PP. 272–282, 2023. https://doi.org/10.5281/zenodo.8271382

Jafar Rezaei, A, systematic review of multi-criteria decision-making applications in reverse logistics, Transportation Research Procedia, Vol. 10, PP. 766–776, 2015. https://doi.org/10.1016/j.trpro.2015.09.030

Jagan, Obbineni, Ilanthenral Kandasamy, W. B. Vasantha and Florentin Smarandache, Combining SWOT analysis and neutrosophic cognitive maps for multi-criteria decision making: A case study of organic agriculture in India, Soft Computing, Vol. 27, PP. 18311–18332, 2023. https://doi.org/10.1007/s00500-023-08097-w

Kala Raja Mohan, R., Narmada Devi, Nagadevi Bala Nagaram, T. Bharathi and Suresh Rasappan, Neutrosophic statistical analysis on gold rate, Neutrosophic Sets and Systems, Vol. 60, PP. 113–123, 2023. https://doi.org/10.5281/zenodo.10224140

Kumar, Nitesh, Tarun Soota, Neetesh Gupta and Sunil Kumar Rajput, Multi attribute outranking approach for supplier selection, IOP Conf. Series: Materials Science and Engineering, Vol. 404, 012008, PP. 1–11, 2018. https://doi.org/10.1088/1757-899X/404/1/012008

Mahammad Nuriyev, Z., Numbers Based Hybrid MCDM approach for energy resources ranking and selection, International Journal of Energy Economics and Policy, Vol. 10, No. 6, PP. 22–30, 2020. https://doi.org/10.32479/ijeep.9950

Martin, Aruldoss, T. Miranda Lakshmi and V. Prasanna Venkatesan, A survey on multi criteria decision making methods and its applications, American Journal of Information Systems, Vol. 1, No. 1, 31–43, 2013. https://doi.org/10.12691/ajis-1-1-5

Mohamed, Abdel-Basset, Rehab Mohamed, Mohamed Elhoseny, Mohamed Abouhawash, Yunyoung Nam and Nabil M. Abdel Aziz, Efficient MCDM model for evaluating the performance of commercial banks: A case study, Computers, Materials & Continua, Vol. 67, No. 3, PP. 2729–2746, 2021 https://doi.org/10.32604/cmc.2021.015316

Nguyen, Anh Tuan, Truong Thi Hue, Luong Thuy Lien, Truong Duc Thao, Nguyen Duy Quyet, Luu Huu Van and Luong Tram Anh, A new integrated MCDM approach for lecturers' research productivity evaluation, Decision Science Letters, Vol. 9, PP. 355–364, 2020. http://dx.doi.org/10.5267/j.dsl.2020.5.001

Pavani, Sirigiri, H. S. Hota and L. K. Sharma, Students performance evaluation using MCDM methods through customized software, International Journal of Computer Applications, Vol. 130, No. 15, PP. 11–15, 2015. https://doi.org/10.5120/ijca2015907171

Rana, Muhammad Zulqarnain, Aiyared Iampan, Imran Siddique and Hamiden Abd ElWahed Khalifa, Some fundamental Operations for multi-polar interval-valued neutrosophic soft set and a decision-making approach to solve MCDM problem, Neutrosophic Sets and Systems, Vol. 51, PP. 205–220, 2022. https://doi.org/10.5281/zenodo.7135277

Regan, Murugesan, Yamini Parthiban, R. Narmada Devi, Kala Raja Mohan and Sathish Kumar Kumaravel, A comparative study of fuzzy cognitive maps and neutrosophic cognitive maps on Covid variants, Neutrosophic Sets and Systems, Vol. 55, PP. 329–343, 2023. https://doi.org 10.5281/zenodo.7832763

Reza, Khalesi and Hamidreza Maleki, A new method for solving fuzzy MCDM problems, The Journal of Mathematics and Computer Science, Vol. 1, No. 4, PP. 435–438, 2010. http://dx.doi.org/10.22436/jmcs.001.04.20Sahaya Sudha, A., Luiz Flavio Autran Monteiro Gomes and K.R. Vijayalakshmi, Assessment of MCDM problems by TODIM using aggregated weights, Neutrosophic Sets and Systems, Vol. 35, PP. 78–98, 2020. https://doi.org/10.5281/zenodo.3951645

Somen, Debnath, Single-valued neutrosophic covering-based rough set model over two universes and its application in MCDM, Neutrosophic Sets and Systems, Vol. 53, PP. 482–507, 2023. https://doi.org/10.5281/zenodo.7536078

Sumarno, A. P., Margono Setiawan, Siti Aisjah Sunaryo, Employees performance evaluation in defense ministry of the republic of Indonesia based on Multicriteria Decision Making (MCDM) and System Dynamic (SD), International Journal of Operations and Quantitative Management, Vol. 27, No. 3, PP. 245–266, 2021. https://doi.org/10.46970/2021.27.3.4

Vakkas, Ulucay, Adil Kılıc, İsmet Yıldız and Memet Sahin, An outranking approach for MCDM-problems with neutrosophic multi-sets, Neutrosophic Sets and Systems, Vol. 30, PP. 213–224, 2019. https://doi.org/10.5281/zenodo.3569696

Wenshuai, Wu, Grey relational analysis method for group decision making in credit risk analysis, EURASIA Journal of Mathematics, Science and Technology Education, Vol. 13, No. 12, PP. 7913–7920, 2017. https://doi.org/10.12973/ejmste/77913

William, Ho, Prasanta K. Dey and Helen E. Higson, Multiple criteria decision-making techniques in higher education, International Journal of Educational Management, Vol. 20, No. 5, PP. 319–337, 2006. https://doi.org/10.1108/09513540610676403

Yavuz, V. Alpagut, An analysis of job change decision using a hybrid MCDM method: A comparative analysis, International Journal of Business and Social Research, Vol. 6, No. 3, PP. 60–75, 2016. https://doi.org/10.18533/ijbsr.v6i3.935

Zhi, Wen, Huchang Liao, Edmundas Kazimieras Zavadkas and Jurgita Antuchevicine, Applications of fuzzy multiple criteria decision making methods in civil engineering: A state-of-the-art survey, Journal of Civil Engineering and Management, Vol. 27, No. 6, PP. 358–371, 2021. https://doi.org/10.3846/jcem.2021.15252

Zon-Yau, Lee, Mei-Tai Chu, Yu-Ting Wang and Kuan-Ju Chen, Industry performance appraisal using improved MCDM for Next generation of Taiwan, Sustainability, Vol. 12, PP. 1–20, 2020. http://dx.doi.org/10.3390/su12135290

24 Predicting Lumpy Skin Disease Using Machine Learning

*Nagadevi Bala Nagaram, R. Narmada Devi,
Kala Raja Mohan, Sathish Kumar Kumaravel,
and Regan Murugesan*

24.1 INTRODUCTION

Lumpy skin disease is a viral infection that primarily affects cattle. It is caused by the lumpy skin disease virus, which belongs to the family *Poxviridae*. This disease is characterized by the development of nodules on the skin, mucous membranes, and internal organs of affected animals. The first clinical signs of the disease are often a fever and decreased milk production. In more severe cases, nodules or lumps will develop on the skin, particularly around the head, neck, udder, and genitals. It is crucial to understand the disease, its transmission, and prevention to effectively manage and control its impact on cattle populations. Lumpy skin disease has been known to affect cattle for centuries, with records dating to ancient times. The disease has evolved in its impact over the years due to changes in livestock management, trading practices, and the globalization of animal diseases (Anwar et al., 2022).

Machine learning (ML) has been used to guide decisions in an extremely wide range of fields. One set of researchers used ML to forecast the variables that influence the ongoing use of mobile meal delivery applications. The scientists analyzed data from 925 users of five well-known meal delivery apps using a statistical model. The primary variables that were discovered to affect usage were app rating, delivery time, and user demographics. The study findings emphasize that it is critical to comprehend these elements in order to raise customer happiness and engagement levels with mobile meal delivery applications (Ahmad Rabaa'I et al., 2022).

(Jae-Ho Han, 2022) studied improving the accuracy and speed of eye disease diagnosis by analyzing medical images using artificial intelligence (AI) algorithms. The author compared artificial neural networks, support vector machines, and decision trees for detecting three common eye diseases: retinitis pigmentosa, diabetic retinopathy, and glaucoma (Bernardes et al., 2011; Shaikh et al., 2022). A research group used medical images and patient data to train ML models and then predict the likelihood of disease development (Shruthi Bhat et al., 2020). A separate group

DOI: 10.1201/9781032635170-24

"

examined the various ML techniques that have been applied to three ophthalmic imaging modalities: fundus photography, optical coherence tomography, and varilux eyeglasses (Yan Tong et al., 2020). In this chapter, we investigate using machine learning to predict lumpy skin disease in cattle.

24.2 CHARACTERISTICS OF LUMPY SKIN DISEASE

Lumpy skin disease (LSD) is known for its nodular lesions on the skin, mucous membranes, and internal organs. These nodules are typically firm to the touch and can vary in size. Understanding the disease's physical characteristics is crucial for its identification and differentiation from other cattle diseases.

The disease can have considerable impacts on cattle health and productivity. Affected animals experience reduced milk production, weight loss, and a decline in overall health. In addition, LSD can lead to significant economic losses for cattle farmers due to decreased productivity and treatment costs. Veterinary professionals play a crucial role in LSD prevention and management via diagnosis, treatment, and control measures to limit the spread of the disease.

24.2.1 LSD TRANSMISSION ROUTES

A. Direct Contact

The direct contact transmission of LSD occurs when healthy cattle come into direct contact with infected animals. This can happen during feeding, grooming, or other close interactions, allowing the virus to spread to susceptible individuals.

B. Indirect Contact

Indirect LSD transmission involves the spread of the virus through contaminated objects or materials such as equipment, feed, or shared spaces. Infected animals can shed the virus onto surfaces, leading to the potential infection of susceptible cattle upon contact with these contaminated areas.

C. Vector

The vector-borne transmission of LSD is facilitated by biting insects, particularly various species of blood-feeding flies and ticks. These vectors can transmit the virus from infected to susceptible cattle during the feeding process, contributing to the spread of the disease.

D. Environmental

Virus particles shed by infected cattle can contaminate the environment, including water sources, grazing areas, and communal spaces, creating opportunities for healthy animals to become infected.

24.2.2 FACTORS INFLUENCING TRANSMISSION

A. Cattle Movement

The movement of cattle within and between farms can significantly influence LSD transmission. Interactions between infected and healthy animals during transportation or grazing increase the likelihood of disease transmission.

B. Environmental Conditions
Environmental factors such as temperature, humidity, and the presence of suitable vectors can impact the spread of the disease. Favorable environmental conditions for virus survival and vector activity can contribute to increased transmission rates.

C. Biosecurity Measures
Robust biosecurity protocols are essential for preventing or controlling the transmission of LSD. Measures such as quarantine, disinfection, and restricted access to farm facilities play a crucial role in minimizing disease transmission.

24.2.3 Prevention and Control Measures

A. Vaccination
Effective vaccination programs can significantly reduce LSD incidence in cattle populations. Vaccines stimulate the immune system to produce protective antibodies, limiting the severity and spread of the disease.

B. Vector Control
Measures to control and manage vector populations are crucial for reducing the risk of vector-borne transmission. This can involve insecticide use, environmental modifications, and/or targeted vector control strategies.

C. Health Monitoring
Regular health monitoring and disease surveillance in cattle populations enable early detection of LSD. Prompt identification of cases allows for immediate intervention and the implementation of control measures to prevent further spread.

D. Educational Programs
Educational initiatives aimed at farmers, agricultural workers, and veterinary professionals are crucial for raising awareness about LSD and promoting effective prevention and control measures. Training programs can empower individuals to implement best practices.

E. Global Spread
LSD has spread across multiple continents, initiating outbreaks in various cattle populations. Factors such as international trade, movement of infected animals and vectors, and climatic conditions have contributed to the extensive dissemination of the disease.

24.3 PROBLEM STATEMENT

Machine learning can be categorized as supervised or unsupervised according to the learning paradigms, problem-solving types, and methodologies employed.

24.3.1 Supervised Education

In supervised machine learning, a labeled data set containing the input feature labels and their associated output labels is used to train the model; the model then maps between inputs and outputs to predict new, unobserved data. Predicting a discrete label for a class is a supervised learning technique known as classification; sentiment analysis, picture categorization, and spam detection are a few examples. Predicting a continuous measurement is known as regression; forecasting the stock market and home price are two examples.

24.3.2 Unsupervised Learning

In this method, a data set without labels is used to train the model, and it searches the data for structures and patterns without providing explicit output labels. For instance, in clustering, the model aggregates comparable data items in groups according to their shared characteristics; customer segmentation and picture segmentation are two examples. Another unsupervised learning technique is dimensionality reduction, lowering the number of input characteristics without sacrificing significant data. This group includes methods such as principal component analysis.

Semi-supervised learning trains models on data that is partially labeled, unlabeled data with supervised learning and labeled with unsupervised. Reinforcement learning teaches an agent how to interact with its surroundings to accomplish predetermined objectives. It entails receiving feedback for its activities in the form of benefits or penalties, and it makes better decisions based on this knowledge.

Deep learning is a branch of machine learning that models complicated relationships in data by employing multiple-layered artificial neural networks, or "deep architectures." Deep learning has made tremendous strides in activities like image identification, natural language processing, and gaming. Transfer learning entails applying skills from one activity or area to either a similar or an unrelated task or domain; less data helps the model perform better.

Ensemble learning is the process of combining many models (such as neural networks and decision trees) to obtain predictions that are more reliable and accurate. Frequent ensemble learning strategies include bagging, boosting, and stacking. Online learning allows an algorithm to evolve and acquire knowledge from streaming data by continually updating it as new information becomes available. Generative models are those that can create new samples that resemble the training set of data by understanding the fundamental pattern of distribution of the data. Variational autoencoders and generative adversarial networks (GANs) are two examples.

Each of the ML approaches we discuss in this overview has advantages and disadvantages; the best course of action will rely on the particulars of the situation as well as the qualities of the data at hand. With respect to this chapter's study, developing a model to assess pertinent characteristics from clinical data and forecast the chance of lumpy skin disease formation is the first step in applying machine learning to predict the disease. This is a suggested method for treating LSD.

24.4 THE MODEL DEVELOPMENT HIERARCHY

The data collection and preprocessing entailed evaluating ML methods for predicting LSD prevalence using geographic and meteorological information to build a data set.

First we compiled a human-based data set containing region, country, reporting date, chronic liver disease, deep tendon reflex, Framingham Risk Score, positron emission tomography, progressive resistance exercise, tuberomammillary nucleus, transmembrane pressure, and relevant ocular health measurements. Subsequently, we cleaned the data and handled missing values appropriately. We encoded the categorical variables and normalized numeric features when necessary.

We used feature selection techniques (e.g., correlation analysis, mutual information, recursive feature elimination) to identify the most relevant features for disease prediction; features like age, family history, and lifestyle habits can play a significant role in predicting diseases. Model selection entailed choosing a suitable ML algorithm for binary classification (here, has disease or does not have disease) based on the data set size, complexity, and computational resources available. Common choices include logistic regression, decision trees, random forests, support vector machine, or gradient boosting models.

After we split the data into training and testing sets to evaluate the model's performance on unseen data, we trained the selected ML algorithms on the training data. Following that, we tested the model and evaluated its performance on the testing data using metrics such as accuracy, precision, recall, F1 score, and area under the receiver operating characteristic curve. We performed cross-validation to further assess the model's generalizability.

Next we fine-tuned the hyperparameters of the chosen model using techniques like grid search or random search to optimize its performance followed by analyzing the model's features to understand which contribute the most to disease prediction; model interpretability is crucial in the medical domain to gain insights into the prediction process. The final step was model deployment in a production environment considering data privacy and ethical matters. An ongoing step would be regularly updating the model using new data to maintain its accuracy and relevance.

It's essential to emphasize that this is a proposed framework, and developing a disease prediction model requires a well-curated data set, domain expertise, and collaboration with healthcare professionals. Medical applications require careful validation, and any model predictions should be used to complement clinical assessments rather than replace them. Additionally, ensuring the privacy and security of patient data is of utmost importance.

24.5 THE LSD PREDICTION MODEL

We built our LSD prediction model by collecting the data, constructing the Model and evaluating it. We describe the model development process here. First, the data set we compiled contained 11 attributes (Table 24.1).

Using a pipeline in ML helps streamline the workflow by combining multiple steps into a single process. Pipelines can be beneficial for tasks like LSD prediction, where data preprocessing and model training need to be performed in a systematic and organized manner. In this example, we create a pipeline to handle data preprocessing and build an LSD prediction model using the random forest classifier.

TABLE 24.1
Model Data Attributes

Attributes	Type
Region	Nominal
Country	Nominal
Reporting date	Nominal
Chronic liver disease	Numeric
Deep tendon reflex	Numeric
Framingham Risk Score	Numeric
Positron emission tomography	Numeric
Progressive-resistance exercise	Numeric
Tuberomammillary nucleus	Numeric
Transmembrane Pressure	Numeric
Dominant land cover	Numeric

Algorithm: lumpy skin disease prediction using pipeline

Step 1: Import required libraries
Step 2: Import lumpy dataset
Step 3: Create pipeline for algorithm giving accuracy
Step 4: Add this pipeline in dataset
Step 5: Fit the pipeline
Step 6: Compare the accuracies
Step 7: Use data to predict and identify the most accurate model

Creating a pipeline enables easy modification or extension of the workflow without the need to repeat individual steps. This flexibility allows for the addition of more preprocessing steps, experimentation with different models, or incorporation of hyperparameter tuning techniques like Grid Search CV or Randomized Search CV seamlessly into the pipeline. Next, the evaluation phase is a crucial step in the predictive modeling process. During this stage, various assessment methods are applied, including metrics such as F1-score, classification accuracy, and confusion matrix analysis.

In ML, classification accuracy is a commonly used evaluation criterion, particularly in binary classification scenarios like lumpy skin disease prediction. It involves calculating the proportion of accurately identified samples (comprising true positives and true negatives) and comparing this to the total number of samples in the data set. The formula for classification accuracy is as follows:

$$\text{Accuracy} = \frac{\text{Number of correctly classified samples}}{\text{Total number of samples number}}$$

Accuracy is an all-encompassing indicator of how well the model predicts both lumpy-positive and lumpy-negative, and the confusion matrix is used to reach conclusions about accuracy based on the degree to which the model's predictions agree with the actual labels. In binary classification tasks such as disease prediction, the confusion matrix comes in particularly handy because the objective parameter has two classes: positive (like a bumpy) and negative (like no lumpy). A matrix of confusion displays the four possible results of a binary categorization:

True positives (TP): the proportion of samples the model correctly diagnosed as positive (lumpy)

False positives (FP): the proportion of samples the model incorrectly diagnosed as positive (lumpy)

True negatives (TN): the proportion of samples the model correctly diagnosed as negative (no lumpy)

False negatives (FN): the proportion of samples the model incorrectly diagnosed as negative (no lumpy).

Table 24.2 is a common representation of a confusion matrix.

Accuracy: The proportion of correct predictions out of the total number of samples (TP + TN)/(TP + TN + FP + FN)

Precision: The ratio of true positive predictions to all positive forecasts is expressed as TP/(TP + FP). It gauges how well the model can steer clear of false positives.

Recall (Sensitivity): TP/(TP + FN) is the ratio of true positive predictions to all real positive samples. It assesses how well the model can recognize positive samples.

F1: A balanced measurement between the two is provided by F1, which is the harmonic mean of precision and recall (2 × Precision × Recall)/ (Precision + Recall).

24.6   RESULTS AND DISCUSSION

We subjected the LSD prediction data to clustering as the ML classifier, yielding two clusters from 40 iterations; see Figure 24.1. Figure 24.2 is a summary of the cluster data, Figure 24.3 and is a summary of the LSD accuracy data.

TABLE 24.2
Confusion Matrix

Predicted Values		Actual Values	
		LSD Positive	LSD Negative
	LSD Positive	TP	FP
	LSD Negative	FN	TN

```
class
   c0           64.3561  3.6439
   c1            3.7609 32.2391
  [total]       68.117  35.883


Time taken to build model (full training data) : 0.45 seconds

=== Model and evaluation on training set ===

Clustered Instances

0       68 ( 68%)
1       32 ( 32%)


Log likelihood: -7.28232
```

FIGURE 24.1   Summary of lumpy skin disease prediction data.

```
Time taken to test model on training data: 0.02 seconds

=== Summary ===

Correctly Classified Instances          87              87      %
Incorrectly Classified Instances        13              13      %
Kappa statistic                          0.704
K&B Relative Info Score                 34.9625 %
K&B Information Score                    32.3351 bits       0.3234 bits/instance
Class complexity | order 0              92.485  bits       0.9249 bits/instance
Class complexity | scheme               59.4453 bits       0.5945 bits/instance
Complexity improvement     (Sf)         33.0397 bits       0.3304 bits/instance
Mean absolute error                      0.3162
Root mean squared error                  0.3492
Relative absolute error                 70.2899 %
Root relative squared error             73.7182 %
Total Number of Instances              100

=== Detailed Accuracy By Class ===

                TP Rate  FP Rate  Precision  Recall  F-Measure  MCC    ROC Area  PRC Area  Class
                0.924    0.235    0.884      0.924   0.904      0.706  0.939     0.962     c0
                0.765    0.076    0.839      0.765   0.800      0.706  0.939     0.862     c1
Weighted Avg.   0.870    0.181    0.869      0.870   0.868      0.706  0.939     0.928

=== Confusion Matrix ===

  a  b   <-- classified as
 61  5 |  a = c0
  8 26 |  b = c1
```

FIGURE 24.2   Summary of cluster data.

```
=== Summary ===

Correctly Classified Instances      13               86.6667 %
Incorrectly Classified Instances    2                13.3333 %
Kappa statistic                     0.7222
K&B Relative Info Score             38.9586 %
K&B Information Score               5.7559 bits      0.3837 bits/instance
Class complexity | order 0          14.7744 bits     0.985  bits/instance
Class complexity | scheme           9.0124 bits      0.6008 bits/instance
Complexity improvement    (Sf)      5.762  bits      0.3841 bits/instance
Mean absolute error                 0.3119
Root mean squared error             0.354
Relative absolute error             66.8367 %
Root relative squared error         71.5964 %
Total Number of Instances           15

=== Detailed Accuracy By Class ===

                TP Rate  FP Rate  Precision  Recall  F-Measure  MCC    ROC Area  PRC Area  Class
                0.889    0.167    0.889      0.889   0.889      0.722  0.898     0.929     c0
                0.833    0.111    0.833      0.833   0.833      0.722  0.898     0.783     c1
Weighted Avg.   0.867    0.144    0.867      0.867   0.867      0.722  0.898     0.871

=== Confusion Matrix ===

 a b   <-- classified as
 8 1 | a = c0
 1 5 | b = c1
```

FIGURE 24.3 Summary of lumpy skin disease accuracy data.

Focusing on precision is a fantastic way to reduce false positives. In fields like medicine, where you really want to reduce the possibility of missing positive instances (predicting false negatives), recall plays a critical role. In most circumstances, the cost of missing a positive example is greater than the cost of incorrectly identifying something as positive. Because both precision and recall must have a fair value, the F-score incorporates both and is effective even in circumstances when the data sets are unbalanced. If the precision or recall of the positive class is poor, the formula will reduce the weight of the metric value.

24.7 CONCLUSION

The integration of machine learning and confusion matrix analysis offers a promising avenue for enhancing the accuracy of lumpy skin disease detection. In this chapter, we identified 86.667% accuracy from the confusion matrix. As technology continues to advance, these tools play a crucial role in improving the efficiency of disease surveillance and control measures in the livestock industry, ultimately mitigating economic losses and safeguarding animal health.

REFERENCES

Ahmad Rabaa' I, Zhu X, Jayaraman JD, Nguyen, TDM. The use of machine learning to predict the main factors that influence the continuous usage of mobile food delivery apps. Model Assisted Statistics and Applications. 2022;17(4):247–258, https://doi.org/10.3233/MAS-220405.

Anwar A, Na-Lampang K, Preyavichyapugdee N, Punyapornwithaya V. Lumpy skin disease outbreaks in Africa, Europe, and Asia (2005–2022): Multiple change point analysis and time series forecast. Viruses. 2022;14(10):2203, https://doi.org/10.3390/v14102203.

Bernardes R, Serranho P, Lobo C. Digital ocular fundus imaging: A review. Ophthalmologica. 2011;226(4):161–181, https://doi.org/10.1159/000329597.

Bhat S, Mosalagi S, Bhalerao T, Katkar P, Pitale R. Cataract eye prediction using machine learning. International Journal of Computer Applications. 2020;176(35), https://doi.org/10.5120/ijca2020920441.

Han J-H. Artificial intelligence in eye disease: Recent developments, applications and surveys. Diagnostics (Basel). 2022, https://doi.org/10.3390/diagnostics12081927.

Shaikh F, Mali P, Birajdar P, Narute S. Eye disease detection using machine learning. International Journal of Advance Research in Science, Communication and Technology. 2022;2(1):781–785, https://doi.org/10.48175/IJARSCT-4659.

Yan Tong WL, Yu Y, Shen Y. Application of machine learning in ophthalmic imaging modalities. Eye and Vision, 2020;7(22), https://doi.org/:10.1186/s40662-020-00183-6.

Author Index

Subject Index